MOBILITY PROTOCOLS AND HANDOVER OPTIMIZATION

MOBILITY PROTOCOLS AND HANDOVER OPTIMIZATION

DESIGN, EVALUATION AND APPLICATION

Ashutosh Dutta

AT&T, USA

Henning Schulzrinne

Columbia University, USA

IEEE PRESS

WILEY

Library of Congress Cataloging-in-Publication Data

Dutta, Ashutosh.
 Mobility protocols and handover optimization : design, evaluation and application / Ashutosh Dutta, Henning Schulzrinne.
      pages cm
   Includes bibliographical references and index.
  ISBN 978-0-470-74058-3 (hardback)
 1. Mobile communication systems. 2. Computer network protocols. I. Schulzrinne, Henning. II. Title.
  TK6570.M6D84 2014
  621.3845'6–dc23

                                                              2013036263

A catalogue record for this book is available from the British Library.

ISBN: 978-0-470-74058-3

Typeset in 10/12pt TimesLTStd by Laserwords Private Limited, Chennai, India

1   2014

*To my parents Ganesh and Pratima Dutta, my in-laws Late Haripada
and Rekha De, my wife Sarmistha, my sons Srijoy and Arijit,
and my family and friends. Their constant inspiration and support
were invaluable while writing this book.*
Ashutosh Dutta

To Carol, Nathan, and Ilta
Henning Schulzrinne

Contents

About the Authors

Dr. Ashutosh Dutta obtained his Ph.D. in Electrical Engineering from Columbia University, an M.S. in Computer Science from NJIT, USA, and a BSEE from NIT, Rourkela, India. As a seasoned mobility and security architect and an accomplished networking and computer science expert with 20-plus years' experience, Ashutosh has directed multiple IT operations, has led research and development for leading global technology corporations and top universities, and has in-depth expertise in developing and implementing research, analysis, and design initiatives.

His career, spanning 25 years, includes positions as LMTS (Lead Member of Technical Staff) at AT&T, New Jersey; CTO Wireless at NIKSUN, New Jersey; Senior Scientist at Telcordia Technologies, New Jersey; Director of Central Research Facilities at Columbia University, New York; and Computer Engineer at TATA Motors, India. Ashutosh's research interests include wireless Internet, multimedia signaling, mobility management, 4G networks, IMS (IP Multimedia Subsystem), VoIP, and session control protocols. He has published more than 80 conference and journal papers and Internet drafts, and three book chapters, and has given tutorials on mobility management at various conferences. Ashutosh has 21 issued security- and mobility-related US patents. Ashutosh serves as the Editor-in-Chief for the Journal of Cyber Security and Mobility published by River Publishers.

Ashutosh is a senior member of the IEEE and ACM. He has served as an IEEE volunteer and leader at the section, region, chapter, society, MGA, and EAB levels. Ashutosh is a recipient of the 2009 IEEE Region 1, IEEE MGA, and 2010 IEEE-USA Leadership Awards.

Professor Henning Schulzrinne, Levi Professor of Computer Science at Columbia University, received his Ph.D. from the University of Massachusetts at Amherst, Massachusetts. He was an MTS at AT&T Bell Laboratories and an associate department head at GMD-Fokus in Berlin before joining the Computer Science and Electrical Engineering departments at Columbia University. He served as Chair of the Department of Computer Science from 2004 to 2009 and as Engineering Fellow at the US Federal Communications Commission (FCC) in 2010 and 2011, and has been Chief Technology Officer at the FCC since 2012.

He has published more than 250 journal and conference papers, and more than 70 Internet RFCs. Some of the protocols codeveloped by him, such as RTP, RTSP, and SIP, are now Internet standards, used in almost all Internet telephony and multimedia applications. His research interests include Internet multimedia systems, ubiquitous computing, and mobile systems.

He is a Fellow of the IEEE; has received the New York City Mayor's Award for Excellence in Science and Technology, the VON Pioneer Award, the TCCC service award, the IEEE Region 1 William Terry Award for Lifetime Distinguished Service to the IEEE, and the UMass Computer Science Outstanding Alumni recognition; and is a member of the Internet Hall of Fame.

Foreword

In today's world, ubiquitous computing and wireless Internet roaming have become the norm. Pervasiveness needs to support secured and seamless mobility among heterogeneous access networks. During a mobility event, the user of a mobile device changes its point of attachment, and the existing communication is degraded because of the need to manage mobility at multiple layers. Though protocols have been proposed to manage these different layers, there is no systematic method for comprehensive analysis of a mobility event. An optimized mobility management scheme would handle mobility efficiently without degrading quality of service.

While numerous mobility protocols for different layers have been designed to support these kinds of handoffs, most optimization techniques are ad hoc and tightly coupled to a specific mobility protocol. It is essential we develop optimization techniques in a systematic way, so that they may be applied to any type of mobility protocol. These would take into account factors such as security, configuration, authentication, quality of service, and the mobile's movement pattern.

By having a common framework and set of abstract functions that define mobility events, it will be easier to analyze any related protocol and derive associated optimization techniques. There is a dire need for a reference tool that details current best practices and provides a common framework to analyze the performance of mobility protocols and establish a set of versatile systems optimization techniques.

This book is intended to fill that void. It provides a theoretical approach to the management mobility events and develops this common framework, based on practical results from case studies of service provider, enterprise, military, and vehicular networks. It provides widely applicable deployment guidelines, and proposes a formal analysis of the mobility event that is unique and has not been presented in any other book. The book also introduces an abstract model that can be used to evaluate various types of optimization methodologies. By having such a model, it is easier to choose or design a set of protocols that can provide an optimized mobility management scheme specific to a customer's requirements.

Both advanced professionals and specialists responsible for designing future wireless networks will find this book useful. Graduate-level students will learn about the theory of mobility management and associated optimization techniques for different mobility protocols. This book describes new research ideas for providing a quality-of-service guarantee in terms of delay, packet loss, and resource utilization. Network designers can use this book to study the fundamental steps associated with a mobility event and to determine the basic principles of systems optimization for the steps of a handoff, and will find principles that can be applied to any mobility protocol to achieve a desired quality of service, even with constrained resource parameters, in support of both real-time and non-real-time applications.

The authors Ashutosh Dutta and Henning Schulzrinne are well versed in the theoretical knowledge of mobility protocols, wireless Internet, and cellular systems, not to mention practical experience in developing and deploying mobile systems and networks. They are highly qualified to explain the details of different types of mobility protocols, handover optimization, and evaluation, and their application in different deployment scenarios.

Ashutosh Dutta is an accomplished networking and computer science professional with over 25 years of experience in directing multiple IT operations, designing and implementing enterprise-level and wide-area-level networks, and conducting research and development for leading global telecom corporations and academic institutions. He brings forth a unique combination of research, development, network performance analysis, deployment, and standards experience that gives him the ability to blend the theoretical aspects with best practices. Many of the results and experiments illustrated in the book are from the mobility test beds that he has designed and implemented. His 80-plus publications and 21 patents in the mobility and security areas make him an ideal contributor to this book.

Henning Schulzrinne, often known as the "father of Internet telephony," has published more than 70 Internet RFCs, 250 publications, and multiple patents. Some of the protocols codeveloped by him are Internet standards today, used by almost all the popular Internet telephony and multimedia applications. He designed the original version of the application layer mobility protocol known as SIP. Henning has more than 25 years of experience as a chaired professor at Columbia University, a researcher at Bell Labs, a leader in the IETF, and a chief technology officer at the FCC. As a Ph.D. student advisor, Henning has guided numerous systems- and mobility-related Ph.D. theses. In fact, many of the book's chapters are based on Ashutosh's doctoral thesis work, conducted under Henning's supervision.

In summary, this book provides a comprehensive view of mobility management and optimization techniques by explaining different mobility protocols for various layers, as well as theory, design, and practical implementation and validation in mobility test beds. I recommend this book not only to networking professionals in charge of deploying enterprise and service provider mobility networks, but also to researchers, graduate students, systems engineers, and mobility architects.

Professor Zvi Galil
John P. Imlay Jr. Dean of Computing
Georgia Institute of Technology
Atlanta, USA

Preface

In a span of less than thirty years, cell phones have become ubiquitous, and wireless voice and data have become one of the most common methods of communication today. Wireless networks have also evolved to support much higher bandwidth and lower end-to-end delay, supporting delay-sensitive applications such as interactive voice and video. For example, the 1G and 2G networks that were deployed in the late 1980s and early 1990s could only support data rates of up to a few tens of kilobits per second, in addition to voice communication, but by the start of the new millennium, they had evolved into 3G networks supporting up to 2 Mb/s data rate. Currently, 4G networks, primarily based on LTE, HSPA+, and, to a lesser extent, WiMAX, are being deployed that support multimedia communication and provide data transfer rates up to 100 Mb/s while reducing access packet delay to 50 ms. As these networks have been improving, there has also been a dramatic growth in the use of mobile devices and bandwidth-intensive applications, primarily in entertainment and interactive video. All of these applications are sensitive to disruptions due to handoffs between different cell sites and networks. Driven by increased data needs and limited spectrum availability, cell sizes are shrinking, leading to further increases in handoff frequency.

Recently, increased indoor usage and the availability of cheap Wi-Fi access points have motivated interest in heterogeneous networks, combining base stations using licensed white space and unlicensed spectrum with multiple operators, making roaming common. These so-called hetnets are considered one of the distinguishing features of the future so-called 5G networks.

Not just networks but also end systems are becoming more diverse. Embedded mobile devices, for example for vehicular applications ranging from traffic safety to logistics, have extended mobile networks beyond human-centric applications into the Internet of Things. Besides commercial mobile radio services such as cellular networks, other mobile networks, such as ad hoc and mesh networks, enterprise networks, and public safety networks, are also growing in size, complexity, and heterogeneity.

Across all of these networks, mobility emerges as a central and common challenge. While numerous papers have covered mobility and handoff issues for specific networks or specific protocol layers, this book provides a unified, systematic, and rigorous treatment of the topic, identifying common events for all wireless networks with mobile or nomadic end systems: discovery, network selection, configuration, authentication, security association, encryption, binding update, and media redirection.

This book introduces mobility protocols for different layers, provides a systematic analysis of the mobility event, and investigates the optimization techniques associated with each of the handoff operations in different layers. It takes into account various kinds of mobility deployment scenarios, including wireless service providers, enterprise networks, ad hoc networks, and vehicular networks, supporting both unicast and multicast traffic, and provides a formal analysis of the mobility event. The framework and abstract model proposed in this book can be used to understand and analyze the usefulness of various optimization methodologies. We present results and performance analysis from mobility test beds and theoretical models that validate these optimization techniques for various scenarios.

Organization of the Book

This book is organized as follows. Chapter 1 introduces the main theme of the book, underscores the importance of systems optimization in mobility management, and highlights the key technical contributions of each of the subsequent chapters. Chapter 2 introduces mobility management in cellular and IP-based networks and discusses the related mobility protocols that are currently available. We provide a systematic analysis of the mobility event and associated handoff components in Chapter 3. In Chapter 4, we introduce a formal systems model that uses Petri Nets to analyze the behavioral properties of a mobility event and the associated optimization techniques. Chapter 5 discusses optimization techniques for layer 2 handoff in the 802.11 environment. Chapter 6 describes some key mobility optimization techniques for IP-based mobility protocols that we have developed for different components of the handoff event and demonstrates their validation through actual experiments. Chapter 7 discusses optimization techniques associated with multilayer mobility protocols. Chapter 8 introduces simultaneous mobility, analyzes the probability of its occurrence, and proposes optimization techniques for the layer 3 and application layer mobility protocols. Chapter 9 describes optimization techniques for multicast stream delivery in a hierarchical scope-based multicast architecture. Chapter 10 highlights mobility optimization by way of cooperative roaming techniques. Chapter 11 evaluates a few handoff systems that we have prototyped using some of the optimization techniques described in the preceding chapters by way of experiments and Petri Net models, investigates behavioral aspects of handoff operations such as deadlocks, and analyzes the trade-off between different schedules for handoff and systems resources. Chapter 12 concludes with a discussion on best current practices and the underlying principles of mobility optimization, a summary of the contributions of the work described in the book, and some possible future research directions.

We have included an alphabetical list of the abbreviations that are used in the book. We have also included two appendices. In Appendix A, we define the RDF (Resource Description Framework) schema for application layer discovery, and in Appendix B, we define many mobility-related terms.

Intended audience

Networking professionals in the following fields in particular will benefit from this book:

- *Architects for wireless service providers.* Many wireless service providers are building LTE networks based on 3GPP specifications. These systems generally suffer from optimization problems during handover and roaming. Systems architects who design LTE networks will learn about the methods available for optimizing different handoff functions in a heterogeneous access network. Architects are responsible for designing a complete system, where mobility and security protocols interact with each other; it is very important to understand their interdependency. This book provides an overview of how mobility optimization is affected by other protocols related to security, configuration, and authentication.

- *Researchers.* This book covers both the theoretical and the practical aspects of mobility optimization and addresses some of the associated research issues, including modeling and cross-layer optimization. Researchers will be able to analyze their results using mobility models and experiments and enhance their own research work.

- *Systems engineers.* Systems engineers are in charge of integrating different parts of a system and ensuring that it is ready for operation. Since this book covers many experimental results from live test beds involving heterogeneous wireless access technologies such as CDMA and 802.11, it will provide some real insight into the operational aspects of mobility optimization in a real-world deployment scenario.

- *Protocol designers.* Since this book covers the basics of mobility management and the associated optimization techniques, protocol designers will get a chance to learn about the fundamental principles of optimization. This will help them to design new protocols or enhance existing protocols to make them suitable for a specific mobility deployment.

- *IT professionals.* IT professionals in enterprise networks are always challenged to design an optimized enterprise network that can provide the desired quality of service to the end users under conditions of resource constraints. For example, they may have to decide whether they need to provide layer 2 or layer 3 mobility management to support handoff, or whether they need any cross-layer feedback during handoff. This book provides an analysis and comparison of different types of mobility protocols that may be useful for the efficient deployment of mobility protocols.

- *Standards professionals.* This book provides an overview of many important mobility- and security-related protocols developed by the IETF, as well as their application in mobile networks as defined by 3GPP. Contributors to standards will benefit from the description of the fundamental principles of mobility optimization, methodologies, and best current practices in the book. This will help them to design new mobility-optimized protocols and mobility framework.

- *Chief technology officers.* CTOs of wireless service providers will get a bird's eye view of mobility management and optimization techniques that can be applied to many different deployment scenarios. Since these techniques take into account other operational aspects such as security and quality of service, as well as roaming, they will be able to use this book as a reference book when they interact with other groups within their organization such as the CSO, network planning, and network deployment groups and study their interdependencies.

Acknowledgements

First and foremost, Ashutosh Dutta would like to express his sincere gratitude to his thesis advisor and coauthor Professor Henning Schulzrinne for insightful guidance, thoughtful coaching, and the rigors of research during his Ph.D. study at Columbia University. Many of the chapters of the book are based on Ashutosh's Ph.D. thesis, written at Columbia University under Professor Schulzrinne's supervision when Ashutosh was working at Telcordia Technologies. Ashutosh would like to thank his Ph.D. thesis committee members Professor Yechiam Yemini, Professor Nick Maxemchuk, Professor Dan Rubenstein, and Dr. Bryan Lyles for their feedback on his thesis research. He would like to acknowledge Professor Prathima Agrawal, Dr. Toshikazu Kodama, and the late Dr. Dave Sincoskie for their inspiration and mentorship during his employment at Telcordia, where he completed his Ph.D.

We would like to express our appreciation to our past and current employers Columbia University, Telcordia Technologies, NIKSUN, and AT&T for providing encouragement, freedom, and a flexible environment during the preparation of this book. In particular, the authors would like to thank the members of the Internet Real-Time (IRT) Lab of Columbia University, which provided a platform for discussing and conducting mobility-related research. We benefited from feedback from fellow IRT members through joint collaboration and regular weekly group meetings. In particular, we would like to highlight the contributions from IRT Lab colleagues Andrea Forte and Sangho Shin in the area of mobility. The authors also appreciate the collaboration with colleagues from Telcordia, Toshiba, Toyota, and KDDI that resulted in many of the mobility test bed experiments illustrated in the book. Some of the research reported in this book was supported by the National Science Foundation (Grant CISE 02-02063), SIPQuest, and FirstHand Technologies. The authors are grateful to the staff of the Computer Science Department of Columbia University, particularly Rosemary Addarich, Susan Tritto, Daisy Nguyen, and Pat Hervey, for computing and administrative support over the years, and would like to thank Mark Hammond, Susan Barclay, Liz Wingett, Anna Smart, and Req Ang at John Wiley and Alistair Smith at Sunrise Setting for their constant support and follow-up during the preparation of the book.

Finally, without the consistent support and enduring patience of our families during many weekends and long evening hours and their constant inspiration throughout, we would not have been able to complete this book.

List of Abbreviations

1G   First-generation cellular network. 1G networks are based on analog systems meant to carry voice only. These were developed around 1980. NMT, AMPS, and TACS are examples of 1G systems.

2G   Second-generation cellular network. 2G networks are an evolution of 1G networks that was introduced during the 1990s. 2G networks are digital in nature and provide a per-user bandwidth of up to 144 kb/s. GSM, IS-54/136, and IS-95 are examples of 2G systems.

3G   Third-generation cellular network. 3G networks can provide a per-user bandwidth of up to 2 Mb/s and can carry multimedia traffic. WCDMA and CDMA2000 are examples of 3G systems.

3GPP   Third Generation Partnership Project. A collaborative effort by a group of telecommunications associations to define the standards for 3G networks and for the development of WCDMA/UMTS.

3GPP2   Third Generation Partnership Project 2. The standards body and organization that coordinates the development of 3G networks based on CDMA2000.

4G   Fourth-generation cellular network. 4G networks are an evolution of 2G and 3G cellular networks; they are being defined as part of IMT-2000 and can provide a per-user bandwidth of up to 100 Mb/s.

AAA   Authentication, Authorization, and Accounting. AAA is a generic model for IP network access control, initiated and developed by the IETF (de Laat et al., 2000).

AH   Authentication Header. The AH is a component of the IPSec protocol suite (Kent and Seo, 2005) that guarantees connectionless integrity and data origin authentication of IP datagrams.

AKA   Authentication and Key Agreement. The AKA process is a challenge – response-based mechanism aimed at mutual network/terminal authentication and security key distribution (Niemi et al., 2002).

AMT   Automatic Multicast Tunneling. AMT allows multicast communication amongst isolated multicast-enabled sites or hosts, attached to a network which has no native multicast support (Thaler et al., 2007).

ANSI   American National Standards Institute. ANSI is responsible for overseeing the development of voluntary consensus standards for products and services in the United States.

ARP   Address Resolution Protocol. This is the process of finding out a host's link layer address when only a network layer address is given (Plummer, 1982).

AuC	Authentication Center. An AuC is a database used to control the authentication process and compare users' identifications with those recognized as valid by the network in a GSM or UMTS network.
AVP	Attribute – Value Pair. The AVP is a fundamental data representation in computing systems and applications. It is a data structure that allows future extension without modifying existing code or data.
B2BUA	Back-to-Back User Agent. A B2BUA consists of two SIP user agents, where one can initiate a call and the other modifies and terminates the call. A B2BUA can act as a third party call controller and can establish a call between two user agents.
BCCH	Broadcast Control CHannel. A BCCH is a point-to-multipoint, unidirectional downlink channel used in the GSM cellular standard.
BCP	Buffer Control Protocol. Using a buffer control protocol (Dutta et al., 2006e), a mobile node communicates with buffering nodes in a network to reduce packet loss during handoff by adjusting the buffer value dynamically.
BN	Buffering Node. A buffering node is a logical entity in a network that allows the buffering of packets during a handoff.
BSC	Base Station Controller. Part of a network that controls one or more base stations, and interfaces with the switching center (e.g., the MSC in a GSM network).
BSS	Base Station Subsystem. The overall system that encompasses the BTS and BSC and takes care of handling traffic and signaling between a mobile phone and the network switching subsystem. A BSS is typically used in 2G and 3G networks.
BSSID	Basic Service Set IDentifier. This uniquely identifies each basic service set. The BSSID is the MAC address of an 802.11 wireless access point.
BTS	Base Transceiver Station. The base station equipment used to transmit and receive signals to and from mobile handsets.
CARD	Candidate Access Router Discovery. A protocol (Liebsch et al., 2005) that provides a network discovery mechanism in layer 3 by way of signaling exchanges between the routers in the previous and target networks.
CDMA	Code Division Multiple Access. A wireless access mechanism defined for 2G and 3G networks.
CDN	Content Distribution Network. A CDN is a system of computers containing copies of data, placed at various points in a network so as to maximize the bandwidth for access to the data from clients throughout the network.
CGMP	Cisco Group Management Protocol. CGMP (Farinacci et al., 1996/1997) is a Cisco proprietary group management protocol that manages the multicast groups in layer 2.
CoTI	Care-of Test Init. In MIPv6, a mobile node uses a CoTI message to initiate the return routability procedure and request a care-of keygen token from a correspondent node.
CS	Circuit Switch. The CS domain is a subset of a 2G/GSM or 3G/UMTS core network domain dedicated to the support of circuit-based services such as voice calls.
CSMA/CA	Carrier Sense Multiple Access/Collision Avoidance. A mobile uses this mechanism to get access to IEEE 802.11-type networks.

CTN   Candidate Target Network. A CTN is one of the possible network attachment points where a mobile might move to.

DAD   Duplicate Address Detection. A process of verifying the uniqueness of a layer 3 identifier, which is an IP address in a subnet. This is often carried out during the layer 3 configuration process (Narten et al., 1998).

DCDP   Dynamic Configuration Distribution Protocol. DCDP is a protocol that works in conjunction with DRCP to configure servers with a block of addresses that can be distributed to the end clients (McAuley et al., 2001).

DEDS   Discrete-Event Dynamic System. A DEDS (Cao and Ho, 1990) may be viewed as a sequence of events. The completion of an activity may initiate one or more new activities. The order of the sequence of events is not necessarily unique.

DMZ   DeMilitarized Zone. In computer security, a DMZ is a physical or logical subnetwork that contains an organization's external services and exposes them to a larger, untrusted network, usually the Internet.

DRCP   Dynamic Rapid Configuration Protocol. A lightweight version of DHCP that reduces the number of messages over the air and the message size, thereby reducing the configuration time (McAuley et al., 2001).

DTTPN   Deterministic Timed-Transition Petri Net. A type of timed-transition Petri net where each of the transitions is associated with a deterministic firing time (Ramamoorthy and Ho, 1980).

EAP   Extensible Authentication Protocol. An authentication framework (Aboba et al., 2004) that supports multiple authentication methods and can run directly over data link layers such as PPP (Point-to-Point Protocol) or IEEE 802 without requiring IP.

EPC   Evolved Packet Core. See the definition of SAE.

EPS   Evolved Packet System.

ESN   Electronic Serial Number. A 32-bit identifier used mainly with AMPS, TDMA, and CDMA phones in the United States; compare with the IMEI numbers used by all GSM phones.

ESP   Encapsulating Security Payload. ESP (Kent and Atkinson, 1998a) is a member of the IPSec protocol suite that provides origin authenticity, integrity, and confidentiality protection of packets.

FA   Foreign Agent. This acts as a decapsulation agent in a Mobile IPv4 network.

FACH   Forward Access CHannel. A downlink access channel that carries control information to terminals known to be located in a given cell in a GSM network.

FEC   Forward Error Correction. A system of error control for data transmission, whereby the sender adds redundant data to its messages. This allows the receiver to detect and correct errors without the need for retransmission.

FMIPv6   Fast Mobile IPv6. An optimized version of the Mobile IPv6 protocol that reduces the handoff delay and packet loss by using layer 3 optimization techniques (Koodli, 2005).

FMS   Flexible Manufacturing System. An FMS is a manufacturing system in which there is some amount of flexibility that allows the system to react to changes, whether predicted or unpredicted.

FOCC   FOrward Control Channel. This is used to send signaling messages from a base station to one or multiple mobiles in a 1G network.

GCoA   Global Care-of Address. The care-of address assigned to a mobile when it first enters a specific domain. It is used for intradomain mobility protocols such as IDMP (Das et al., 2002).

GFA   Gateway Foreign Agent. A hierarchical mobility agent that helps to reduce binding-update signaling delay.

GGSN   Gateway GPRS Support Node. This acts as a gateway between an external packet data network and a GSM network that supports GPRS.

GIST   General Internet Signaling Protocol. An IETF protocol (Schulzrinne and Hancock, 2008) being developed by the NSIS (Next Steps in Signaling) working group to define a common messaging layer that will provide a common service for diverse signaling applications.

GPRS   General Packet Radio Service. An evolution of GSM whereby packet switching rather than circuit switching is used to provide increased bandwidth.

GPSK   Generalized Pre Shared Key. The EAP-GPSK protocol is lightweight and seeks to minimize round trips (Clancy and Tschofenig, 2009). Hence, it is well suited to all types of device, especially those with processing-power, memory, and battery constraints.

GRE   Generic Routing Extension. A tunneling protocol that can encapsulate a variety of network layer protocol packet types inside IP tunnels.

GSM   Global System for Mobile Communication. A 2G cellular standard based on a digital TDMA system operating on a 200 kHz RF channel.

GTP   GPRS Tunneling Protocol. A group of IP-based communication protocols used to carry GPRS in GSM and UMTS networks. GTP is used to carry user data between an SGSN and a GGSN and between a RAN and a core network.

HA   Home Agent. An encapsulating agent in a home network (Perkins, 2002a).

HAWAII   Handoff-Aware Wireless Access Internet Infrastructure. A micromobility protocol designed to take care of mobility when a mobile's movement is confined within a domain.

HIP   Host Identity Protocol. A mobility protocol that acts as a shim layer between the network layer and the transport layer to provide the desired mobility functions (Moskowitz and Nikander, 2006).

HLR   Home Location Register. A central database within a GSM or CDMA network that contains information about the mobiles that subscribe to that particular network.

HoTI   Home Test Init. As part of the return routability procedure in Mobile IP, a mobile node sends an HoTI message to the correspondent node (via the home agent) to acquire the home keygen token.

HSDPA   High-Speed Downlink Packet Access. A high-speed enhancement of a 3G/UMTS network for network-to-terminal transmission.

IDEN   Integrated Dispatch Enhanced Network. A mobile telecommunication technology developed by Motorola which provides its users with the benefits of a trunked radio and cellular telephone. IDEN places more users in a spectral space compared with cellular systems and uses TDMA.

IDMP   Intra Domain Mobility Protocol. IDMP (Das et al., 2002) is a mobility protocol that optimizes mobility within an administrative domain.

IGMP   Internet Group Multicast Protocol. IGMP is used between a host and a router and is used to manage multicast groups (Cain et al., 2002).

IKE	Internet Key Exchange. The protocol (Harkins and Carrel, 1998) used to set up a security association in the IPSec protocol suite. IKE uses a Diffie–Hellman key exchange to set up a shared session secret, from which cryptographic keys are derived.
IMC	Internet Multimedia Client. An end client capable of receiving multicast traffic in a hierarchically scoped multicast environment (Dutta and Schulzrinne, 2001).
IMEI	International Mobile Equipment Identity. A 56-bit number that is unique to every GSM phone. Used by a GSM network to identify valid mobile devices.
IMS	IP Multimedia Subsystems. IMS is a 3GPP framework designed for delivering IP multimedia services to end users over 3G and 4G networks.
IMSI	International Mobile Subscriber Identity. A serial number (normally contained in a SIM) that identifies the subscriber in a GSM or UMTS network. An IMSI is usually 15 digits long. The first three digits are the mobile country code, and the following digits are the mobile network code.
IPCP	Internet Protocol Control Protocol. The IPCP (Dutta and Schulzrinne, 1992) is responsible for configuring, enabling, and disabling the IP protocol modules at both ends of a PPP link.
IS-95	Interim Standard Number 95. IS-95 was the first CDMA-based digital cellular standard and was pioneered by Qualcomm. It is also known as CDMAOne and TIA (Telecommunication Industry Associations)-EIA (Electronic Industries Alliance)-95.
ISAKMP	Internet Security Association and Key Management Protocol. The ISAKMP (Maughan et al., 1998) defines the procedures for authenticating a communicating peer, creation and management of security associations, key generation techniques, and threat mitigation (e.g., denial-of-service and replay attacks).
LAI	Location Area Identity. An internationally unique identifier composed of a three-decimal-digit country code, a two-to-three-digit mobile network code, and a location area code. The LAI is broadcast regularly by the BCCH. A mobile station stores its LAI in the subscriber identity module (SIM).
LCoA	Local Care-of Address. A new LCoA is assigned to a mobile when it moves between subnets within a mobility domain in both MIPv4 and MIPv6 networks.
LDAP	Lightweight Directory Access Protocol. Developed by the IETF (Wahl et al., 1997), this is an application protocol for querying and modifying directory services running over TCP/IP.
LMA	Local Mobility Agent. The home agent for the mobile node in a Proxy Mobile IPv6 domain (Gundavelli et al., 2008).
LTE	Long Term Evolution. LTE is the fourth generation of radio technologies that has resulted from enhancements to UMTS. It is being defined as part of the 3GPP standards process. LTE uses OFDM for the downlink, that is, from the base station to the terminal. In the uplink, LTE uses a precoded version of OFDM called single-carrier frequency division multiple access (SC-FDMA).
MA	Mobility Agent. An MA is an anchor agent closer to the mobile and provides similar functionality to the home agent in MIP.
MAG	Mobility Access Gateway. The MAG is a functional component of an access router that manages the mobility-related signaling for a mobile node that is attached to its access link.

MAHT   Maximum Acceptable Handoff Time. This defines the amount of media interruption that a mobile can withstand during handoff; it depends upon the type of application.

MAP   Mobile Application Part. MAP is the application layer of the SS7 (Signaling System Number 7) protocol that is used for communication between several components of a GSM network, such as the HLR, VLR, MSC, and AuC.

MICS   Media Independent Command Service. As one of the key functional components of MIHF (Das et al., 2009), MICS commands are used to gather information about the status of the connected links and to execute higher-layer mobility and connectivity decisions in the lower layers. MICS is useful for optimizing handover among heterogeneous access networks.

MIHF   Media Independent Handover Function. This defines a set of abstract functions that help mobility protocols to achieve seamless handover between heterogeneous access networks. MIHF was defined by the IEEE 802.21 Working Group.

MIIS   Media Independent Information Service. This is a key component of MIHF, and provides a framework and corresponding mechanisms by which an MIHF entity can discover and obtain network information within a geographic area. MIIS was developed by the 802.21 Working Group as part of MIHF.

MIN   Mobile Identification Number. A 10-digit identifier for a mobile subscription. The MIN is stored in a database managed by the cellular provider.

MIP   Mobile IP. A standard developed by the IETF that takes care of session continuity of IP-based sessions for mobile users (see 2002b, Section 2.6.1.)

MIP-LR   Mobile IP with Location Registers. A modified version of Mobile IP (Jain et al., 1999) that provides survivability and eliminates triangle routing of the data path.

MIP-RO   Mobile IP with Route Optimization. A modified version of Mobile IP that reduces the data transfer delay by finding a direct path (see Perkins (2002b), Section 2.6.1).

MIP-RR   Mobile IP with Regional Registration. An optimized version of Mobile IP that reduces the binding update delay by confining the binding update (Perkins, 2002c).

MME   Mobility Management Entity. The MME is a part of an EPC network that is in charge of session and user-mobility management.

MMP   Micro-mobility Management Protocol. MMP includes a suite of protocols that take care of mobility within an administrative domain (Wong et al., 2002).

MOM   MObile Multicast (Williamson et al., 1998). MOM is a home subscription-based multicast mobility protocol to reduce tunnel convergence.

MPA   Media Independent PreAuthentication. This defines a framework that allows a mobile to preauthenticate itself with the target network when the mobile is still in the serving network.

MSC   Mobile Switching Center. The switching center where a mobile network interfaces to the public telephone system.

NAMONC   Network-Assisted MObile and Network-Controlled. A method (Malki, 2004) where a mobile host is assisted by the network with an impending layer 2 handoff.

NAT   Network Address Translation. NAT is the process of modifying network address
                   information in datagram packet headers while in transit across a traffic-routing
                   device for the purpose of remapping a given address space into another (Egevang
                   and Francis, 1994).

NIMOT   Network-Initiated, MObile-Terminated. A network node initiates a handoff by
                   sending handoff triggers to the mobile node.

NMT   Nordic Mobile Telephony. NMT is based on an analog technology
                   (first-generation standard) specified by the Nordic countries and was
                   commercially deployed in 1979.

NUD   Neighbor Unreachability Detection. NUD is a process of figuring out the
                   existence of neighbors on a link in an IPv6 network. Usually, a mobile uses
                   neighbor solicitation (Narten et al., 1998) to find out about the existence of
                   neighbors. If a neighbor is not found, it is deleted from the neighbor cache entry.

OWL   Web Ontology Language. The OWL Web ontology language (McGuinness et al.,
                   2004) is designed for use by applications that need to process the content of
                   information instead of just presenting information to humans.

PANA   Protocol for carrying Authentication to Network Access. An application layer
                   protocol (Jayaraman et al., 2008) defined by the IETF that is used to authenticate
                   a user independently of the access network.

PBU   Proxy Binding Update. A request message sent by a mobile access gateway to a
                   mobile node's local mobility anchor to establish a binding between the mobile
                   node's home network prefix(es) assigned to a given interface of the mobile node
                   and its current care-of address.

PC   Pilot Channel. This acts as a timing beacon for the system and is used by mobile
                   stations in neighboring cells to assess the suitability of a cell for handover. Pilot
                   channels are used in CDMA networks.

PCF   Packet Control Function. An entity in a radio access network that controls the
                   transmission of packets between the base station and the packet data serving
                   node in a CDMA2000 network.

PDIF   Packet Data Interworking Function. This acts like an access router in an IP-based
                   noncellular network.

PDN-GW   PDN-GW (where PDN stands for Packet Data Network) is part of EPC and
                   terminates the SGi interface towards the packet network.

PDSN   Packet Data Serving Node. The layer 3 point of attachment in a CDMA network.

PHT   Proactive Handover Tunnel. This is a tunnel between a mobile node and the
                   router in a target network to complete a proactive handover. Traffic flows
                   through a proactive handover tunnel even before the handover is complete.

PS   Packet Switch. The PS domain is a subset of the 2G/GSM and 3G/UMTS core
                   network domain dedicated to the support of packet-based services.

PTK   Pairwise Transient Key. A PTK is a collection of four keys, referred to as
                   temporal keys, that are derived from the pairwise master key (PMK) every time a
                   mobile device is associated with an access point. These keys are used to encrypt
                   data and protect it from modification.

RA   Router Advertisement. The ICMP router discovery messages are called router
                   advertisements. Each router periodically multicasts an RA from each of its
                   multicast interfaces, announcing the IP address(es) of that interface.

RAN	Radio Access Network. This controls and terminates radio signals in 3G and 4G networks.
RAS	Radio Antenna Server. Radio antenna servers (Dutta and Schulzrinne, 2001) are content servers towards the edge of a network that are capable of translating streaming content from a globally scoped multicast address to a locally scoped multicast address.
RAT	Radio Access Technology. This refers to the type of cellular access that is in use. GSM, WCDMA, and CDMA2000 are all different types of radio access technology.
RBMOM	Range-Based Mobile Multicast (Lin and Wang, 2000). RBMOM improves upon the performance of MOM by dynamically assigning a designated multicast service provider.
RDF	Resource Description Framework. A general method for the conceptual description or modeling of information that is implemented in Web resources, using a variety of syntax formats. RDF is a family of W3C specifications (Lassila et al., 1999).
RDFS	RDF Schema (McBride, 2004).
RDQL	RDF Data Query Language. RDQL is an SQL-like RDF query language derived from Squish.
RHT	Reactive Handover Tunnel. This is a tunnel set up between the router in the previous network and a mobile's new care-of address so that the transient in-flight traffic can be forwarded to the mobile after the handover. Reactive handover tunnels are used to reduce the packet loss due to handoff.
RNC	Radio Network Controller. The RNC is the network element responsible for the control of the radio resources of UTRAN (Bauer et al., 2003).
RSC	Radio Station Client. The RSCs are the multicast traffic sources (Dutta and Schulzrinne, 2001) in a hierarchically scoped multicast environment.
S-GW	Serving GateWay. The S-GW is part of EPC. It is the functional network entity that forwards and routes packets between eNodeB and the packet data network.
SA	Security Association. An SA establishes shared security information between two network entities to support secure communication.
SAE	System Architecture Evolution. SAE defines the packet core of a 4G network, where LTE is the RAN. SAE is also known as EPC.
SAT	Supervisory Audio Tone. One of three tones, at 5970, 6000, or 6030 Hz, that is transmitted by the base station and repeated back by a mobile.
SC	Sync Channel. This allows a mobile station to achieve time synchronization with the base station and the network.
SCTP	Stream Control Transport Protocol. SCTP (Stewart et al., 2000) is a transport layer protocol that serves a similar role to TCP and UDP. SCTP places messages and control information in separate chunks (data chunks and control chunks), each identified by a chunk header.
SGSN	Serving GPRS Support Node. This keeps track of location information and security information associated with mobile stations within a service area.
SID	System IDentifier. The SID is used by a mobile in a cellular network to ensure that it is on the correct network.

SMR	Specialized Mobile Radio. A two-way radio system in which two mobile transceivers are linked by a single repeater.
SMS-SC	Short Message Service–Service Center.
SNR	Signal-to-Noise Ratio. The SNR is the ratio of the signal to the background noise and is used as an indication of signal quality. It can be used as a measure of receiver performance.
SPARQL	SPARQL Protocol and RDF Query Language. This is an RDF query language and is considered a key Semantic Web technology.
SPI	Security Parameter Index. The SPI is a 32-bit value used to distinguish between different SAs terminating at the same destination and using the same IPSec protocol (Kent and Atkinson, 1998b).
SRTP	Secure RTP. A profile of the Real-time Transport Protocol (RTP), which can provide confidentiality, message authentication, and replay protection for RTP traffic and the control traffic for RTP (defined by the Real-time Transport Control Protocol (RTCP)) (Baugher et al., 2004).
SSID	Service Set IDentifier. This identifies a particular 802.11 wireless LAN. The SSID can be up to 32 characters long. The SSID is defined as a sequence of 1–32 octets, each of which may take any value.
SSM	Source Specific Multicast. A multicast protocol used mostly for multicasting content from one source to many users.
TACS	Total Access Communication System. TACS is a 1G cellular standard based on the analog system used in European countries. TACS was first used in the United Kingdom in 1985 and then later used in other countries, including Hong Kong and Japan. In Japan, it is called JTAC.
TDMA	Time Division Multiple Access. This allows several users to share the same frequency channel by dividing the signal into different time slots.
TELEMIP	Telecommunications-Enhanced Mobile IP. A mobility management protocol to take care of mobility within a domain (Das et al., 2000).
TLS	Transport Layer Security. The TLS protocol (Dierks and Allen, 1999) allows client/server applications to communicate across a network in a way designed to prevent eavesdropping, tampering, and message forgery. TLS is usually implemented on top of the transport layer protocols, encapsulating application-specific protocols such as HTTP, FTP, SMTP, NNTP, SIP, and XMPP.
TLV	Type–Length–Value. Any information in any communication protocol can be encoded in the form type–length–value, where the type and length are of fixed size, whereas the value is of variable size.
TMSI	Temporary Mobile Subscriber Identity. A temporary identifier assigned to a subscriber by a network.
UMTP	UDP Multicast Tunneling Protocol. This protocol enables a host without multicast connectivity to make an ad hoc connection to the MBone by tunneling multicast UDP datagrams inside unicast UDP datagrams.
URI	Uniform Resource Identifier. The URI is an addressing technology that consists of a string of characters used to identify or name a resource on the Internet. URIs were originally defined as two types: Uniform Resource Locators (URLs), which

	are addresses with network locations, and Uniform Resource Names (URNs), which are persistent names that are address-independent (Hansen et al., 2006).
UTRAN	UMTS Terrestrial Radio Access Network. UTRAN represents the access network of a 3G UMTS network.
VCC	Voice Call Continuity. A set of 3GPP standards to maintain call continuity between cellular and IP-based networks.
VLR	Visitor Location Register. A VLR is a database that holds information about all the mobiles that are visitors and are under one visited MSC (V-MSC) in a GSM network.
VPN	Virtual Private Network. The concept of a VPN is that a private network is superimposed on a public network so that one can obtain the advantages of a dedicated network. VPNs are often installed by organizations to provide remote access to a secure organizational network.
WCDMA	Wideband Code Division Multiple Access. A system based around a direct-sequence spread spectrum that enables multiple users to access a cellular channel simultaneously.
WPA	Wi-Fi Protected Access. This is built around the RSN (Robust Security Network) standard. It was an interim standard before IEEE 802.11i was standardized.
XML	Extensible Markup Language. XML is a textual data format with strong support via Unicode for the languages of the world. It is a subset of SGML (Standard Generalized Markup Language) that can be served, received, and processed on the Web as easily as HTML.

1

Introduction

Wireless connectivity to communications and information has advanced the world towards ubiquitous computing. In the space of less than thirty years, cell phones have become ubiquitous and wireless data access has become common. However, this access has brought with it a variety of technical problems. Radio physics and power constraints, the need to reuse spectrum, economic constraints on facility placement, and service balkanization due to competitive and political factors force us to implement wireless systems as cells of limited range. Furthermore, cells may use very different wireless technologies or provide fundamentally different services, such as VoIP (Voice over IP), streaming, or direct short-range communications for telematics. We then need handoff mechanisms, often in multiple protocol layers, to allow a mobile terminal to move from cell to cell and maintain service continuity.

Mobility can be described as movement of a terminal, resulting in the release of the terminal's binding to the current cell (point of attachment to the network) and the establishment of bindings to the new cell being entered, while preserving the existing sessions associated with higher-level services. The cellular telephony community has long implemented service- and technology-specific mobility protocols that hand off voice sessions as the user moves from cell to cell. Because voice service quality is highly sensitive to service interruptions, cell-to-cell handoffs in a cellular environment have been highly optimized and are not noticed by the public. Tripathi et al. (1998) summarized some of the handoff technologies associated with cellular mobility. Pollini (1996) discussed some of the trends in handover design in cellular networks that may affect the optimization of handover performance.

For IP traffic, the IETF has defined mobility protocols for both IPv4 (Perkins, 2002b) and IPv6 (Johnson et al., 2004). However, IP traffic is dramatically more diverse than cellular voice in the range of link layer technologies used to support IP traffic, the number of economic units supplying IP services, and the authentication protocols and services running above IP. This diversity has meant that the IETF could not easily design access-specific handoff optimization techniques such as soft handoff (Chen and Mary, 2003; Wong and Lim, 1997), often seen in cellular voice, into the mobility standards. As a result, unoptimized Mobile IP handoffs can take a few seconds to perform, and degrade the quality of service in the process.

IP's transformation from a service supporting email and file transfer to the base layer for network convergence means that the constraints on handover performance are becoming much more stringent. Handovers cannot interrupt real-time services. The mechanisms and design principles needed for

Mobility Protocols and Handover Optimization: Design, Evaluation and Application, First Edition.
Ashutosh Dutta and Henning Schulzrinne.
© 2014 John Wiley & Sons, Ltd. Published 2014 by John Wiley & Sons, Ltd.

building optimized handovers in the context of mobile Internet services are poorly understood and need better analysis. To the best of our knowledge, none of the existing work has attempted to model the systems aspects of a mobility event nor was intended to systematically analyze the elementary operations involved in a mobility event. This body of work is also lacking in the ability to predict the performance of a mobility event. Some of the existing work has focused on optimizing only parts of a mobility event in an ad hoc manner, specific to a mobility protocol, without providing a comprehensive approach to solving the optimization problem in all layers or functional modules. We provide an overview of this related work for each of these mobility functions along with a detailed description of the proposed techniques in Chapter 6.

This book is intended to contribute to a general theory of optimized handover, especially with respect to the mobility of Internet-based applications. The contributions fall into four categories:

1. Identification of fundamental properties that are rebound during a mobility event. Analysis of these properties provides a systematic framework for describing mobility management and the operations that are intrinsic to handover.
2. A model of the handover process that allows one to predict performance both for an unoptimized handover and for specific optimization methodologies under conditions of resource constraints. This model also allows one to study behavioral properties of the handoff system such as data dependency and deadlocks.
3. A series of optimization methodologies, experimental evaluations of them, and optimization techniques that can be applied to the link, network, and application layers and preserve the user experience by optimizing a handover.
4. Application of the model to represent optimizations, and comparison of the results with experimental data.

1.1 Types of Mobility

There are several types of mobility, such as terminal mobility, personal mobility, session mobility, and service mobility. Schulzrinne and Wedlund (2000a) introduced several different types of mobility to support multimedia traffic for an IP-based network. We briefly review each type of mobility.

1.1.1 Terminal Mobility

Terminal mobility allows a device to move between networks while continuing to be reachable for incoming requests and maintaining existing sessions during movement. It allows an established call or session to continue when an MS (mobile station) moves from one cell to another without interruption in the call or session.

Terminal mobility may also arise from a change in the network condition, whereby a mobile may switch between two neighboring networks even without any movement. We describe here the types of terminal mobility that arise from different types of handoffs. Handoff, often also known as handover, is a process that results when a mobile disconnects from one point of attachment in a network and reconnects to another point of attachment in the same or a different network.

The handoff process can be either hard or soft. With hard handoff, the link to the prior base station is terminated before or as the user is transferred to the new cell's base station. Thus, the mobile is linked to no more than one base station at a given time. Initiation of the handoff may begin when the signal strength at the mobile received from the target base station is greater than that from the current base station. As the mobile moves into a new cell, its signal is abruptly handed over from its current cell (or base station) to the new one. In the old analog systems, hard handover could be heard as a click or a very short beep. In digital cellular systems, this is not noticed. However, in an IP-based handoff scenario, hard handover contributes 4–15 s of delay (Dutta et al., 2005c). With soft handoff (Wong and Lim, 1997), the MS continues to receive and accept radio signals from the base stations that are part of the previous as well as the new cell for a limited period of time. The MS signal is also received at multiple base stations. In order to ensure the layer 2 independence requirement of mobility management schemes, a maximum acceptable handoff time (MAHT) is required, which will vary based on the access type.

In an end-to-end wireless IP environment, four logical levels of handoff procedures can be defined:

- *Layer 2 handoff.* This allows an MS to move from one layer 2 point of attachment to another layer 2 point of attachment that belongs to the same subnetwork. Each layer 2 point of attachment may be equipped with same or a different type of radio access technology. One subnetwork may consist of multiple layer 2 radio access networks. The IP address of the mobile host remains the same during this handoff.
- *Subnet handoff.* This allows an MS to move from a radio access network within a subnet to an adjacent radio access network within another subnet that belongs to the same administrative domain. The IP address of the mobile may or may not change.
- *Domain handoff.* This allows an MS to move from one subnet within an administrative domain to another in a different administrative domain. Domain handoff can take place between two administrative domains that belong to the same operator or different operators.
- *Interoperator handoff.* This allows a mobile to move from one wireless service provider to another wireless service provider. Interoperator handoff involves additional steps such as authentication and authorization with the mobile's home networks and may be subject to various roaming agreements among wireless operators. In most cases, interoperator handoff is not as frequent as subnet handoff or layer 2 handoff.

In the following section, we define several different combinations of handoffs in a heterogeneous environment.

We have analyzed, modeled, and experimented with all three of these types of handoff in heterogeneous access environments such as CDMA (code division multiple access) and IEEE 802.11. We focus on the systems optimization aspects of terminal mobility in this book. Systems optimization techniques minimize the delay and packet loss contributed by several handoff components under certain conditions of resource constraints on the mobile and the network, namely constraints on battery power, CPU cycles, and network capacity.

1.1.1.1 Heterogeneous Handover

An access network is defined as the backhaul network that provides the first-hop or last-hop access to a mobile in an end-to-end communication system. When a mobile moves from one access network to another access network during an active communication session and changes its point of attachment, it is subject to disruption in the continuity of service. During the handover, as the mobile changes its point of attachment in the network, the mobile terminal may end up communicating using a second interface with the new network, or change its subnet or the domain it is connected to. Heterogeneous handover is a type of handover that requires authorization for the acquisition or modification of resources assigned to a mobile, and the authorization needs interaction with a central authority in a domain. In many cases, the authorization procedure in a heterogeneous handover follows an authentication procedure that also requires interaction with a central authority in a domain. Based on the type of handover and access technology, the following heterogeneous handover scenarios can be defined. These can be categorized primarily as (A) intersubnet and (B) intrasubnet.

An intersubnet handover can comprise the following combinations of handovers that fall into category A:

1. intertechnology, interdomain;
2. intertechnology, intradomain;
3. intratechnology, interdomain;
4. intratechnology, intradomain.

An intrasubnet handover can comprise the following types of handovers that fall into category B:

1. intratechnology, intradomain;
2. intertechnolgy, intradomain.

Intrasubnet

When a mobile moves between two radio access networks that are part of the same subnet and does not change broadcast domain, this is called an intrasubnet handover. During an intrasubnet handover, the mobile may be subject to an intertechnology handover as well if both of the access networks, with different radio access technologies (e.g., CDMA and 802.11), belong to the same subnet. However, during an intrasubnet and intertechnology handover, the effective layer 3 identifier of the terminal may change if the terminal starts to communicate using an interface that is part of a different access network, since the IP addresses associated with each interface will be different even though these addresses belong to the same subnet.

Intersubnet

A mobile is subject to an intersubnet handover when it moves between two radio access networks that belong to two different subnets. As a result, its layer 3 identifier (e.g., IP address) is changed, thus giving rise to a need for a mobility management protocol such as Mobile IP (Perkins, 2002b), Mobile IPv6 (Johnson et al., 2004), SIP-Mobility (Wedlund and Schulzrinne, 1999), and HIP (Moskowitz and Nikander, 2006) that can take care of the continuity of the existing application. An intersubnet handover can be viewed as orthogonal to intrasubnet handoff scenarios and may include intradomain, interdomain, intertechnology or intratechnology handover. Intersubnet handover potentially gives rise to packet loss and jitter because of the delay associated with transitions in layers 2 and 3.

Intertechnology

A mobile may be equipped with multiple interfaces, where each interface can support a different access technology (802.11 or CDMA). A mobile may prefer to communicate with only one interface at any time in order to conserve power. During the handover, the mobile may move out of the footprint of one access technology (e.g., 802.11) and into the footprint of a different technology (e.g., CDMA). This will warrant switching of the communicating interface on the mobile as well. This type of intertechnology handover is often called vertical handover, since the mobile makes a movement between two different cell sizes. A vertical handover can be termed an upward or a downward vertical handover based on the direction of movement, namely from a smaller cell to a larger cell or vice versa as described by Stemm and Katz (1998). A mobile moving from an 802.11 network to a cellular network can be viewed as an upward vertical handover. An intertechnology handover may affect the quality of service of multimedia communication, since each access network may offer a different bandwidth.

Intratechnology

An intratechnology handover is defined as event when a mobile moves between two examples of the same type of access technology, such as between 802.11[a, b, n] and 802.11[a, b, n] or between CDMA1XRTT and CDMA1EVDO. In this scenario, a mobile may be equipped with a single interface (with multiple PHY types of the same technology) or with multiple interfaces. An intratechnology handover may involve intrasubnet or intersubnet movement and thus may need to change its layer 3 identifier, depending upon the type of movement. An intratechnology handover may involve networks with different access characteristics and thus may very well belong to the heterogeneous-handover category. However, a handover between two 802.11b networks, if these belong to the same subnet or domain, need not be termed a heterogeneous handover.

Interdomain

A domain can be defined in several ways. But, for the purposes of roaming, we define a domain as an administrative domain that consists of networks that are managed by a single administrative entity which authenticates and authorizes a mobile for accessing those networks. An administrative entity may be a service provider, an enterprise, or any organization. As a mobile moves between two administrative domains, it is also subject to intersubnet handover, as the two different domains exclusively have two different subnets. Thus, an interdomain handover will by default be subject to intersubnet handover and, in addition, it may be subject to either intertechnology or intratechnology handover. Interdomain handover will be subject to all the transition steps that a subnet handover goes through in addition to an authentication process. These extra steps contribute to additional delay on top of the delays due to regular subnet handover.

Intradomain

When a mobile's movement is confined within an administrative domain, it is called intradomain movement. An intradomain movement may involve intrasubnet, intersubnet, intratechnology, and intertechnology handovers as well.

1.1.2 Personal Mobility

The concept of personal mobility was initially introduced as part of Universal Personal Telecommunications (UPT), as described by Zaid (1994). Personal mobility removes the fixed association between the terminal and the user, thereby allowing an additional degree of mobility over and above terminal mobility in mobile networks.

Figure 1.1 illustrates some fundamental differences between personal mobility and terminal mobility by showing the relationships between path identification, terminal identification, and user identification.[1]

---

[1] These terms are defined in Appendix B.

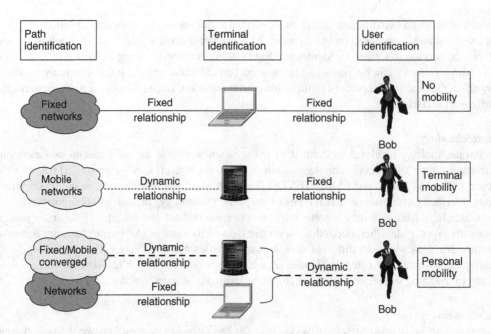

Figure 1.1   Personal and terminal mobility

For multimedia communication, personal mobility is a form of mobility by which a user can be reached at different terminals using the same logical address or resource identifier (Schulzrinne et al., 1996), regardless of the point of attachment of the mobile and the identifier that it obtains when it attaches to a network. As the mobile changes its point of attachment, it acquires a new identifier in the new network and updates this identifier by means of registration with a central authority, either in the home network or in the visited network. The central authority is often a SIP (Session Initiation Protocol) registrar (Rosenberg et al., 2002) that keeps a binding between the new terminal identifier and the Uniform Resource Identifier (URI) that is assigned to the mobile user and is unique for each mobile. A URI is typically a SIP URI and is of the form sip:alice@xyz.edu as defined by IETF RFC 2396 (Berners-Lee et al., 1998). Personal mobility can involve both 1-to-n mappings, where one address can be associated with many potential terminals, and m-to-1 mappings, where multiple addresses map to one device. Thus, by having a mapping between the terminal identifier and the resource identifier, it is possible to direct the data to one or more interfaces, where the interfaces can be part of the same device or multiple devices.

1.1.3   Session Mobility

Session mobility allows a user to continue an existing multimedia session or part of a session as the user moves from one device to another. For example, a user who is part of a multimedia session including voice and video on a cell phone can transfer the video to another device such as a TV, thus splitting the existing multimedia session over two devices. Similarly, an existing audio session can be transferred from a cell phone to a desktop phone. The MEGACO (Cuervo et al., 2000), third-party call control (Rosenberg et al., 2004), and REFER (Sparks, 2003) mechanisms are some of the approaches that can be used to implement session mobility. Most recently, 3GPP (the Third Generation Partnership Project) has been following an approach called Voice Call Continuity (VCC) (Third Generation Partnership Project 2 and Telecommunications Industry Association 2007) that allows a user to move between an IP network and a cellular network. Shacham et al. (2008) and Shacham et al. (2007) discussed the technical details of session mobility and described how multimedia sessions can be transferred between different terminals.

1.1.4   Service Mobility

Service mobility allows a user to maintain their access to services even when changing devices and network service providers. In a VoIP (Voice over IP) environment, the typical services that users may wish to maintain include speed dial lists, address books, call logs, media preferences, buddy lists, and call-handling instructions. However, in order to be able to obtain the same service independently of the service provider, the mobile device may often require cross-provider relationships.

1.2   Performance Requirements

In order to provide the desired quality of service (QoS) for interactive VoIP and streaming traffic, one needs to limit the end-to-end delay, network jitter, and packet loss to an acceptable level. The performance requirements will vary based on the type of application and its characteristics such as delay and loss tolerance. Various standards organizations have defined limits for these metrics. For example, 3GPP TS23.107 (Greis, 2001) defines four application classes, namely conversational, streaming, interactive, and background (e.g., file transfer or email), each with different sets of end-to-end delay and QoS requirements. Based on the type of application (e.g., interactive, streaming, or data), these values may vary. For example, for the one-way delay, ITU-T G.114 (Time, 2000) recommends 150 ms as the upper limit for VoIP applications and considers 400 ms as a generally unacceptable delay. Similarly, the streaming class has a tolerable packet (SDU) error ratio ranging from 0.1 to 0.00001 and a transmission delay limit of less than 300 ms.

In general, the handoff process contributes to packet loss and network jitter and adversely affects the overall throughput of data traffic because of interruption and retransmission of data due to the change in the point of attachment to the network. A mobility event contributes to two kinds of delays that affect performance, namely the *handoff delay* and *one-way delay* of the packet. The handoff delay is defined as the time between when the last packet is received at the old point of attachment and when the first packet is received at the new point of attachment. The end-to-end delay (or one-way delay) consists of several components, namely the transmission delay, propagation delay, network delay, operating system delay, codec delay, and application delay. Wenyu and Schulzrinne have done a complete analysis of these delays (Jiang and Schulzrinne, 2000). Handoff contributes to the network delay component of the end-to-end delay.

During a mobile's handoff process, in-flight transient traffic cannot reach the mobile. In-flight packets are defined as packets that are in transit during a mobile's movement from one network point of attachment to another. Network jitter contributes to the variation in interpacket arrival time of consecutive packets at the receiver. This is caused by variation in the one-way transmission delay of the consecutive packets. These in-flight packets can be either lost or buffered. If the in-flight packets are lost, this contributes to the interpacket arrival delay between the last packet before handoff and the first packet after handoff. If the packets are buffered, packet loss is minimized, but there is additional jitter for the in-handoff packets when they are flushed after the handoff. Buffering during handoff avoids packet loss, but at the cost of additional one-way delay. A trade-off between one-way delay and packet loss is desired, based on the type of application. For real-time communication, if a packet is received after a certain delay threshold, it is also considered lost.

We have verified experimentally that in the absence of any optimization technique, a mobile can be subject to a handoff delay of between 4 and 17 s (Dutta et al., 2005c), resulting in transient service interruption and packet loss. This value varies depending upon the type of mobility protocol used; the type of handover, such as vertical (i.e., handover between different network types) or horizontal (i.e., handover between networks of the same type); and the type of access network, namely 802.11 or CDMA. Thus, it is desirable to conduct a formal analysis of the discrete events that constitute the handoff process; build a system model that can predict handoff performance; develop relevant optimization techniques for these operations; and analyze the dynamic behavior of the system during handoff, including resource utilization.

While several mobility protocols have been defined for different layers, to the best of our knowledge, there has been no formal analysis or system model that can study the basic operations associated with mobility events and various systems optimization techniques.

1.3 Motivation

The following are the key mobility issues that are covered in the book:

1. The existing mobility protocols affect the performance of real-time communication because of the sequence of discrete events associated with the handoff event.
2. The existing mobility optimization mechanisms are tightly coupled to the corresponding mobility management protocols and do not provide a generalized approach to optimization. For example, it is impossible to apply mobility optimization mechanisms designed for Mobile IPv4 (Perkins, 2002b) or Mobile IPv6 (Johnson et al., 2004) to MOBIKE (Eronen, 2006). Thus, it is desirable to develop a set of formal methodologies with specific design criteria to help formulate system optimization techniques that can be applied to any mobility management protocol and access technology.
3. The available mobility management techniques do not provide any systematic framework to formalize the different states and transition processes involved with a mobility event. Thus, it becomes difficult to study the behavior of a handoff event and evaluate the performance of any mobility protocol or to devise any improvements.
   (a) As far as we know, there has not been a systematic mobility system model that can analyze behavioral characteristics of a handoff event such as deadlock and help formulate systems optimization techniques for cellular or IP-based mobility management protocols.
   (b) The existing work does not provide a generalized mobility optimization framework that can support horizontal and vertical handovers across administrative domains. [2]
   (c) There has not been any formal analysis of how a specific mobility optimization technique might affect other system resources in the network.

This book addresses the above issues. We analyze the basic operations associated with a mobility event in detail. We develop a formal systems model of mobility by representing the basic operations

---

[2] "Administrative domain" is defined in Appendix B.

associated with a handoff as a series of discrete events. We formalize the associated states and transitions in the form of a discrete-event dynamic systems (DEDS) model. We then analyze this DEDS-oriented mobility model using discrete timed-transition Petri nets (DTTP). Based on an analysis of the properties associated with a mobility event, we propose several systems optimization techniques for the basic operations associated with the handoff event. We demonstrate these techniques with models, experiments, and numerical analysis using a few network layer and application layer mobility protocols. We then apply these optimization techniques to a timed Petri net model and compare the results with experimental results. We perform a trade-off analysis between the utilization of systems resources and the handoff performance metrics obtained using these optimization techniques. Finally, we also use this model to study behavioral properties of the handoff such as deadlock and data dependency under conditions of systems and network resource constraints.

1.4   Summary of Key Contributions

The highlights of the key contributions of Chapters 2–12 are summarized in Table 1.1. For each chapter, we summarize the technical problems that are addressed, the details of the proposed mechanisms, and the key benefits, with experimental results.

Table 1.1   Summary of key contributions

No.	Chapter title	Summary of key contributions
2	Analysis of mobility protocols for multimedia	Comprehensive analysis and comparison of several generations of mobility protocols (e.g., 1G, 2G, 3G, and 4G) to extrapolate the common abstract functions for a mobility event.
3	Systems analysis of mobility events	Development of a new synthesis that derives a fundamental taxonomy of handover functions and their relationships. This taxonomy provides a basis for describing and characterizing optimization in each layer. Experimental analysis of the handoff delays for the application layer and network layer mobility protocols based on this handover taxonomy.
4	Modeling mobility	• Data dependency analysis and resource analysis of the handover components based on the mobility taxonomy. • Design of the first mobility system model for handoff processes using deterministic timed-transition Petri nets based on data dependency and resource dependency. • Development of Petri-net-based mechanisms to predict the systems performance and behavior of a handoff system. • New mechanisms to investigate the opportunity for parallelism based on resource modeling.
5	Layer 2 optimization	• This chapter introduces a new handoff procedure that reduces the MAC layer handoff latency, in most cases to a level where VoIP communication becomes seamless. This new handoff procedure reduces the discovery phase using a selective scanning algorithm and a caching mechanism. • Using the selective scanning mechanism, it is possible to reduce the total handoff latency to an average value of 129 ms, and by using the caching mechanism it is possible to reduce the handoff latency to 3 ms.

(*continued overleaf*)

Table 1.1   (*continued*)

No.	Chapter title	Summary of key contributions
	Mobility optimization techniques	Proposed proactive, reactive, and cross-layer mechanisms to optimize several handoff components as determined in Chapter 3.
6	Discovery (Section 6.3)	Application layer discovery mechanism that discovers the network elements of the target networks in an access-independent manner. By discovering these elements proactively and caching some of these at the mobile, network discovery latency is reduced to 4 ms.
	Authentication (Section 6.4)	Network-layer-assisted layer 2 preauthentication mechanism that bootstraps the layer 2 authentication process in the neighboring networks by deriving preshared keys prior to the handover of the mobile. This mechanism reduces the authentication delay to 16 ms for both intersubnet and interdomain handover.
	Layer 3 configuration (Section 6.5)	• Proactive IP address acquisition scheme that reduces the signaling exchange by obtaining the IP address from the target network over a secured tunnel before layer 2 handover. • Reactive router-assisted duplicate address detection mechanism, where the router multicasts the ARP cache at periodic intervals so that the mobile avoids ARP checking for duplicate address detection.
	Layer 3 security association (Section 6.6)	• Anchor-agent-assisted layer 3 mechanism that maintains the layer 3 security context by hiding the change of network layer identifier address of the mobile and reduces the delay by avoiding the rekeying process. • Preregistration-based mechanism that establishes the security context prior to handoff by generating security keys. The handover delay due to layer 3 security association is completely eliminated.
	Binding update (Section 6.7)	• Reactive hierarchical binding update mechanism that uses a two-level hierarchy of addresses and an anchor agent to limit the global signaling update during mobility of the mobile within a domain. This mechanism achieves a reduction of about 70% in global signaling overhead for a 10 subnets/domain scenario. • Proactive binding update mechanism over a secured tunnel that eliminates the binding update delay completely at the cost of maintaining a proactive tunnel between the mobile and the target network.
	Media rerouting (Sections 6.8 and 6.9)	• Reactive forwarding mechanism that redirects the in-flight data from the previous network using an application layer mobility proxy in the previous network. • Mobile-controlled buffering mechanism that controls the buffering period dynamically based on handoff duration during proactive handoff. • Proactive multicasting mechanism that multicasts the in-flight data to the neighboring networks and reduces in-flight packet loss during handoff.
	Route optimization (Section 6.10)	• Packet-interceptor-assisted mechanism that modifies the source and destination addresses of the packets at the end hosts to maintain a direct path between the communicating hosts. This mechanism reduces transport delay by 50% for large packets. • Proxy-assisted packet interceptor that eliminates the trombone routing delay and reduces the signaling-related delay by 60% in an IMS (IP Multimedia Subsystem) environment.

Table 1.1 (*continued*)

No.	Chapter title	Summary of key contributions
		• Binding-cache-based mechanism that minimizes the end-to-end media transport delay by a factor of 5 by localizing the media traffic for a localized mobility protocol (e.g., ProxyMIPv6).
	Media-independent cross-layer triggers (Section 6.11)	• First set of cross-layer triggers based on abstract primitives that can pass information across layers and expedite handoff-related operations independently of the access mechanism (e.g., CDMA or 802.11). These proposed cross-layer triggers were standardized in IEEE 802.21.
7	Optimization with multilayer mobility protocols	Multilayer mobility management scheme that uses cross-layer triggers from data link layers and application layers and optimizes several handoff operations, namely address configuration, layer 3 binding update, and media traversal. The proposed mechanism increases the data throughput by 50% in a high-mobility scenario by reducing the binding update traversal.
8	Optimization for simultaneous mobility	• Proposal of an analytical framework for simultaneous mobility that can predict the probability of simultaneous mobility based on interhandoff time and binding update latency.
		• Outlines solutions to the simultaneous mobility problem for network layer and application layer mobility protocols based on timer-based retransmission, forwarding, redirecting mechanisms, and simultaneous bindings.
9	Handoff optimization for multicast streaming	• Hierarchical scope-based multicast streaming architecture and implementation that offer local and global program management, and localized advertisement insertion using control information of real-time traffic (e.g., RTCP).
		• Reactive and proactive fast-handoff mechanisms using application layer triggering that reduce the join latency by a factor of 10 during subnet handover.
10	Cooperative roaming	• A novel approach, namely cooperative roaming, in which mobile nodes can collaborate with each other and share useful information about the network in which they move.
		• This achieves seamless L2 and L3 handoffs regardless of the authentication mechanisms used and without any changes to either the infrastructure or the protocol.
11	System evaluation	• Verification of experimental results from a few mobility systems that we built using optimization techniques for several handoff components that we developed, and validation of these results with results from corresponding Petri net models.
		• Behavioral analysis to study deadlocks and the effect of concurrency on handoff operations.
12	Conclusions and future work	We infer the best current practices for designing a mobility protocol based on the results obtained from our mobility taxonomy and systems optimization mechanisms.

2

Analysis of Mobility Protocols for Multimedia

Mobility management consists of two components: *location management* and *handoff management*. *Location management* enables the network to discover the current point of attachment of the mobile user so that a new connection can be established when a new multimedia call arrives. *Handoff management*, often known as terminal mobility, allows the network to maintain the user's connection binding as the mobile node moves from one attachment point to another in the network. We focus on *handoff management* in this book.

2.1 Summary of Key Contributions and Indicative Results

Over the last three decades, several generations of mobility protocols have evolved without any systematic design approach, and these protocols have used ad hoc mechanisms to optimize the handoff performance. Without any systematic analysis of the handover components and optimization mechanisms, it is difficult to predict the systems performance of these mobility protocols or design any new mobility protocol for the next generation of networks.

Here, we analyze the system architecture of each of the available mobility protocols (e.g., 1G, 2G, 3G, 4G, and several IP-based mobility protocols), describe the respective handoff mechanisms, and then compare the handoff mechanisms in terms of their common mobility functions. For example, we extrapolate how discovery, configuration, authentication, and media routing functions are performed for each of the cellular and IP-based mobility protocols and then map the respective network parameters for these mobility protocols to each of the common mobility functions.

There is no prior work that extrapolates these common mobility functions from the existing cellular and IP-based mobility protocols. A comparative analysis and extrapolation of the abstract primitives can determine the handoff functions that are needed to design a new mobility protocol with given resource parameters and design optimization mechanisms for these functions.

Mobility Protocols and Handover Optimization: Design, Evaluation and Application, First Edition.
Ashutosh Dutta and Henning Schulzrinne.
© 2014 John Wiley & Sons, Ltd. Published 2014 by John Wiley & Sons, Ltd.

2.2   Introduction

As a mobile goes through a handover process, it is subject to connection disruption because of the rebinding of its association in several layers of the protocol stack. The delays incurred because of rebinding in each of these layers affect the ongoing multimedia application and data traffic within the client. Several basic operations are associated with the reestablishment of the binding process across these layers. These operations can be affected by several factors, such as access characteristics (e.g., bandwidth or channel characteristics), the access mechanism (e.g., CDMA, CSMA/CA (carrier sense multiple access/collision avoidance), or TDMA (time division multiple access)), reconfiguration of identifiers, reauthentication, reauthorization, and rebinding of security associations in all layers.

Mobility protocols have evolved over the last three decades. Based on access characteristics and bandwidth, these can be classified into five main categories: 1G cellular, 2G cellular, 3G cellular, 4G, and IP-based mobility. The List of Abbreviations section defines 1G, 2G, 3G, and 4G. These mobility protocols exhibit some similar functionalities during their handoff operations. We shall highlight these similarities when we describe these protocols later in the chapter. Figure 2.1 shows the evolution of access technologies with the generations of mobility standards. Table 2.1 shows details of the access characteristics, frequency, and data rate of these protocols. In this section, we briefly discuss how mobility operations are performed in 1G, 2G, 3G, and 4G access networks, including IP-based mobility protocols. We analyze the associated abstract functions in Chapter 3.

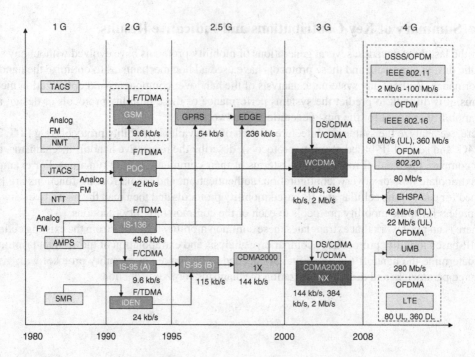

Figure 2.1   Evolution of wireless access technologies

Table 2.1 Access characteristics of cellular protocols

Generation	System	Channel spacing	Access type	Uplink data rate
1G	AMPS	30 kHz	FDMA	N/A
	TACS	25 kHz	FDMA	N/A
	NMT	25 kHz	FDMA	N/A
	NTT	25 kHz	FDMA	N/A
2G	GSM	200 kHz	TDMA	9.6 kb/s
	PDC	30 kHz	TDMA	42 kb/s
	IS-136	30 kHz	F/TDMA	48 kb/s
	IS-95 (A)	1.25 MHz	F/CDMA	14.4 kb/s
	iDEN	25 kHz	F/TDMA	24 kb/s
2.5G	GPRS	200 kHz	TDMA	45 kb/s
	EDGE	200 kHZ	TDMA	236 kb/s
	IS-95 (B)	1.25 MHz	F/CDMA	115 kb/s
	CDMA2000 1X	1.25 MHz	CDMA	144 kb/s
3G	UMTS/WCDMA	5 MHz	CDMA/TDMA	2 Mb/s
	CDMA2000 1xEV-DO	1.25 MHz	CDMA	2 Mb/s
4G	LTE	20 MHz	OFDMA	50 Mb/s
	WiMAX	2.5 GHz	OFDM	40 Mb/s
	UMB	5 MHz	OFDMA	75 Mb/s

2.3 Cellular 1G

1G refers to the first generation of wireless telephone technology. 1G cellular is based on analog technology; it was introduced in the 1980s and continued until it was replaced by 2G digital telecommunications. Several 1G cellular standards were developed in different countries. One such standard is NMT (Nordic Mobile Telephone), used in the Nordic countries, Eastern Europe, and Russia. Others include AMPS (Advanced Mobile Phone System), used in the United States; TACS (Total Access Communications System), used in the United Kingdom; JTACS, used in Japan; C-Netz, used in West Germany; Radiocom 2000, used in France; and RTMI (Radio Telefono Mobile Integrato), used in Italy. In 1979, the first analog cellular system, the Nippon Telephone and Telegraph (NTT) system, was started. In 1981, Ericsson Radio Systems AB deployed the Nordic Mobile Telephone 900 system, and in 1983 ATT deployed the Advanced Mobile Phone Service as a trial in Chicago. However, the two most important 1G systems deployed in the world are AMPS and TACS. All the 1G systems use frequency division multiple access but each system works in a different frequency range. For example, AMPS operates on the 800 MHz frequency band, whereas TACS and NMT450 operate on the 900 MHz and 450 MHz frequency bands, respectively. We now briefly introduce the 1G architecture and describe how handoff is performed in a 1G network.

2.3.1 System Architecture

Figure 2.2 shows the simple system architecture of a 1G system. The main components of the system are the mobile host (MH), base station (BS), base station controller (BSC), and mobile

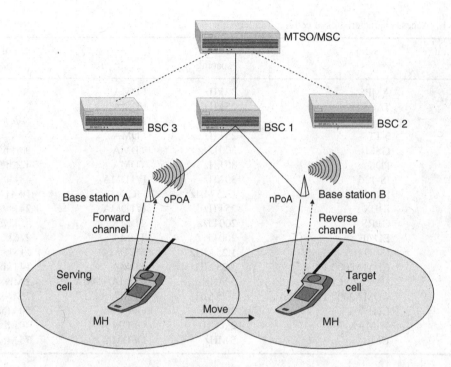

Figure 2.2   First-generation cellular architecture

switching center (MSC) (often known as the mobile telephone switching office (MTSO) in AMPS). Base stations are considered to be the point of attachment (PoA) for the mobile host in a specific radio cell. The MSC acts like a mobility anchor agent in the network. Definitions of these terms are given in the List of Abbreviations and Appendix B.

A mobile is assigned a mobile identifier number (MIN) that is equivalent to its home address. This includes an area code identifying the home address area, a three-digit exchange number, and a four-digit subscriber identification number. Each mobile also has an electronic serial number (ESN) that is permanently assigned by the manufacturer. This is equivalent to a device identifier and helps to secure the mobile. While voice traffic is transmitted using a traffic channel, signaling among the network components is done by control channels that are used to initiate calls and for handoff-related operations. There are two kinds of control channel: forward and reverse. For AMPS, the forward channel is known as the forward control channel (FOCC), and the other is the reverse control channel (RECC). The FOCC is transmitted continuously by the base station so that it can be received by all of the mobiles. A mobile uses the FOCC to associate with the network and perform a handover. The mobile uses the RECC to register with the network.

The network and base stations are provided with identification codes. For example, in the USA, the system identifier (SID) is assigned by the FCC along with the carrier frequencies for each geographical area. The mobile host uses the system identifier to ensure that it is on the correct network. The MSCs are linked together to provide a fully integrated network. Base station controllers typically control a small number of base stations and are linked either by a wire land line or, often, by a short-range microwave link.

2.3.2   Handoff Procedure

First-generation cellular networks follow network-initiated handover. During the call, the base station network entity measures the signal strength from a mobile device as received at the serving cell and passes this to the MSC. As the signal-to-noise ratio begins to fall below a certain threshold, the nearby cells are requested to perform signal strength measurements. The scanning receiver in the BS measures the signal strength of the MSs in the neighboring cells and reports it to the MSC. If the signal received is better in another base station, and if the MSC decides that a handover is necessary, it instructs the neighboring BS to allocate a channel to the mobile. A second voice path is then set up and bridged across to the existing one in preparation for the handoff. Once this is complete, the system or network generates a handoff instruction and sends it over the FOCC. This handoff instruction carries information that includes the new channel to be used, along with the SAT (supervisory audio tone). The mobile accepts the information and sends an ST (signaling tone) for 50 ms. It then turns off the reverse channel transmission and retunes to the new channel. Once it reaches the new channel, the mobile turns its voice transmitter channel transmission on and retransmits the SAT of the new base station. This acts as a confirmation that the new link has been established. Accordingly, when the new base station receives this, it informs the mobile switching center that the handoff has been successfully accomplished. The new base station then informs the old base station that it should release the channel that the mobile was previously using and make it available for further use. Thus, the original voice path via base station A is rerouted via base station B as shown in Figure 2.2.

2.4   Cellular 2G Mobility

A variety of second-generation digital cellular systems were developed in different parts of the world during the early 1990s. These include the GSM/DCS1800/PCS1900 standard in Europe, the PDC standard in Japan, and the IS 54/136 and IS-95 standards in the USA. In this section, we briefly discuss two second-generation mobile systems, namely GSM (Global System for Mobile communication) and IS-95, and highlight their handover procedures.

2.4.1   GSM

GSM is considered to be a 2G cellular system that is a natural evolution from the TACS (Total Access Communication System) mostly used in European countries.

2.4.1.1   System Architecture

Figure 2.3 shows a generalized architecture for GSM. The main elements of the system are the base transceiver station (BTS), the base station controller, the mobile switching center, and the registration and authentication components. The BTS and BSC form the radio access network (RAN) part of the system. Registration functionalities are provided by the home location register (HLR) and the visitor location register (VLR). The equipment identity register (EIR) checks the validity of a mobile by checking its International Mobile Equipment Identity (IMEI). The home location register contacts the authentication center (AuC) in order to authenticate the mobile. These terms are defined in Appendix B.

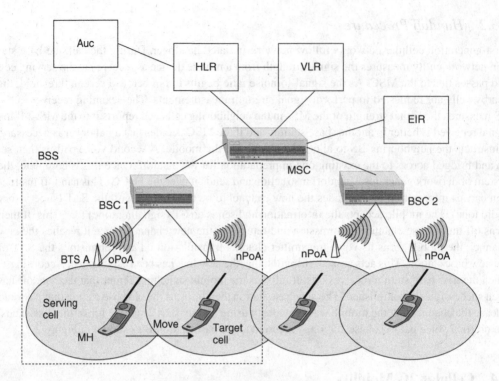

Figure 2.3 GSM-based mobility

A mobile node can be associated with several types of identifiers, each with its own function. The SIM (subscriber identity module) is a small memory card that provides information about the subscriber. The IMEI is a 15-digit number used to identify the equipment and is used by the EIR to decide if the mobile is properly authenticated. The IMSI is like the home address of the mobile. It enables the operator to link the phone number and the subscriber. The TMSI (temporary mobile subscriber identity) is like a care-of address for the mobile that is assigned when the mobile visits a network. The TMSI changes whenever the mobile changes its network. In order to provide proper authentication to the mobile, an authentication key is stored on the SIM card and is used to compute a cipher key (Kc). The cipher key is used in an encryption algorithm to prevent unauthorized listening to the mobile message.

There are a variety of control channels that are used to provide the required functionality to enable mobiles and the BTS to communicate, set up, and manage calls. These can be split into groups, namely three broadcast channels for initial synchronization, three common control channels for initiating calls, and three dedicated control channels to manage calls. The BCCH (broadcast control channel), synchronization channel (SCH), paging channel, slow associated control channel (SACCH), and fast associated control channel (FACCH) are mostly used for handoff management. The SCH provides the BTS identification and allows the mobile to become associated with the BTS. The BCCH is continually broadcast on the downlink channel by the BTS and contains information including the base station id and frequency allocations. The paging channel is used to locate the mobile when there is an incoming call. The downlink SACCH provides beacon frequencies of the neighboring cells, and the uplink SACCH includes a measurement report that gives strength measurements for signals received from beacon transmissions from neighboring cells.

We discuss the handoff procedure associated with GSM in the next section.

2.4.1.2 Handoff Procedure

Unlike handoff in 1G cellular systems, handover (handoff) in GSM can be either network-initiated and mobile-assisted or mobile-initiated and network-controlled. Network-initiated-handoff is triggered by the network based on radio subsystem criteria such as RF (radio frequency) signal-to-noise ratio and distance from the base station, as well as network-directed criteria such as the current traffic load per cell. In order to support a handover that is based on signal-to-noise ratio criteria, the MS takes radio measurements from neighboring cells and reports these to the serving cell on a regular basis. When the network determines that there is a need for a handover, appropriate handoff procedures are followed.

Connections in GSM may be handed off between radio channels in the same cell, between channels in different cells with the same base station subsystem (BSS) coverage, or between cells under the coverage of different BSSs or even different MSCs. Based on the type of movement, handoffs can be categorized into two types, namely *internal-connection handoff* and *external-connection handoff*. As part of the internal-connection handoff technique, the BSS may handle the connection handoffs autonomously in the same cell or between cells under its own coverage. The MSC is involved in managing handoffs that need to take place between cells under the coverage of two different BSSs. This process is called external-connection handoff. In general, when the BSS indicates that an external handover is required, the decision about when and whether an external handover should occur is taken by the MSC. The MSC uses the signal quality measurement information reported by the mobile stations, which is preprocessed at the BSS for determination of the external handover. The details of connection handoff were discussed by Rahnema (1993).

Figure 2.4 shows an example of GSM-based mobility. It illustrates intra-BSC, inter-BSC, and three different types of inter-MSC handover, namely anchor-to-relay, relay-to-relay, and relay-to-anchor.

The MSC acts like an anchor agent during the duration of a call. During an external handover, the original MSC handling a call keeps control of the call when the mobile is handed over to the target MSC or even to a subsequent MSC. When the BSS performs an internal-connection handoff, it informs the MSC when the process is completed. Either the mobile or the BSS indicates an impending need for connection handoff by way of the FACH (forward access channel). The BSS usually monitors the quality of the radio signal received and transmits the results to the MSC, which keeps a more global view of the radio channels belonging to its BSSs. The MSC may also initiate a need for a connection handoff in an attempt to balance out traffic load in the network.

Table 2.2 shows how different mobility-related tasks such as radio channel measurement, requesting a handover, and confirming a handover are taken care of by the network elements, namely the MS, BS, and MSC, in a distributed manner.

2.4.2 IS-95

Interim Standard Number 95 (IS-95) was developed by Qualcomm, and was rebranded as CDMAOne in 1997. IS-95 is based on a spread-spectrum technology platform that enables multiple users to

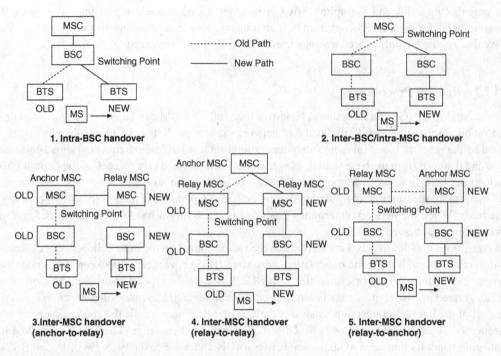

Figure 2.4   Types of GSM handover

Table 2.2   Distributed mobility functions for GSM networks

Task	MS	Old base station	New base station	MSC
Radio channel measurements	Make periodic measurements on current and neighboring channels. Send results to BS.	Monitor backwards channels. Give measurement order to MS.		
	Start measurements. Send results to BS.			
Issue handover request		Send measurement results to MSC. Request handover.		Evaluate handover request. Inform new BS
		Evaluate handover requests. Request handover.		Inform new BS
Confirm/disconfirm handover			Accept, block, delay handover request	Permit, drop, delay handover

occupy the same radio channel or frequency spectrum at the same time. IS-95 has two distinct versions, IS-95A and IS-95B. IS-95B provides additional ISDN-like data rates.

2.4.2.1   System Architecture

Figure 2.5 shows the system architecture of the IS-95 system. This is based on the TR-45 (Telecommunication Industry Association 2013) reference model. The main components of this architecture are the mobile station, base station, mobile switching center, home location register, visited location register, authentication center, and equipment identity register. The base station consists of a base transceiver system and base station controller. Besides the switching functions, the MSCs behave as anchor agents during handoff and help to direct voice traffic. In terms of their functionality, handoff-related MSCs can be categorized as anchor MSCs, border MSCs, candidate MSCs, originating MSCs, target MSCs, serving MSCs, and tandem MSCs. An anchor MSC is the first MSC that gives radio contact in a call, a candidate MSC could possibly accept a call or a handoff, a serving MSC currently provides service to a call, a tandem MSC provides a trunk connection for a call in which a handoff has occurred, and a target MSC is the MSC selected for a handoff.

The channels in CDMAOne can be split into a forward link channel and a reverse link channel. For the forward link channel, there are four types of coded channels originating from the base station: pilot, sync, paging, and traffic. For the reverse link channel, there are two types of coded channels originating from the MS: access and traffic. The pilot, sync, paging, and access channels carry the necessary control data, while the traffic channels carry digital voice. The pilot channel (PC) provides

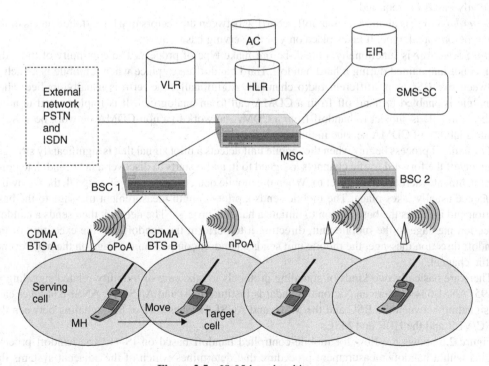

Figure 2.5   IS-95-based architecture

the MS with a beacon, a timing and phase reference, and the signal strength for power control. The pilot channel is transmitted continuously by each sector of a base station. A mobile uses the pilot channel to identify a cell and can identify the strongest sector within a cell based on a measurement of the signal-to-noise ratio of the pilot signal. This channel is useful for initiating a handoff operation. The sync channel (SC) provides the MS with critical time synchronization data. The message on the SC contains information necessary for the MS to align its timing with the pilot channel. A mobile uses the sync channel to discover the network and its parameters. The paging channel contains messages with parameters that the MS needs for access and paging. The messages convey system parameters, access parameters, and the neighbor list. This channel is used to communicate with the MS when there is no call in progress. The paging channel can be used to locate a mobile.

For the reverse link, there are two basic channels, namely the access channel and the reverse traffic channel. A mobile uses the access channel to communicate with the base station when no traffic channel has been set up, and the reverse traffic channel to send date and control signals to the base station.

2.4.2.2 Handoff Procedure

CDMAOne supports three main types of handover: soft handover, softer handover, and hard handover.

Soft handover involves an intercell handoff and is a make-before-break connection. The connection between the mobile and the network is maintained by several base stations during the process. A soft handoff occurs only when the old base station and the new base stations are operating on the same CDMA frequency channel. The mobile communicates simultaneously with these BSs until it is clear that only one BS is required.

Softer handover is an intracell handoff occurring between the sectors of a base station and is of the make-before-break type. It takes place only at the serving base station.

Hard handover is functionally a break-before-make type of process. The continuity of the radio link is not maintained during a hard handoff. Hard handoff takes place when a mobile is switched between two BSs using different radio channels. Additionally, the term is usually applied when a mobile is enabled to hand off from a CDMA call to an analog call. It is implemented in areas where a mobile is subject to handoff from a CDMA network to a non-CDMA network because of nonavailability of CDMA service in that area.

The handoff process begins when the mobile unit detects a pilot signal that is significantly stronger than any of the forward traffic channels assigned to it, and it starts to discover a new candidate point of attachment where it can connect to. When the mobile detects a stronger pilot signal, the following sequence usually takes place. The mobile sends a pilot strength measurement message to the base station and instructs the base station to initiate a handoff process. The network then sends a handoff direction message to the mobile unit, directing it to perform the handoff. On the execution of the handoff direction message, the mobile unit sends a handoff completion message on the new reverse traffic channel.

There are basically two kinds of signaling protocols to take care of mobility-related signaling in IS-95: ANSI-634 (American National Standards Institute-634) and ANSI-41. ANSI-634 takes care of signaling between the BSC and the MSC, and ANSI-41 takes care of the signaling between the MSC/VLR and the HLR and MSCs.

Figure 2.6 shows the flow for mobile-controlled handoff based on IS-41. The handoff process begins with a handoff measurement procedure that determines which of the adjacent systems the

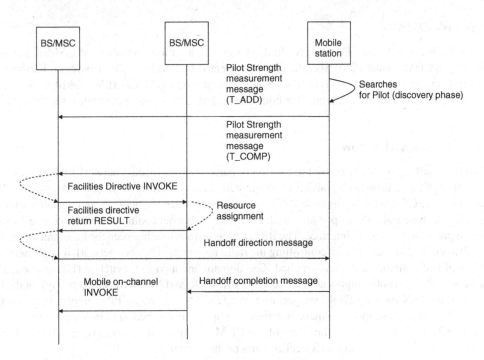

Figure 2.6   IS-41-based mobility

mobile should switch to. IS-41-based intersystem handoff comprise two types of handoff, namely *handoff forward* and *handoff back*. Handoff forward defines a type of intersystem handoff where a mobile is handed off from one mobile switching center (MSC A) to another (MSC B), where MSC A is the serving MSC. Handoff back is the process of handing back to the serving MSC.

IS-95 also offers an automatic roaming functionality, by which it allows a mobile to originate a call in a visited system and receive a call destined for a roaming subscriber. Three basic processes, namely registration, call origination, and call delivery, constitute the roaming function.

2.5   Cellular 3G Mobility

The main vision of third-generation mobile networks (3G networks) is to provide ubiquitous wireless network access that can support voice, multimedia, and high-speed data communication. The IMT-2000 (International Mobile Telephone 2000) standards were developed by the International Telecommunications Union Radio Communication sector (ITU-R) and the International Telecommunication Union Telecommunications sector (ITU-T) to define the operations of 3G networks. One of the main attributes of IMT-2000 is the introduction of wireless wideband packet-switched data services for wireless access to the Internet with speeds of up to 2 Mb/s. IMT-2000 proposed ten different multiple access schemes. Two of these schemes were based on TDMA and the remaining eight on CDMA. We describe two of these access schemes, namely WCDMA and CDMA2000.

2.5.1   WCDMA

WCDMA (Wideband Code Division Multiple Access) is a natural evolution of GSM-based systems. The WCDMA networking specification was created by 3GPP (Third Generation Partnership Project). Compared with the 2G version of cellular systems and CDMA2000, WCDMA uses a much wider spectrum of one 5 MHz channel for both data and voice, providing data rates of up to 2 Mb/s.

2.5.1.1   System Architecture

WCDMA-based 3G networks consist of two major parts: the radio access network and the core network. The radio access network consists of a base station and a radio network controller (RNC). The base station in 3G (also known as node B) is an interface between the network and the WCDMA air interface. The base station is responsible for taking care of channel coding, interleaving, rate adaptation, and processing of the air interface. The RNC acts as an interface between the base station and the core network. It is responsible for controlling the radio resources. The core network in a 3G network consists of two domains: a circuit-switched (CS) domain and a packet-switched (PS) domain. The core network consists of components such as the VLR, HLR, and MGW (media gateway) on the CS side and the SGSN (serving GPRS support node) and GGSN (gateway GPRS support node) on the PS side. A detailed description of these functional components can be found in Tachikawa (2002).

Figure 2.7 shows the system architecture of a WCDMA network. It shows the components of both the packet-switched and the circuit-switched parts of the network.

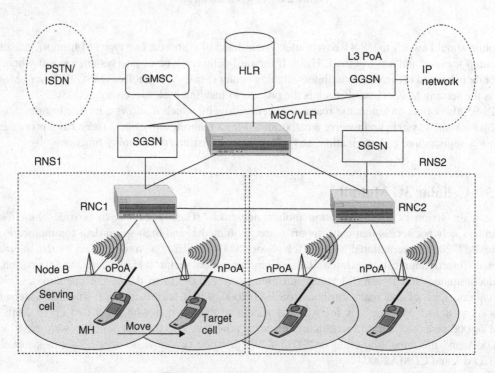

Figure 2.7   WCDMA architecture

2.5.1.2 Handoff Procedure

In this section, we describe the handoff procedure associated with WCDMA. We briefly discuss the mobility management of a WCDMA system, for both the circuit-switched and packet-switched domains. Mobility management for the circuit-switched domains in WCDMA consists of call origination, termination, and handover control. In WCDMA, the mobile node is also called the UE (user equipment). As part of the location-updating process during the *attach* procedure, a temporary mobile subscriber identity (TMSI) is assigned to the UE, which is identified by an International Mobile Subscriber Identity (IMSI). The IMSI is allocated permanently to each UE. However, every time a mobile changes its network, it obtains a new TMSI and it binds it to the IMSI. This process can be defined as the general binding update process of the handoff procedure.

The handover decisions are generally handled by the RNC (radio network controller). The RNC continually monitors information regarding the signals being received by both the mobile node and node B. When the quality of a particular link falls below a given level and another, better radio channel is available, the mobile initiates a handover. As part of this monitoring process, the UE measures the received signal code power (RSCP) and received signal strength indicator (RSSI), and information is then returned to node B and hence to the RNC on the uplink channel.

There are several cases of handovers in WCDMA based on the movement pattern of the UE with respect to the cell and other nodes (e.g., node B and the RNCs) within a UTRAN (UMTS Terrestrial Radio Access Network). WCDMA-based handoff can be categorized into three types: soft handover, softer handover, and hard handover.

Soft handover
Soft handover takes place when the operating frequency remains the same between neighboring cells. Several types of soft handover are discussed here.

1. *Intra-node B/intra-RNS (radio network subsystem).* This type of handover is done if the UE moves from one cell to another cell where both belong to the same node B.
2. *Inter-node B/inter-RNS/intra-SGSN.* This type of handover is done if the UE moves from a cell in one node B to a cell in another node B that belongs to a different RNS.
3. *Inter-node B/inter-RNS/inter-SGSN.* In this case, the UE moves from a cell in one node B to a cell in another node B that belongs to a different RNS. The node Bs are connected to different RNCs, and these two RNCs are also connected to different SGSNs. In this case, the UE may even move between two SGSNs.
4. *Inter-GGSN.* The GGSN is the layer 3 point of attachment that acts as the default router. Unless the mobile moves between GGSNs, the layer 3 identifier of the mobile does not change.

Softer handover
When the UE moves between two different sectors on the same base station, this is called softer handover or handoff. Since the processing elements are shared, this enables a softer handoff to be accomplished more easily than a soft handoff.

Hard handover
Hard handover (or interfrequency handover) is only needed if the mobile needs to change its frequency after the handover or an interface does not exist between two RNCs in the case of a soft handover (inter-node B/inter-RNS). A frequency change may take place as a result of change of

W-CDMA cell level, that is, from a macrocell to a satellite, or another change of the radio access technology (inter-RAT handover), for example from a UMTS to a WLAN or GSM network. A hard handover occurs quite rarely and differs a lot from the soft handover types described above.

2.5.2 CDMA2000

CDMA2000 builds on CDMAOne to provide an evolution path to 3G. Like WCDMA, it supports two kinds of access: DS/CDMA, based on FDD (frequency division duplex), and T/CDMA, based on TDD (time division duplex). There are several versions of CDMA systems available, namely CDMA2000 1X, CDMA 1xEV-DO (Evolution Data Only), and CDMA 1xEV-DV (Evolution Data and Voice).

2.5.2.1 Architecture

In contrast to WCDMA, CDMA2000 uses a spectrum of 1.25 MHz per channel. CDMA2000 1X doubles the voice capacity of the CDMAOne system and also supports high-speed data services. It can support a peak data rate of up to 153 kb/s. CDMA 1xEV-DO is an evolution of CDMA 2000 that is designed for data-only use, and it provides a peak rate capability of over 2.45 Mb/s on the downlink. CDMA2000 1xEV-DV is an evolution of CDMA2000 that can simultaneously transmit voice and data. The peak data rate is 3.1 Mb/s on the forward link, and the reverse link is limited to 384 kb/s.

In order to be able to support packet-based services more efficiently, CDMA2000 upgrades some existing network elements of IS-95, namely the BTS and BSC, and adds many new network elements, such as a packet data serving node (PSDN); authentication, authorization, and accounting (AAA); and a home agent (HA). The BSC is equipped with additional IP routing functionality and the new PDSN establishes, maintains, and terminates point-to-point sessions with the subscriber and initiates AAA for the mobile station client to the AAA server. In addition, it also augments the functionality of the HLR and VLR. For an architecture based on CDMA2000 in conjunction with legacy network components such as MSCs and HLRs, signaling between the BSC and MSC is taken care of by the TIA-2001 standards (http://www.tiaonline.org), and signaling between the MSC, VLR, and HLR and between MSCs is taken care of by the ANSI-41 standards. TIA-2001 describes the overall system functions, including the services and features required for interfacing a base station with the MSC, with other base stations, and with the packet control function (PCF), and for interfacing the PCF with the PDSN. The mobile node usually initiates a data call via the BTS, which is the first point of attachment to the network. The BSC responsible for this BTS forwards the call to the associated PCF. The PCF selects a PDSN based on certain unique characteristics of the mobile and establishes a GRE (generic routing extension) tunnel with the PDSN. At that point, the mobile node initiates a PPP session that is terminated at the PDSN. Thus, there is single-hop IP connectivity between the mobile and the PDSN.

2.5.2.2 Handoff

Compared with IS-95, CDMA2000 introduces additional levels of thresholds to limit the frequency of handovers. By limiting unnecessary handovers, it increases the capacity of the system for data traffic. Like WCDMA, CDMA2000 supports three types of handoff operations: hard handoff, soft handoff, and softer handoff. Figure 2.8 shows a hierarchical network architecture with all the

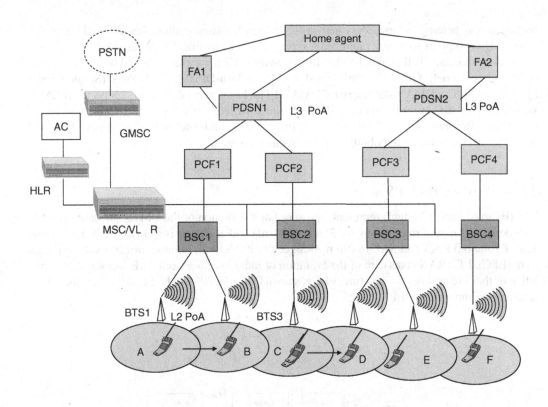

Figure 2.8   CDMA2000 architecture

networking elements of a CDMA2000 network. A node's mobility is taken care of at each level of the hierarchy, namely the BTS, PCF, and PDSN levels. There are several movement scenarios that can take place within this hierarchy. When the mobile moves between two BTSs, as long as the movement is confined to the PDSN the mobile does not need to set up a new PPP session. But if the mobile moves between two BTSs that are controlled by two different PCFs, each PCF may choose a new PDSN in the hierarchy. This will require the PPP session to be terminated and a new one to be started. When the mobile chooses a new PDSN to reestablish a new PPP session, it obtains a new layer 3 identifier such as an IP address. During packet-based communication, a mobile node is identified using a simple IP addressing scheme or Mobile IP. In the case of a simple IP addressing scheme, the mobile obtains an IP address from a DHCP server, which is usually co-located with the PDSN. This identifier changes as the mobile moves between PDSNs. However, if the mobile uses a Mobile IP-based approach, this IP address does not change. If the IP address changes, layer 3 mobility can be taken care of by mobility protocols such as Mobile IP in the PDSN layer as discussed in Section 2.4.1.

2.6 4G Networks

4G refers to the fourth generation of cellular wireless and is an evolution of 3G networks. 4G networks provide additional features such as higher data rates of up to 100 Mb/s, seamless mobility support across heterogeneous access networks, secured IP-based communication, and quality of service (QoS) support for many multimedia services such as mobile TV, MMS (Multimedia Messaging Service), and DVB (Digital Video Broadcasting). 4G networks are based on several access technologies, namely OFDMA (Orthogonal FDMA), Flash-OFDM, SC-FDMA (Single-Carrier FDMA), and MC-CDMA (Multicarrier CDMA). LTE (Long Term Evolution) supports OFDMA, WiMAX (IEEE 802.16) supports Flash-OFDM, and IEEE 802.20 supports MC-CDMA as access techniques. However, we shall describe two specific 4G candidate networks, namely LTE and WiMAX, and the associated mobility features in this section.

2.6.1 Evolved Packet System

EPS (Evolved Packet System) represents the very latest evolution of the UMTS standard. EPS has been defined as part of release 8 by the 3GPP standards bodies. It consists of two parts: LTE (Long Term Evolution) RAN and SAE (System Architecture Evolution), often known as Evolved Packet Core (EPC). LTE RAN takes care of the evolution of radio interfaces and SAE focuses on the evolution of the core network architecture. EPS is shown in Figure 2.9. We briefly describe some of the functional components of EPS.

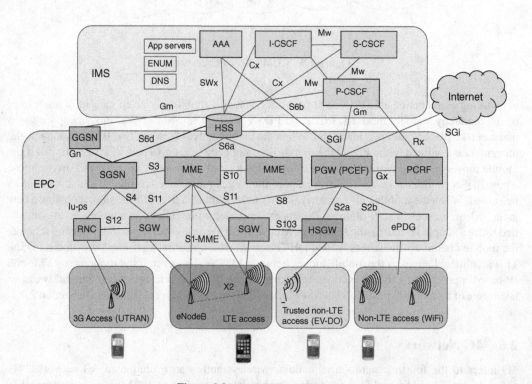

Figure 2.9   SAE/LTE architecture

2.6.1.1 System Architecture

LTE's radio access network, which is also known as Enhanced UTRAN (E-UTRAN), defines a packet-optimized access network that can efficiently support IP-based real-time and non-real-time services, providing performance comparable to circuit-switched networks. Unlike UTRAN, E-UTRAN relies on a fully shared radio resource allocation scheme that allows resource usage to be maximized by combining all radio bearers on a shared high-bit-rate radio channel.

The goals of E-UTRAN are to provide reduced latency, higher data rates, improved system capacity, and improved coverage, as well as reduced cost for the operator and packet-optimized radio access technology. The most important component of E-UTRAN is eNode-B (eNB). Unlike UMTS, there is no separate RNC, but RNC functionality has been integrated into eNode-B.

SAE also defines a simplified core network composed of one packet domain that can support all packet-switched services and interworking capabilities with traditional PSTN. EPS represents a migration from the traditional hierarchical system architecture to a flattened architecture that minimizes the number of communication hops and distributes the processing load across the network. SAE is also known as EPC. EPC is composed of several functional entities, namely the MME (mobility management entity), the serving gateway (S-GW), the PDN-GW (packet data network) gateway, and the PCRF (policy and charging rules function). The MME is in charge of all the control plane functions and provides features such as security procedures, terminal-to-network session handing, and idle-terminal location management. The serving gateway is the termination point of the packet data interface towards E-UTRAN and serves as a local mobility anchor supporting intra-E-UTRAN mobility and mobility with other 3GPP technologies such as 2G-based GSM and 3G-based UMTS. The PDN-GW is the termination point of the packet data interface towards the packet data network. The PDN-GW supports policy enforcement features, packet filtering, and enhanced charging support. The PDN-GW also acts like a home agent.

EPS provides much better performance than the previous releases of UMTS. A peak date rate of up to 100 Mb/s downlink and 10 Mb/s uplink can be obtained with E-UTRAN. It offers a radio access network latency of 5 ms in the user plane and 100 ms in the control plane. Latencies for inter-RAT (radio access technology) are restricted to 300 ms for real-time traffic and 500 ms for non-real-time traffic.

2.6.1.2 Handoff Procedure

LTE supports mobility in both the radio layer and the network layer. It supports UE-assisted network-controlled hard handover when the UE is in active mode. There are two types of handoff defined for LTE: intra-MME and inter-MME. During intra-MME handoff, the mobile moves between two eNBs that are connected to the same MME. Inter-MME defines a handover when the eNBs are controlled by two different MMEs. Two eNBs can be served by two different serving gateways.

LTE supports mobility with both 3GPP and non-3GPP access systems. Mobility across 3GPP access systems is called *local mobility* and mobility across non-3GPP access systems is called *global mobility*. Non-3GPP access systems include wireless systems such as IEEE 802.11 and IEEE 802.16. Both mobility based on GTP (GPRS Tunneling Protocol) mobility and Mobile IP-based mobility are used to support local mobility. However, only Mobile IP-based mobility is used to support global mobility.

We describe in the following the specifics of a handoff process involving E-UTRAN. Handoff management within an LTE network involves several stages, namely discovery, measurement, handover preparation, and, finally, execution of the handover. The UE, which is equivalent to a mobile host, can be identified using a list of identifiers that are used to assist handoff. The IMSI and IMEI provide the subscriber ID and equipment ID, respectively. The M-TMSI (M-temporary mobile subscriber identity) is used to identify the UE within the MME. The GUTI (globally unique temporary UE identity) is allocated by the MME. The GUTI identifies a globally unique MME and the UE within the MME. The UE is assigned an S-TMSI (S-temporary mobile subscriber identity), which is a shortened version of the GUTI and is used to locate the UE. The UE is also assigned an ID that uniquely identifies the UE in a tracking area. An eNodeB sends a tracking area identifier (TAI) on the broadcast control channel. Upon power on, the UE performs a cell search, discovers the EPS/LTE access system, and performs access system and network selection. After the cell is discovered, the UE attaches to the radio network and is authenticated with the MME. After a successful authentication, the MME registers with the HSS (home subscriber server) as the serving UE in the HSS. Network layer configuration is done by the PDN-GW so that the mobile can configure itself with an IP address and the default router.

Figure 2.10 describes an intra-MME handover process where the UE is handed off from eNB1 to eNB2, which are part of the same MME. It shows the signaling between the eNBs, MME, and S-GW.

Based on a certain policy, the UE sends a measurement report to the source eNB or serving eNB. The source eNB makes a decision based on the measurement report and radio resource management (RRM) function to hand over the UE to the target eNB. The source eNB sends a handover request to the target eNB and passes the necessary QoS information that it presently supports for the UE. This ensures that the target eNB configures the required resources for the UE when it arrives in the target network. A GTP-U tunnel is also set up between the source eNB and target eNB during

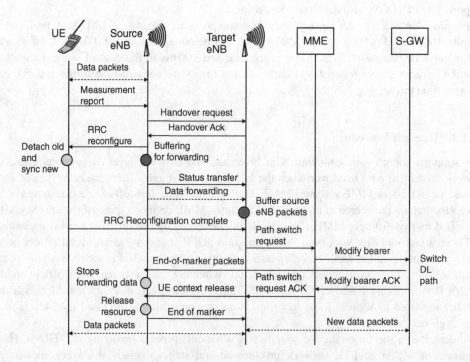

Figure 2.10 Call flow for intra-MME handoff

this time. The mobile is then assigned radio resources by the target eNB. During this time, the data destined for the UE is buffered in the source eNB. Once all the details of the UE, such as the UL (uplink) PDCP sequence number receiver status and DL (downlink) sequence number transmitter status, are known to the target eNB, the target eNB receives the buffered data from the source eNB and buffers it locally for further delivery to the UE at a later time. Once the UE has connected to the target eNB successfully, the target eNB informs the MME that it should do a path switch. On receiving the path switch request, the MME communicates with the S-GW to modify the bearer through a "modify bearer" request. The S-GW transmits end-of-marker packets to the source eNB before starting transmission on the new path. This helps the source eNB to stop forwarding packets to the target eNB. The target eNB delivers the buffered packets before delivering packets on the new downlink path. After the new path has been established, the source eNB releases its resources. Thus, the new data is delivered to the UE via the target eNB. Buffering at the source eNB and target eNB helps to reduce the packet loss during handoff.

2.6.2   WiMAX Mobility

The WiMAX NWG reference model defines three main network components, namely the mobile station (MS), access service network (ASN), and connectivity service network (CSN), that participate in the handoff process. The MS provides user access to the WiMAX network. The ASN consists of base stations and ASN gateways (ASN-GWs). The CSN provides IP connectivity to the radio equipment and acts like a gateway to the Internet. There are two types of handover scenario in WiMAX, namely ASN-anchored mobility and CSN-anchored mobility. Figure 2.11 illustrates ASN-anchored and CSN-anchored mobility scenarios. ASN-anchored mobility is defined

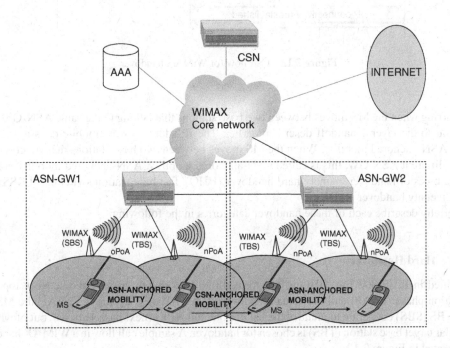

Figure 2.11   WiMAX handover

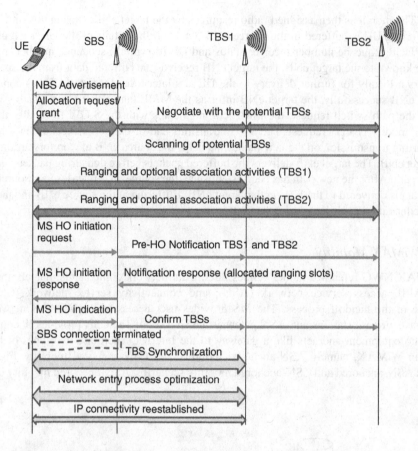

Figure 2.12   Call flow for WiMAX handover

as occurring when the MS moves between two base stations that belong to the same ASN-GW. This is similar to the layer 2 handoff described earlier. The IP address of the mobile does not change during ASN-anchored mobility. When the MS moves between two base stations that are controlled by two different ASN-GW, this is called CSN-anchored mobility. ASN-anchored mobility consists of three types of handover, namely hard handover (HHO), fast base station switching (FBSS), and macrodiversity handover (MDHO).

We briefly describe each of these handover categories in the following.

2.6.2.1   Hard Handover

As defined in IEEE 802.16e, HHO is categorized into two phases, namely the network topology acquisition phase (NTAP) and the actual handover phase (AHOP). During the NTAP, the MS and service BS (SBS) cooperate to identify the list of potential NBSs (next base stations), out of which a particular target base station (TBS) is chosen for handover. A sample call flow in a WIMAX handover is illustrated in Figure 2.12.

1. *Network Topology Acquisition Phase*. The following is the sequence of messages during the NTAP phase:
   (a) *Network topology advertisement*. The serving base station uses a MOB_NBR-ADV (mobile neighbor advertisement) signaling message to broadcast information about the state of the NBSs.
   (b) *MS scanning procedure*. The MS scans the advertised BSs and then selects suitable BSs for the handover. This procedure is carried out with the help of scanning interval allocation request and response messages (MOB_SCN-REQ and MOB_SCN-RSP).
   (c) *Ranging and optional association activities*. The scanning is followed by contention/ noncontention ranging activities, through which the MS gathers further information about the PHY channel related to the selected TBSs. The MS uses a ranging request (RNG_REQ) and ranging response (RNG_RSP) for this purpose.
2. *Actual handover phase*. During this phase, the MS switches location from the SBS to the TBS. The following are the various steps associated with the actual handover phase:
   (a) *Handover initiation*. During the process of deciding on the TBS, the MS decides the final TBS to hand over to. This can be initiated by the MS or the SBS. If it is initiated by the MS, then the MS communicates a MOB_MSHO-REQ message containing the list of selected TBSs to the SBS and the SBS replies with a MOB_BSHO-RSP message. If the decision is made at the SBS, a MOB_BSHO-REQ message is used. Once a specific TBS has been selected from the list of suitable candidate TBSs, the MS informs the current SBS about the beginning of handover activity by sending a MOB_HO-IND (mobile handover indication) message.
   (b) *TBS synchronization and ranging process*. During this process, appropriate synchronization and ranging activities take place with the TBS.
   (c) *Authentication, authorization, and registration phase*. As part of the next step, authentication, authorization, and registration processes take place with the TBS. After this process is complete, the mobile is reconnected with the new base station and IP connectivity to the new network is restored.

2.6.2.2 Soft Handoff

In the case of the two optional handover approaches of MDHO and FBSS, the MS communicates simultaneously with multiple BSs using the concept of the diversity set (DS) and anchor BS (ABS). Based on the signal strength, the DS includes the most active NBSs that could be involved during handover. The most active BS, which has the most powerful signal strength, is chosen as the ABS. In the case of MDHO, the MS communicates with all the BSs within a DS. In the case of FBSS, each MS communicates with the ABS. As the mobile moves, based on the signal strength of the neighboring BSs, the DS and ABS are continuously updated. In the case of MDHO, during the handover process, the MS simultaneously receives and transmits data from the multiple BSs that are part of the DS. In the case of FBSS, the MS and the current ABS jointly select the target ABS. During the BS switching, the MS remains connected to the current and target ABSs.

Compared with HHO, the MDHO and FBSS approaches are designed to provide full, seamless mobility at higher speeds and reduced handover delay, and to contribute to obtaining almost zero packet loss. However, these two optional handover mechanisms need a large number of WiMAX BSs within a specified area. These two techniques also face deployment challenges such as sharing the same carrier frequency among multiple BSs, perfect synchronization of the active set BSs, and increased deployment cost.

2.7   IP-Based Mobility

IP-based mobility management techniques can be implemented in several layers of the protocol stack, such as the network layer, transport layer, and application layer. IP-based mobility protocols can be used to take care of mobility for 3G- and 4G-based systems. MIPv4 (Mobile IPv4) (Perkins, 2002b) and its several variants, namely MIP-RO (MIP with Route Optimization), MIP-RR (MIP with Regional Registration) (Perkins, 2002c), MIP-LR (MIP with Location Registers) (Jain et al., 1999), MIPv6 (Johnson et al., 2004), and MOBIKE (S. Eronen, 2006), are a few of the network layer mobility protocols that were defined by the IETF. Cellular IP (Campbell et al., 2000), HAWAII (Handoff Aware Wireless Access Internet Infrastructure) (Ramjee et al., 2000), Proxy MIPv6 (Gundavelli et al., 2008), and IDMP (Intra Domain Mobility Protocol) (Das et al., 2002) are the network layer micromobility protocols suitable for intradomain mobility. Intradomain mobility refers to a movement scenario in which the mobile's movement is confined to one administrative domain. MSOCKS (Maltz and Bhagwat, 1998), TCP-Migrate (Snoeren and Balakrishnan, 2000), and SCTP (Stream Control Transport Protocol) (Koh et al., 2003) have been designed to take care of mobility in the transport layer. SIP-based mobility (Schulzrinne and Wedlund, 2000a) takes care of mobility by means of application layer signaling, such as by SIP (Session Initiation Protocol) (Rosenberg et al., 2002). HIP (Host Identity Protocol) (Moskowitz and Nikander, 2006) defines a shim layer between the network layer and transport layer to provide terminal mobility in a way that is transparent to both the network layer and the transport layer.

We have provided a survey of the related mobility protocols and issues (Dutta et al., 2001, 2002a). We have also experimented with several mobility protocols, namely MIPv4, MIPv6, SIP-based mobility, MIP-LR, and ProxyMIPv6, and verified that these mobility protocols in their current form are not adequate to meet the delay and packet loss performance requirements for real-time traffic (Dutta et al., 2005c, 2007d), and hence these protocols will benefit from overall systems optimization.

We now briefly describe some of these IP-based mobility protocols and categorize them into network layer macromobility, network layer micromobility, application layer mobility, and transport layer mobility.

2.7.1   Network Layer Macromobility

Network layer mobility can be categorized into two types: *macromobility* and *micromobility*. The macromobility mechanism takes care of global mobility where the mobile moves between administrative domains. We describe two types of network layer macromobility, namely Mobile IPv4 and Mobile IPv6.

2.7.1.1   Mobile IPv4

Mobile IP is a mechanism developed for the network layer to support mobility (Perkins, 2002a). Originally it was intended for travelers with laptops to provide portability, and was later adopted by the wireless community. It supports transparency above the IP layer, including the maintenance of active TCP connections and UDP port bindings. A mobile host is identified by a node identifier such as a fixed IP (home IP address). When the mobile host connects to a visited network that is different from the one that its IP address belongs to, its home network forwards packets to the mobile. A router (or an arbitrary node), which is usually known as the home agent, on the user's home network forwards the packets. There are two different methods to deliver packets to a mobile host when it is on a foreign network. With the first method, the mobile host adopts a second (temporary) IP address known as the care-of address (CoA) and registers it with its home agent. When the home agent receives a packet for this user, it encapsulates the packet in another IP packet with the care-of address as the destination address and sends it to the foreign network (Perkins, 1996a,b). Encapsulating a packet within another packet until it reaches the care-of address is known as tunneling. Note that encapsulation adds between 8 and 20 bytes of overhead, which can be significant for voice packets of this size.

The care-of address in the first method is said to be co-located, and it can be acquired via services such as DHCP (Dynamic Host Configuration Protocol) (Droms 1997) or an optimized version such as DHCP with rapid commit (Park et al., 2005) in a local area network, or via PPP (Simpson, 1994) in a point-to-point networking environment. With the second method, the mobile host first registers with a foreign agent (FA) in the network it is visiting. The foreign agent sends (registers) its address to the mobile host's home agent as the care-of address of the mobile host. Packets that are intended for the mobile host are sent to the foreign agent after the home agent has encapsulated them with the IP address of the foreign agent. After decapsulating these packets, the foreign agent delivers them to the mobile host.

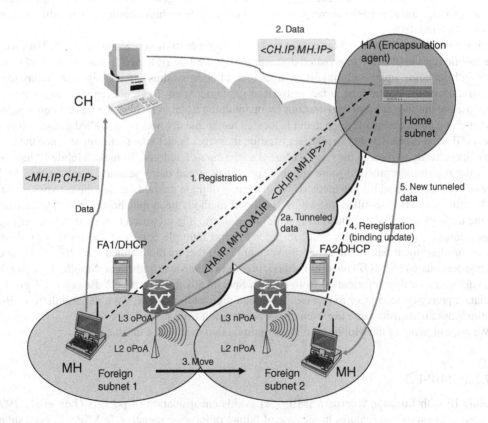

Figure 2.13 Mobile IPv4

Figure 2.13 shows the functional details of Mobile IPv4. In this specific figure, the mobile node moves from subnet 1 to subnet 2 and in the process changes its layer 2 and layer 3 points of attachment and either reconfigures itself with a new care-of-address from a DHCP server or uses the FA's address as its new CoA.

For the methods outlined above to be able to work, a mobile host needs to be able to learn that it has moved from its home network to a foreign network. For this purpose, home agents and foreign agents advertise their presence periodically in their own broadcast domains. A mobile host can also solicit agent advertisements if these advertisements are absent. Packets from the correspondent host must first travel to the home agent and then be later forwarded to the mobile host either by way of the foreign agent or directly. Packets from the mobile host do not have to traverse the home agent; the mobile host sends them as usual with its home IP address as the source address, which is known as triangular routing.

Routing of all incoming packets via the home network may cause additional delays and waste of bandwidth. However, if the correspondent host knows where the mobile host is, it can send packets directly to the care-of address of the mobile host. This is achieved by a route optimization (Perkins, 2002a) process that enables the mobile to send mobility binding updates directly to the corresponding host. Binding updates are sent from the home agent upon request from the mobile, or can be sent upon receiving a warning message from a foreign agent if the mobile host changes location during a communication session. In the second case, the former foreign agent will keep forwarding packets to the new foreign agent until the correspondent host updates its mobility binding cache (this is known as smooth handoff).

Another optimization that has been proposed is regional registration (Calhoun et al., 2003b), where the mobile registers locally in a visited domain. In base Mobile IP, a mobile host is required to register with its home agent each time it changes care-of address, thus causing signaling delay to the registration if the mobile host is far away from its home agent. Regional registration attempts to decrease the number of home registrations by maintaining a hierarchical structure of foreign agents. As long as a mobile host's foreign agent is located hierarchically under a so-called gateway foreign agent (GFA), it is unnecessary to relay registration messages back to the home agent, since the home agent has already registered the GFA's address as the care-of address. To make Mobile IP handoffs (i.e., the registration process) more suitable for real-time and delay-sensitive applications, Malki (2004) proposed two additional methods. With the first of these methods, called the network-assisted mobile and network-controlled (NAMONC) handoff method, the mobile host is informed (assisted) by the network that a layer 2 handoff is anticipated. Here, it is proposed to use simultaneous bindings (multiple registrations at a time) in order to send multiple copies of the traffic to potential points of attachment before the actual movement. The other method, called the network-initiated, mobile-terminated (NIMOT) handoff method, proposes extensions to the base Mobile IP so that foreign agents can utilize information from layer 2. Specifically, foreign agents use layer 2 triggers to initiate a preregistration prior to receiving a formal registration request from the mobile host. Both methods assume considerable involvement of information from layer 2.

We present many of the Mobile IP-related optimization techniques in Chapter 6.

2.7.1.2 MIP-LR

Mobile IP with Location Registers (MIP-LR) avoids encapsulation of packets (Jain et al., 1999) and provides survivable features in the case of failure of location registers. In MIP-LR, each subnet may contain a host that functions as a visitor location register and/or a host that functions as a home location register. Each mobile host can be served by multiple HLRs. Each VLR advertises its presence on its local subnet using agent advertisement messages similarly to Mobile IP. When a mobile host is located on its local subnet, it is not registered at either the HLR or the VLR. When the mobile moves to a foreign network, it obtains a care-of address from the pool of addresses that the VLR has. The mobile host registers with the foreign VLR using the CoA it has obtained, which in turn relays the registration to the mobile host's HLR. The HLR returns a registration reply containing the allowed lifetime for this registration; the VLR records the mobile host's CoA and the lifetime and forwards the reply to the mobile host. A correspondent host wishing to send a packet to the mobile host for the first time issues a query to the HLR, which returns the mobile host's CoA as well as the remaining registration lifetime. The correspondent host then sends the packet directly to the mobile host's CoA. The correspondent host caches a binding for the mobile host's CoA and uses this binding for subsequent packets destined for the mobile host. The correspondent host must refresh its binding cache by querying the HLR again before the mobile host's remaining registration lifetime expires. In MIP-LR, unlike Mobile IP, the HLR can be geographically distributed anywhere. We have implemented an extension of MIP-LR using an application layer module that does not require any kernel changes. Having an application layer module, this allows a mobile to use a policy-based approach to trigger MIP-LR for certain types of application. Figure 2.14 shows the functionalities of Mobile IP with Location Registers when the mobile moves from one subnet to another and in the process changes its layer 2 and layer 3 points of attachment. In this case, there is no foreign agent in the visited network and it is also not a requirement that the location register needs to be in the home network.

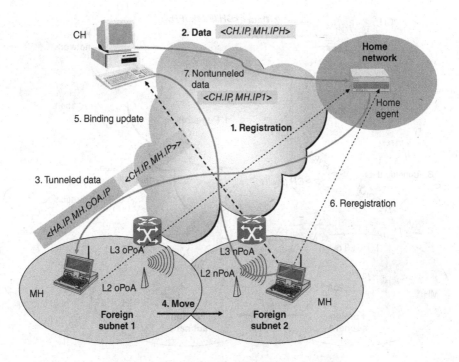

Figure 2.14   Mobile IP with Location Registers

After a mobile host moves, if the mobile host was previously registered at some other foreign VLR, the new VLR deregisters the mobile host at the old VLR. This deregistration is required so that the mobile host's old CoA can eventually be released for use by some other mobile host. If a VLR runs out of CoAs temporarily, it can still issue its own IP address as a CoA and, when a mobile host registers using this CoA, inform the HLR accordingly.

2.7.1.3 Mobile IPv6

IPv6's (Deering and Hinden, 1998) increased address space and inherent support for security and autoconfiguration have made it an attractive candidate to support mobility for the next-generation Internet. Mobile IPv6 is the protocol to support mobility for IPv6 nodes. Since address autoconfiguration is a standard part of MIPv6, the MH will always obtain a CoA routable to a foreign network. Thus, there is no need to have a foreign agent in MIPv6. When the mobile node moves to a new foreign network, it acquires a temporary care-of-address using stateless autoconfiguration (Thomson and Narten, 1998) or via DHCPv6 (Droms et al., 2003).

Figure 2.15 shows the functional components of Mobile IPv6. Unlike Mobile IPv4, the visited networks do not have any foreign agents. MIPv6's route optimization feature also enables direct data delivery from the correspondent host (CH) to the mobile node.

Although Mobile IPv6 is defined as a network layer approach and one needs to install an MIPv6 stack so as to support mobility in an IPv6 space, any standard operating system will in future come with inherent Mobile IPv6 support.

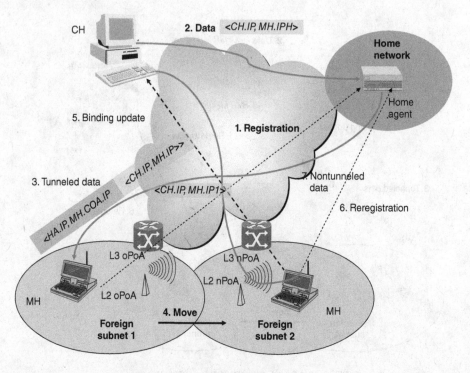

Figure 2.15   Mobile IPv6

While Mobile IPv6 provides a way of making sure of the uniqueness of an address as a mobile moves to a new router space, it also adds delay to the binding update and binding acknowledgement as in Mobile IPv4. However, compared with regular Mobile IP, there are inherent advantages to MIPv6. Route optimization is a standard feature of MIPv6, and thus there is no need for the CH to be equipped with additional software like MIP-RO. The MH sends a binding update directly to the CH and makes use of the home address destination option as part of the binding update. This allows the correspondent host to keep a binding cache that maps the care-of address of the mobile to the mobile's home address. For the ongoing traffic, this avoids triangular routing, and thus packets from the CH to the MH need not be encapsulated but are sent directly to the MH with its CoA as the source route. However, when a new CH needs to communicate with the mobile for the first time, the packets from the CH need to travel to the home agent and be tunneled to the mobile host. As the mobile moves during the packet transfer process, the subsequent packets are tunneled directly to the mobile host without being routed via the home agent.

2.7.1.4 Fast Mobile IPv6

While Mobile IPv6 takes care of session continuity during handoff, by itself it lacks the ability to provide the low-latency handoff and reduced packet loss that are essential for many interactive applications such as Voice over IP, gaming, and conferencing. Most of the handoff delays observed in Mobile IPv6 are due to IP address configuration and binding update delay when the home agent is far away. Fast Mobile IPv6 (FMIPv6) (Koodli 2008) proposes mechanisms to reduce the hand-off delay by way of localizing the binding updates to the edges of the network, reducing the delay due to IP address acquisition, and buffering at the edge routers. This involves additional protocol exchange between the mobile host, the current router (pAR), and the next access router (nAR). These mechanisms can be categorized into two types of handover, namely predictive and reactive. The FMIPv6 protocols work in conjunction with the existing MIPv6 stack. Figure 2.16 shows the inter-action among several network elements. For brevity, it does not reflect the MIPv6-related signaling, however. Figure 2.17 and Figure 2.18 describe the call flows for predictive and reactive handovers, respectively.

We now briefly describe the predictive operation of FMIPv6. The mobile host sends a router solicitation for a proxy (RtSolPr) message to its default access router (pAR) in order to obtain information related to the link layer addresses of the neighboring access points discovered during the layer 2 scanning process, and the prefixes associated with the neighboring access router (nAR). The current access router (pAR) communicates with the nAR using protocols such as Candidate Access Router Discovery (CARD) (Liebsch et al., 2005) to obtain the relevant information about the neighboring network elements. The pAR serving the user responds with a proxy router advertisement (PrRtAdv) containing the requested information, thus allowing the mobile host to perform address autoconfiguration prior to its movement to the new network. The host, after formulating a prospective new CoA, sends a fast binding update (FBU) to its default router instructing it to tunnel packets addressed to its old CoA (oCoA) towards its new CoA (nCoA). The access router currently serving the host (pAR) starts buffering newly arriving packets with the oCoA as their destination and exchanges handover initiate (HI) and handover acknowledge (HAck) messages with the nAR to initiate the process of the MH's handover. This HI/HAck message exchange can also serve for validation of the nCoA already formed by the host. The pAR responds to the MH with a fast binding acknowledge (FBack) message on both links (old and new) and starts the tunneling of buffered and arriving data to the nCoA. These

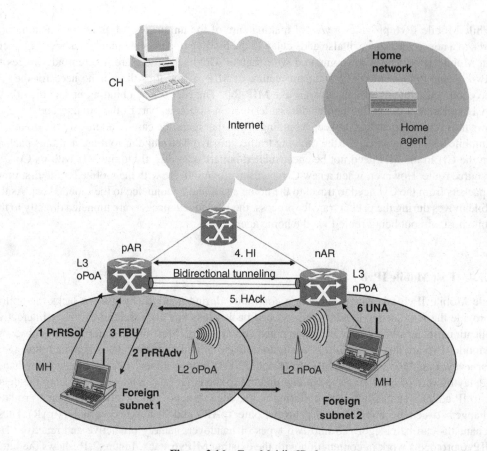

Figure 2.16   Fast Mobile IPv6

packets are also buffered at the nAR until the mobile arrives at the new point of attachment. The MH, as soon as it attaches to the new link, transmits a UNA (unsolicited network advertisement) to inform the nAR of its presence. Buffered packets at the nAR can be delivered immediately to the MH on the new link.

In the reactive mode of FMIPv6, the FBU is sent after the mobile connects to the new network. Thus, the FBU is routed through the nAR but is processed at the pAR. Unlike predictive handoff, the packets destined for the previous address of the mobile are forwarded to the nAR instead of being buffered at the nAR. Packet loss is minimized in predictive handover owing to buffering at the nAR.

2.7.2   Network Layer Micromobility

There are several network layer micromobility protocols that are meant to optimize mobility when the mobile's movement is confined within a domain. These protocols avoid the overhead associated with tunneling over the air, and reduce the signaling overhead when the mobile's movement

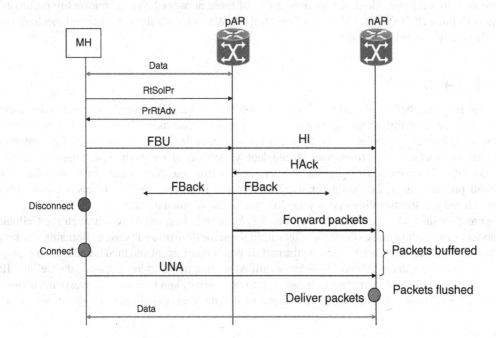

Figure 2.17   Call flow for predictive handoff

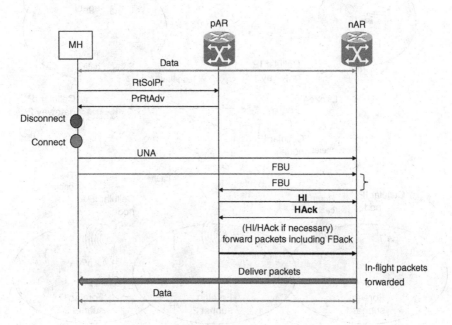

Figure 2.18   Call flow for reactive handoff)

is confined to a domain. Here, we describe a few of these network layer micromobility protocols, namely Cellular IP, HAWAII, IDMP, and ProxyMIPv6. We focus mainly on the general mechanisms that they use to optimize mobility.

2.7.2.1 Cellular IP

Cellular IP (Campbell et al., 2000) is a micromobility management protocol. It separates local and wide-area mobility, adopts a domain-based approach, and uses Mobile IP for interdomain (wide-area) mobility. Cellular IP isolates the wireless access network from the core of the Internet via a gateway that acts like a foreign agent, and deploys network elements (base stations) specialized for mobility management. Isolating the wireless access from the core is necessary since Cellular IP itself provides an IP forwarding engine that replaces IP in the wireless access network. This approach reduces the signaling updates and localizes them within a domain.

Figure 2.19 shows how the packets destined for the mobile host are routed through the Cellular IP nodes as the mobile moves between cells within the same domain and between domains. Packets from a correspondent host are first sent to the mobile host's home agent and then tunneled to the gateway, where they are decapsulated. Hosts are identified by their home addresses inside the Cellular IP cloud. Packets generated by mobile hosts are sent to the gateway and later to the correspondent host. Each Cellular IP domain is equipped with a gateway router that periodically broadcasts a beacon.

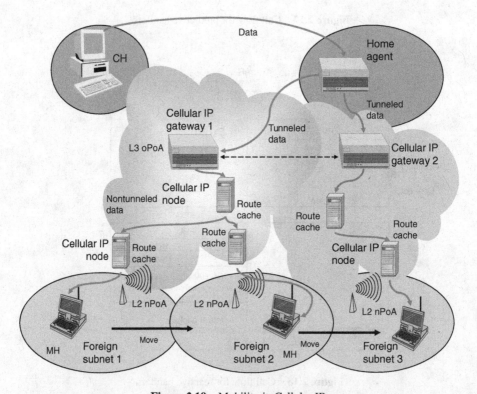

Figure 2.19   Mobility in Cellular IP

This beacon is broadcast to all the Cellular IP nodes within the domain, and thus each node between the mobile and the gateway learns of its neighbor's address in the uplink and uses that information to route any data traffic destined for the gateway. Cellular IP base stations snoop actual data packets sent from mobile hosts to the gateway to cache the path taken by them (in fact, base stations record only the host IP number and the neighboring base station from which the packet was received). To route packets from the gateway to the mobile host, base stations use the reverse of this path. Hosts that have not transmitted packets for a while are removed from the routing cache of the base stations. The location-tracking method for mobile hosts depends on whether the host is active or idle. An idle host is one that has not received or transmitted a packet for a specific time. It is adequate to maintain the position of idle hosts in a distributed paging cache. To achieve this, a technique known as passive connectivity in cellular telephone systems is mimicked in the Cellular IP layer. Base stations are grouped geographically into paging areas, where a paging area may include more than one base station. Idle hosts send infrequent paging-update packets to the gateway. For active hosts, a distributed routing cache maintains the exact location of each host. Once a mobile host has moved to another base station during a call, it sends a route-update packet back to the gateway. The new base station(s) record this path accordingly.

2.7.2.2 HAWAII

Handoff-Aware Wireless Access Internet Infrastructure (HAWAII) is another effort to compensate for Mobile IP's inefficiency in supporting intradomain mobility (Ramjee et al., 2000). In that sense its objective is similar to that of Cellular IP, but unlike Cellular IP, HAWAII defines separate control messages to set up host routes on the intermediate routers to route packets between the domain root router and the mobile host. HAWAII operates on the basic assumption that most user mobility occurs within a domain, and therefore optimizing routing and forwarding for efficient support of intradomain mobility will complement Mobile IP's interdomain mobility support. Another assumption is that base stations are capable of IP routing. HAWAII segregates the network into a hierarchy of domains. On the top of the hierarchy for each domain, there is a domain root router. Packets addressed to a mobile host in a specific domain first reach the domain root router and are then sent to the mobile host. As long as the mobile host is moving within a domain, it retains its IP address. Once the mobile host moves into another domain, it is assigned a co-located care-of address, and the home agent in the home domain tunnels packets to this address. Figure 2.20 shows a sample architecture for HAWAII.

The path (route) between the mobile host and the domain root router is specific to that host. It is established during power-up and updated during movement of the mobile host in the domain root router and pertinent intermediate routers. This information is refreshed periodically by the mobile host, which allows the routers to maintain the path state. This idea is similar to Cellular IP and the regional registration in Mobile IP: the involvement of a physically distant home agent is not desirable each time the mobile host moves. Four path setup methods that can be used to reestablish path states when the mobile host moves within a domain are proposed in this scheme. Two of them forward packets from the old base station to the new base station for a short period (i.e., until the relevant routers update their entries for the specific host). The other two methods do not forward packets during a handoff. Rather, they either bicast the packets to two base stations or unicast them for hosts that can simultaneously listen to two base stations. Obviously, HAWAII requires all routers in a domain to be augmented with mobility support so that they are able to handle host-specific path setup messages.

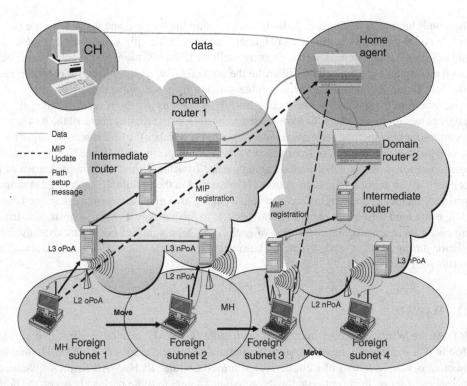

Figure 2.20   Mobility in HAWAII

2.7.2.3   TeleMIP

TeleMIP (Telecommunications-Enhanced Mobile IP) is an intradomain mobility framework which uses two layers of scoping within a domain and is based on IDMP (Intra Domain Mobility Protocol) (Das et al., 2002). Figure 2.21 shows the basic architecture for TeleMIP. By specifying an intradomain termination point called a mobility agent (MA), it helps to reduce the signaling updates during movement within a domain. The mobility agent has similar functionality to a foreign agent but is placed hierarchically at a higher level. This reduces the signaling traffic due to frequent handoffs within a domain. The mobility agent acts like a domain-wide anchor point similar to a gateway foreign agent in MIP-RR. Unlike other proposed intradomain mobility management schemes, IDMP uses two dynamically autoconfigured care-of addresses for routing packets destined for mobile nodes. The global care-of address (GCoA) identifies the mobile node's attachment to the current domain, while the local care-of address (LCoA) changes with subnet changes and identifies the mobile's attachment to a specific subnet.

Intradomain updates are sent only up to the MA, which provides a globally valid care-of address to the mobile host. TeleMIP reduces the frequency of global update messages since the MA is located at a higher level in the hierarchy than subnets, and global updates (to the home agent, correspondent hosts, etc.) only occur for interdomain mobility. It also reduces the requirement for additional public addresses (as in IPv4) by adopting a two-level addressing scheme.

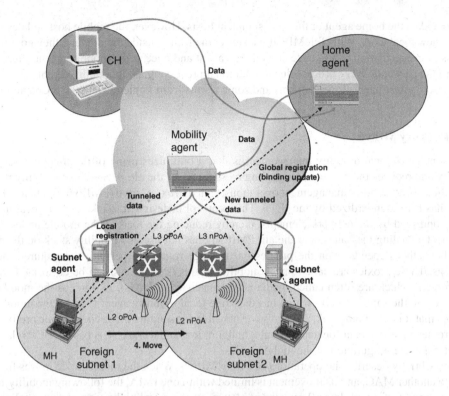

Figure 2.21   Mobility in TeleMIP

In TeleMIP, the network is divided into domains as in Cellular IP and HAWAII. Each domain is identified by the mobility agent's IP address. Domain identifiers are broadcast as part of agent advertisements in each subnet. Each mobility agent is identified with a unique domain identifier. The MA's address can also serve as a domain identifier. The network retains control of MA assignment. When a mobile host first moves into a domain, it obtains a global care-of address (the MA's address) as well as a local care-of address. The MA's global CoA is sent in the registration message to the HA. The mobile host also registers itself (using its local CoA) with its MA. There are foreign agents and DHCP servers, organized hierarchically at the subnet level under an MA. These provide the mobile host with a locally scoped address (LCoA), which identifies the location of the mobile within the domain. The MA provides the mobile host with a global care-of address that stays constant as long as the mobile stays within the domain.

Each domain is equipped with at least one mobility agent. However, multiple MAs can be provisioned for load balancing and redundancy within the domain. The mobile host's location is known only up to the MA-level granularity. A mobile host retains the same MA (global care-of address) within the same domain. All packets from the global Internet are tunneled to the MA, which acts as a single point of enforcement. The MA forwards packets to the mobile host, using regular IP routing, by using the local CoA (co-located or at a foreign agent) as the destination. On subsequent movement within the domain, the mobile host obtains only a new local CoA. At that point there is no need to update the home agent or the correspondent hosts. However, the mobile host updates its MA with its new local CoA. With TeleMIP, if packets come from outside the domain, they go through a process of encapsulation twice, once at the home agent and once at the mobility agent, thus adding delay to the packet delivery process due to additional processing at the mobility agent.

We describe fast-handoff mechanisms and some results from implementations in Chapter 6.

2.7.2.4   Proxy MIPv6

An advantage of local mobility management is that it optimizes many of the functions related to mobility and reduces the number of signaling messages over the air. A candidate mobility protocol to optimize local mobility management operations, called Proxy MIPv6 (PMIPv6) (Gundavelli et al., 2008), has been standardized by the IETF. This protocol is designed to take care of local mobility and is controlled by the network elements, thereby reducing the load on the mobile nodes and the number of signaling messages over the air. PMIPv6 does not use any mobility stack on the mobile node, but rather depends upon the proxies on the edge routers to perform the required mobility functions. These proxies are called proxy mobility agents (PMAs) and can be co-located with the edge routers, which are often called media access gateways (MAGs). As long as the mobile node moves within the same domain, which has the same local mobility agent (LMA), the mobile node assumes that it is in a home link. A PMA is responsible for sending the proper mobile prefix as part of the router advertisement for stateless autoconfiguration, and it can also act as a DHCP relay agent for stateful autoconfiguration of the mobile nodes.

We now briefly describe the operations of Proxy MIPv6. When the mobile node moves from one MAG to another MAG, and its movement is limited within one LMA, the following mobility-related operations are performed: layer 2 movement, detection of a new link by way of router solicitation, access authentication, profile verification, proxy binding update, and address reconfiguration.

Figure 2.22 illustrates the network elements associated with the operation of Proxy MIPv6. After the mobile node connects to a new point of attachment as part of the initial bootstrapping process or after movement to a new domain, access is authenticated with the designated AAA server. During this process, the PMA sends a binding update to the LMA with the address of the PMA that is specific to the home prefix of the mobile node. In the absence of a preexisting tunnel, this process helps to set up a tunnel between the LMA and the respective PMA. The mobile node configures its address using the prefix included in the router advertisement and interface-id, which can be assigned by the PMA or created by the mobile node itself. After movement to the new access network, if the same prefix is advertised by the new PMA, then the IP address of the new mobile does not change. A tunnel is not desirable on the mobile node, because it adds extra processing and bandwidth constraints to the wireless hop. A Proxy MIPv6-based mobility protocol is preferred when mobility is confined within a domain and wireless service providers do not want to overload the mobile node's stack by setting up a tunnel between the mobile and the HA. Figure 2.23 shows the sequence of events as the mobile attaches from the previous point of attachment and connects to a new point of attachment.

2.7.3   NETMOB: Network Mobility

A mobile network is a network segment or subnet that can move and attach to arbitrary points in the routing infrastructure. A mobile network can only be accessed via specific gateways, called mobile routers, that manage its movement. Mobile networks have at least one mobile router serving them. A mobile router does not distribute the mobile network routes to the infrastructure at its point of

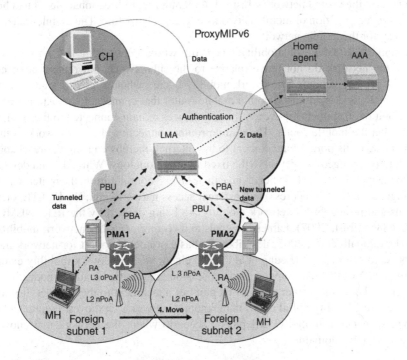

Figure 2.22   Proxy Mobile IPv6

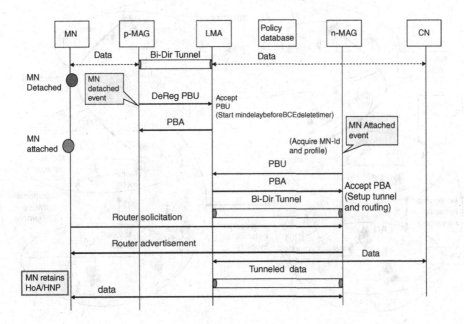

Figure 2.23   Flow for Proxy Mobile IPv6 handoff

attachment (i.e., in the visited network). Instead, it maintains a bidirectional tunnel to a home agent that advertises an aggregation of mobile networks to the infrastructure. The mobile router is also the default gateway for the mobile network.

Network mobility is a type of mobility where the whole network moves while the end users connected to the network do not move relative to the subnetwork, but remain connected to the subnetwork. A typical example of network mobility occurs when a moving train, a moving bus, or a tank in a battlefield is equipped with a mobile router that connects to an external network and changes its point of attachment. However, the end devices remain connected to the subnetwork and communicate via the mobile router. The mobile router connecting the subnetwork to the Internet dynamically changes its point of attachment to the Internet, thereby causing the reachability of the said network to be changed in relation to the fixed Internet topology. While the end devices that are connected to the mobile network do not change their layer 3 configuration, the router may change its layer 3 configuration (i.e., IP address) and external access network (e.g., Wi-Fi, LTE, or WiMAX). Requirements and protocols for network mobility are being defined by the IETF NEMO working group. RFC 4886 (Ernst, 2007) defines the goals and requirements for network mobility support. RFC 3963 (Devarapalli et al., 2005) describes the basic protocol support for network mobility for IPv6 networks, and RFC 5177 (Leung et al., 2008) elaborates on network mobility extensions for Mobile IPv4. The NEMO basic support ensures session continuity for all the nodes in the mobile network, even as the mobile router changes its point of attachment to the Internet. It also provides connectivity and reachability for all nodes in the mobile network as it moves. Figure 2.24 illustrates a typical type of network mobility. It depicts a scenario where the network may move within a domain or between two domains.

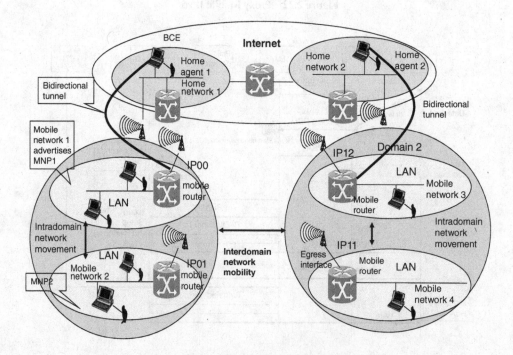

Figure 2.24   Network mobility

We briefly describe the signaling aspects of NEMO here. NEMO signaling is performed with extended MIP messages. Specifically, messages contain an additional router (R) flag to signal a mobile router instead of a mobile node. These messages are sent using the mobility extension header in IPv6 and UDP control messaging in IPv4. The two major messages used by NEMO are binding updates (BUs) and binding acknowledgements. Mobile routers use BUs to notify their HAs of a new CoA, and thus of a new Internet attachment point. BUs contain the new CoA, the router flag, and an optional mobile network prefix which is used to update the network's prefix information. Upon receipt of a BU, the HAs send a binding acknowledgement.

The NEMO Basic Support Protocol defines two operating modes for NEMO: implicit and explicit. In implicit mode, BUs do not contain a mobile network prefix option. Instead, the HA must determine the mobile router's network prefix by some means outside of the NEMO Basic Support Protocol. When in explicit mode, all BUs contain one or more mobile network prefix options. The HA is then able to use the network prefix option to set bindings for the mobile router.

Lastly, the NEMO Basic Support Protocol specifies the routing of packets to and from the mobile network. Packets sent to an MN (mobile node) from a correspondent node are routed over the Internet using standard routing until they reach the HA. The HA intercepts packets and encapsulates them in a tunnel to the mobile router. The mobile router decapsulates the packets and routes them to the MN. Reverse traffic must be reverse tunneled to the HA before being routed to the correspondent node. The Basic Support Protocol specifies bidirectional tunneling so that only mobile routers and HAs need to be aware of the network's mobility, and also to prevent firewalls from dropping packets. Some firewalls will drop packets to prevent spoofing when the source address of the packets (the home address) does not match the network from which they are being sent.

2.7.4   Transport Layer Mobility

There are several transport layer mobility solutions. TCP-Migrate (Snoeren and Balakrishnan, 2000) proposes a new set of migration options for TCP which provide a pure end-system alternative to network layer solutions. With this extension, established TCP connections can be suspended by a TCP peer and reactivated from another IP address without a third party except for the involvement of dynamic DNS updates to locate the mobile host. However, this approach requires modifying the transport protocol at the end terminals. MSOCKS (Maltz and Bhagwat, 1998) is another transport layer solution, which introduces a proxy in the middle of a network and is built on top of the SOCKS protocol (Leech et al., 1996) often used for firewall traversal. Upon movement of the mobile and a change in its address, the intermediary proxy helps splice the TCP connection. The transport protocol SCTP (Stream Control Transport Protocol) (Stewart et al., 2000) also adds support for mobility. It has a built-in ADDIP feature (Koh et al., 2003) that provides continuity support when the mobile's IP address changes. Figure 2.25 shows the splicing of a TCP connection when there is a break in the communication due to a mobile's movement.

2.7.5   Application Layer Mobility

Application layer mobility uses the Session Initiation Protocol (SIP) (Rosenberg et al., 2002) as the underlying signaling mechanism to provide a host-based mobility solution. This mechanism does not depend on a home agent or foreign agent in the network, nor does it require additional mobility software in the end hosts. Figure 2.26 shows the functional components of SIP-based mobility.

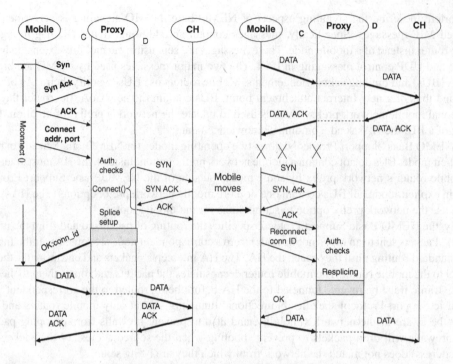

Figure 2.25 MSOCKS-based mobility

Just like other mobility protocols, SIP-based mobility refers to a similar scenario where the mobile moves between two different subnets and, in the process, changes its layer 2 and layer 3 points of attachment. Just as with direct binding updates in the case of MIPv6 and MIP-LR, SIP-based mobility management uses a re-INVITE signal to the correspondent host as a way of performing a direct binding update that helps to update the binding cache in the correspondent host. Thus, the data is delivered using an optimized direct path. Application layer mobility based on SIP can support a variety of mobility mechanisms such as personal mobility, service mobility, and terminal mobility. As part of the research presented in this book, we have experimented with SIP-based application layer mobility protocols for both RTP- and TCP-based applications and designed several optimization techniques for SIP-based application layer mobility protocols. These optimization techniques for SIP-based mobility management are described in Chapter 6.

2.7.6 Host Identity Protocol

HIP (Moskowitz and Nikander, 2006) defines a new protocol layer between the inter-networking layer and the transport layer to provide terminal mobility in a way that is transparent to both the network layer and the transport layer. The proposed host identity namespace takes care of some of the important gaps between the existing IP and DNS namespaces. The host identity namespace consists of host identifiers (HIs). A host identifier is cryptographic in its nature and is the public key of an asymmetric key pair. Each host identity uniquely identifies a single host. By this means, IP addresses

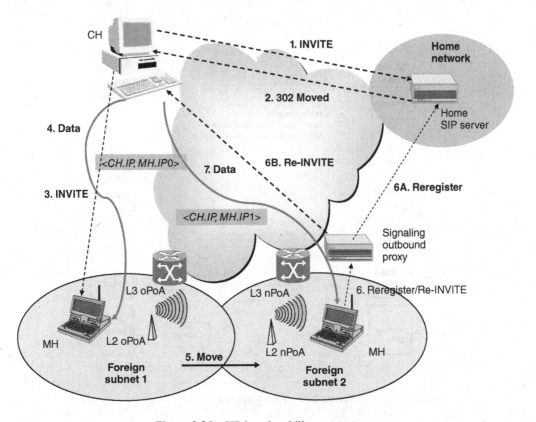

Figure 2.26   SIP-based mobility management

can be decoupled from the higher-layer applications in a secure manner and the end hosts can be authenticated by a public key as they move around. The public key is one component of an asymmetric cryptographic key pair and is used as a publicly known identifier for cryptographic identity authentication. By decoupling the network and transport layers, applications and transport connections can be made independent of any underlying IP address changes, thereby enabling alternative solutions for host mobility, host multihoming, and network address translation. A potential benefit of HIP is that it can be integrated directly with the IP security protocol IPSec. Figure 2.27 shows how the end-point address and the locator are separated whenever the IP address changes.

Since IPSec security associations (SAs) are bound to host identifiers, not addresses, an IP address change does not break the existing transport connections, nor does it trigger a reestablishment of IPSec SAs. HIP-enabled mobility provides an optimization technique similar to the route optimization technique for Mobile IPv6 by sending a direct update to the correspondent host. However, unlike MIPv6, HIP does not have any home network. HIP uses a *Readdress* packet that is similar to the binding update for MIPv6. However, unlike MIPv6, it inherently secures the readdressing process. In the HIP architecture, the end-point names and locators are separated from each other. IP addresses continue to act as locators. The host identifiers play the role of end-point identifiers. It is important to understand that the end-point names based on host identities are slightly different from interface names; a host identity can be reachable through several interfaces simultaneously.

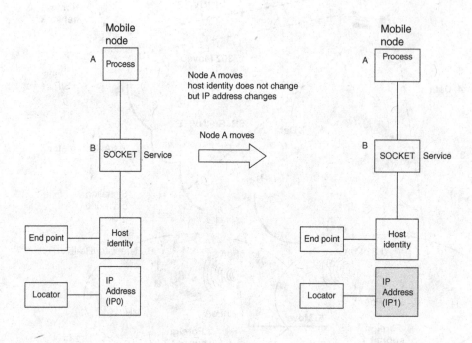

Figure 2.27   Host Identity Protocol

2.7.7   MOBIKE

MOBIKE (IKEv2 Mobility and Multihoming) (S. Eronen, 2006) is an extension to IKEv2 (Kaufman, 2005) that provides mechanisms so that mobile clients with VPN (virtual private network) connectivity do not need to tear down the existing security association during their layer 3 handoff. In the base IKEv2 protocol, the IKE SAs and tunnel mode IPSec SAs are created implicitly between the IP addresses that are used when an IKE SA is established. When the mobile moves, its IP address changes, giving rise to the need for a new IKE process and a new security association between the mobile's new IP address and the VPN gateway. This process results in suboptimal operation and media interruption. However, MOBIKE allows the IP addresses associated with IKEv2 and tunnel mode IPSec security associations to be modified by initiating additional signaling messages such as UPDATE_SA_ADDRESS. As a practical application, a mobile VPN client could use MOBIKE to keep the connection with the VPN gateway active while the mobile itself changes its address owing to a change in the network point of attachment. Similarly, a multihomed host could use MOBIKE to move traffic to a different interface if, for instance, the one currently being used stops working. MOBIKE is probably the most preferred mobility protocol for taking care of mobility for clients in mobile VPN environments, where the mobile does not need any additional home agent. In Chapter 6, we propose optimization techniques to improve the handoff delay performance for mobiles in a VPN environment.

Table 2.3 Qualitative comparison of IP-based mobility protocols

Mobility protocol	Intradomain encapsulation	Interdomain encapsulation	End system changes	Triangle routing	Network change	Fast-handoff technique	Mobility layer	Mobility type
Mobile IPv4*	X	X	X	X	–	–	Network	Macro
MIPv6*	–	–	X	–	–	X	Network	Macro
MIP-RO	X	X	X	–	–	–	Network	Macro
MIP assisted by FA	X	X	–	X	X	X	Network	Macro
IDMP*	X	X	–	X	X	X	Network	Micro
MIP with LR*	–	–	X	–	–	X	Network	Macro
Cellular IP*	–	X	–	X	X	X	Network	Micro
HAWAII	–	X	X	X	X	X	Network	Micro
MSOCKS	–	–	X	–	X	–	Transport	Macro
TCP-Migrate	–	–	X	–	X	–	Transport	Macro
SIP*	–	–	–	–	–	X	Application	Macro

X, Yes; – , No.

Table 2.3 shows a qualitative comparison of some of the available IP-based mobility management protocols for the wireless Internet. Only network layer, transport layer, and application layer mobility protocols are considered in this comparison. The mobility protocols with an asterisk next to them are the candidate protocols that we experimented with and are described in this book. The metrics against which these protocols were compared can be defined briefly as follows. "Intradomain encapsulation" involves an extra level of encapsulation due to additional tunneling in the network. "Interdomain encapsulation" involves tunneling when the mobile moves between domains. "End system changes" refers to whether additional software is needed to make a specific mobility protocol work. "Fast-handoff technique" refers to whether the handoff delay can be reduced further to support real-time communication.

2.7.8 IAPP

IEEE 802.11f defines a recommended practice that describes a service access point (SAP), service primitives, a set of functions, and a protocol called Inter Access Point Protocol (IAPP) that allows APs to interoperate on a common DS (distribution system) using the backbone network over TCP/IP or UDP/IP to carry IAPP packets between the APs. 802.11f also makes use of a RADIUS server that maps the BSSID of an AP to its IP address on the distribution system and distributes keys to the APs to allow secure communication between the APs. IAPP helps the exchange of information between APs when the mobile station roams from one AP to another. IAPP also facilitates the creation and maintenance of the ESS (Extended Service Set), supports the mobility of STAs (stations), and enables APs to enforce the requirement of a single association for each STA at a given time. When a station moves from one base station subsystem to another within the same ESS, IAPP enables the transfer of data relevant to that mobile station from the old AP to the new AP. The primary objective is to reduce the amount of signaling that occurs over the wireless medium every time the station moves. This also speeds up the handoff. There are two main IAPP procedures, namely association and reassociation. IAPP procedures are initiated by the Access Point Management Entity (APME). The IAPP association process consists of the primitives MLME-Associate, Add.request, L2 Update Frame,

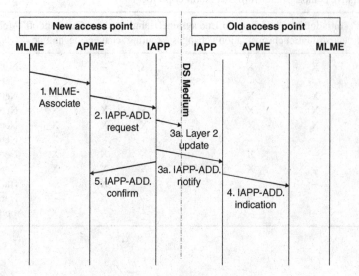

Figure 2.28   IEEE 802.11f association

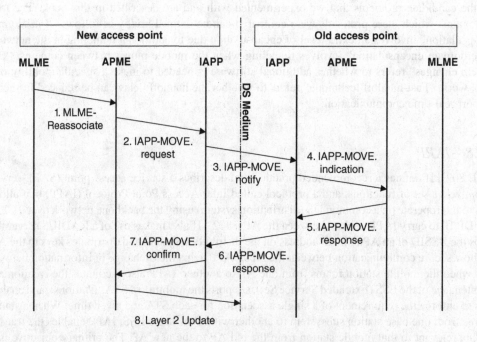

Figure 2.29   IEEE 802.11f reassociation

Add-notify, ADD.indication, and ADD.confirm. The IAPP reassociation process consists of the following primitives: the MLME.Reassociate, MOVE.Request, MOVE-notify, MOVE.indication, MOVE.response, and MOVE.confirm messages. There is a third procedure that is initiated by invoking an IAPP-CACHE.request to cache context in neighboring APs to facilitate fast roaming. We now briefly describe the various steps in the association process. The MLME-Associate message is sent by the MLME when a station has successfully been associated with the AP. The ADD.request contains the MAC address of the station and the sequence number of the association request frame. L2 Update Frame is sent to the distribution system to update all layer 2 devices, such as bridges. Subsequent to this message, the ADD-notify message is transmitted to all IAPP multicast IP address. The ADD.indication primitive informs the older AP of the new association, so that if the station was formally associated with an AP, that AP dissociates the station. Figure 2.28 shows the interaction of the protocols during the association process and Figure 2.29 shows what happens when the client reassociates with the access point.

IAPP supports three protocol sequences. One is initiated by invoking the IAPP-ADD.request after the APME receives an MLME-ASSOCIATE.indication. The second is initiated by invoking the IAPPMOVE.request after the APME receives an MLME-REASSOCIATE.indication. The third is initiated by invoking the IAPP-CACHE.request to cache context in neighboring APs to facilitate fast roaming.

2.8   Heterogeneous Handover

Supporting terminal handovers across heterogeneous access networks, namely 802.11, CDMA2000, UMTS, and LTE networks, is a challenge, as each access network has different QoS, security, and bandwidth characteristics. A mobile may be equipped with multiple interfaces, where each interface can support a different access technology (e.g., 802.11 or CDMA). It may be preferable for a mobile to communicate with one interface at a time in order to conserve power. During the handover, the mobile may move out of the footprint of one access technology (e.g., 802.11) and into the footprint of a different technology (e.g., CDMA, UMTS, or LTE). This will require switching of the communicating interface on the mobile as well. This type of intertechnology handover is often called vertical handover, since the mobile moves between two cells of different sizes. A vertical handover can be termed an upward or a downward vertical handover based on the direction of movement, such as from a smaller cell to a larger cell or vice versa (Stemm and Katz, 1998). For example, the movement of a mobile from an 802.11 network to a cellular network can be viewed as an upward vertical handover. An intertechnology handover may affect the quality of service of multimedia communication, since each access network will offer a different bandwidth and each of the access-specific handoff operations may require a different amount of resources. We provide two illustrative examples of intertechnology handover involving 802.11, UMTS, and LTE networks.

2.8.1   UMTS–WLAN Handover

Release 6 of 3GPP's description of interworking with WLAN systems (3GPP, 2005) describes the mechanisms associated with handoff scenarios involving UMTS and WLAN networks. In particular, it highlights several fundamental operations such as network advertisement, network selection, authentication, authorization, IP address allocation, and tunneling with the home agent in the home network. Figure 2.30 depicts a scenario involving handover between WLAN and UMTS networks.

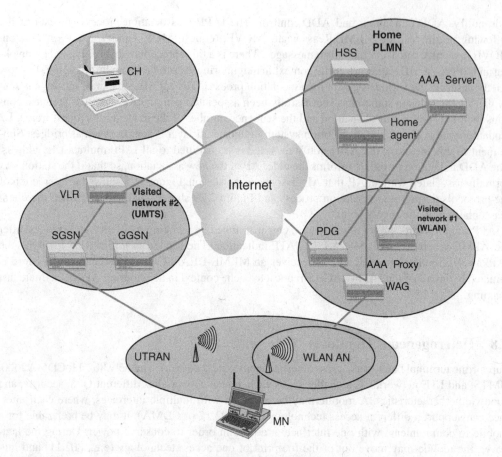

Figure 2.30   UMTS–WLAN handover scenario

In this specific scenario, the mobile node is equipped with both an 802.11 interface and a WCDMA interface.

Figure 2.30 shows three different networks, namely visited network 1 (a WLAN network), visited network 2 (a UMTS network), and the home network of the service provider (labeled "Home PLMN"). When a subscriber is in roaming mode and is away from the home network, it could potentially move between the two access networks. In this specific example, the mobile uses Mobile IPv6 as the mobility protocol to maintain session continuity during an intertechnology handover involving UMTS and 802.11 networks (3GPP, 2005; Kwon et al., 2005).

When the mobile is connected to the UMTS network, it follows the standard 3G access mechanism as defined in 3GPP and communicates with the correspondent host (CH). As the mobile moves to a hotspot area (WLAN), it may be preferable for it to communicate via its WLAN interface. However, before it can communicate with the correspondent host, it will need to reestablish the authentication and authorization with the home network so that the data is rerouted via the home agent to the new access network. Figure 2.31 shows the steps associated with a handoff where a mobile moves from a WLAN network to a UMTS network and then back to the WLAN network during a specific media session.

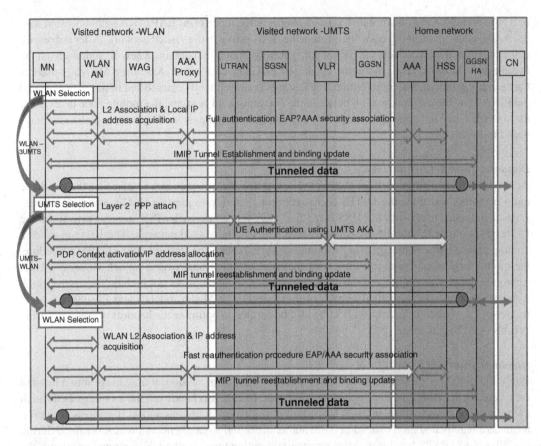

Figure 2.31   Flow during UMTS–WLAN handover

During the handover from one access network to another, the mobile needs to go through several steps as described in Chapter 3, namely discovery of the new access network, acquisition of a new layer 3 identifier (e.g., an IP address), authenticating with the home network using a VLR/AAA proxy with the HSS in the home network, informing the home agent about its new IP address by way of a binding update, and restablishing the tunnel (in the case of MIP) so that data can be redirected to the mobile in its new access network. This will ensure that session continuity is maintained as the mobile moves from the WLAN network to the UMTS network and then back again. We now briefly describe the steps involved during the mobile's WLAN roaming and handover to and from the UMTS network (Kwon et al., 2005).

WLAN Roaming
First we assume that the multihomed mobile moves into a WLAN coverage area such as a Wi-Fi hotspot. Although the hotspot area has wide-area network (UMTS) coverage, the mobile prefers to connect to the Wi-Fi network as per the the mobile's connectivity preferences. As the mobile powers on the interface, it detects the WLAN signal, selects the access point, and associates in layer 2. After layer 2 association, the mobile obtains a local IP address from the DHCP server. This address could

be a NAT address. Once the mobile has basic connectivity, it performs EAP-AKA authentication with the AAA server in the home network through the AAA proxy in the visited network. After successful authentication, the UE selects the packet data gateway (PDG) and obtains a new IP address from the home agent. In most cases, the home agent and PDG are co-located. Assuming MIPv6 is used as the mobility protocol, a secured bidirectional tunnel is subsequently established between the UE and the PDG. Now the communication between the correspondent host and the mobile node takes place through the home network. Some mobility optimization techniques such as route optimization have been omitted here.

Handover from WLAN to UMTS

As the mobile moves away from the WLAN coverage area, it prepares to discover the available radio networks and finds the UMTS access network. The UE requests a UMTS attach procedure, and authenticates itself with the HSS using the AKA procedure. After PDP context activation, the UE receives a new IP address allocated by the GGSN in the visited network. The UE then sends a binding update to the home agent in the home network. After a successful binding update procedure, the mobile continues communication with the correspondent host over a tunnel. However, while session continuity is maintained during this handover, these steps do not include the optimization techniques that could be applied to each of these steps. The Media Independent Preauthentication framework described in Section 11.3.1 can also be applied to optimize the handoff operations during WLAN–UMTS handover.

Handover from UMTS to WLAN

While the mobile is connected to the UMTS network and is in session, it may move into a hotspot area, detect a WLAN signal, and decide to hand over. During this handover, the mobile goes through a similar set of steps to those in the WLAN roaming situation. It obtains a local IP address; however, it can do a reauthentication instead of a full authentication if a reauthentication id is obtained during the WLAN-to-UMTS handover. After a successful authentication, the UE communicates with the PDG and obtains a remote IP address that is used for a binding update. Once the binding update is complete, the UE continues to communicate with the correspondent host via the home agent.

2.8.2   LTE–WLAN Handover

In order to provide high-speed connectivity and the desired quality of service and end-user experience, operators are deploying LTE networks. Intratechnology handoff involving LTE networks has been described earlier in this chapter. However, LTE and WLAN interworking will gain in significance, considering the large deployment of Wi-Fi hotspots. Wi-Fi hotspots will complement LTE by providing cheaper roaming, better indoor coverage, and offloading of data-intensive traffic. Figure 2.32 illustrates how two types of WLAN network, namely a *trusted Wi-Fi network* operated by 3GPP operators and a *nontrusted Wi-Fi network* operated by aggregators, can interoperate with an LTE core network. The untrusted model includes any type of Wi-Fi access that either is not under the control of the operator or does not provide sufficient security. Trusted access refers to operator-built Wi-Fi access with over-the-air encryption and secure authentication. The figure shows three different types of UE, namely (1) a cellular UE connected to an LTE network through eNodeB, (2) a Wi-Fi UE connected through an operator-owned hotspot (a non-3GPP secured network), and (3) a Wi-Fi UE connected through a public hotspot (non-3GPP nonsecured network).

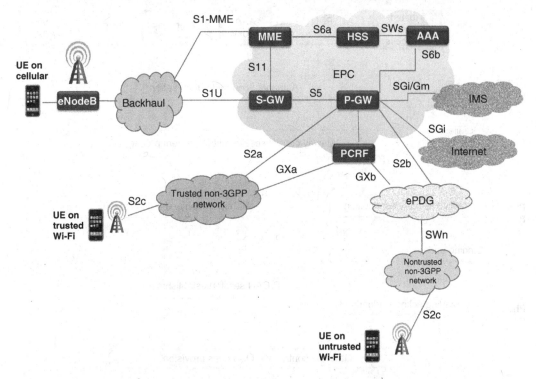

Figure 2.32   LTE–WLAN interworking

First we discuss the scenario where the UE is connected to the trusted non-3GPP access network. In this case, the UE can access the PGW (PDN gateway) using either an S2a or an S2c interface. Figure 2.33 and Figure 2.34 show the detailed call flows for these two cases (3GPP, 2009), depicting the initial attachment procedure when the mobile is in the WLAN network. In both of these figures, the trusted non-3GPP access network represents the Wi-Fi network inclusive of the WAG (wireless access gateway).

The S2c interface is based on the dual-stack Mobile IP version 6 (DSMIPv6) and requires that the UE supports it. DSMIPv6 sets up a tunneled connection between the UE and the PGW. This tunnel is used to forward traffic to and from the user equipment. Figure 2.34 shows the three phases of attachment as defined by 3GPP. During phase A, the mobile attaches to the network. As part of phase B, a DSMIPv6 tunnel is established with the PGW. Phase C confirms that the session is active. The S2c-based approach requires changes in the UE.

Alternatively, the UE can communicate with the PGW using an S2a interface. The S2a interface is based on the Proxy Mobile IPv6 (PMIPv6) protocol, which sets up a tunnel between the WAG and the PGW. In this case the mobility function resides on the WAG and provides Mobile IP functions transparently for the client. The mobile client does not have to change its IP address. The S2a-based approach eliminates the need for client software.

Figure 2.35 shows the call flow in a network-assisted S2a-based handover when the UE hands over from the WLAN network to the LTE network. We now briefly describe the steps involved in the handover process. Initially, the mobile is connected using the WLAN interface and communicates

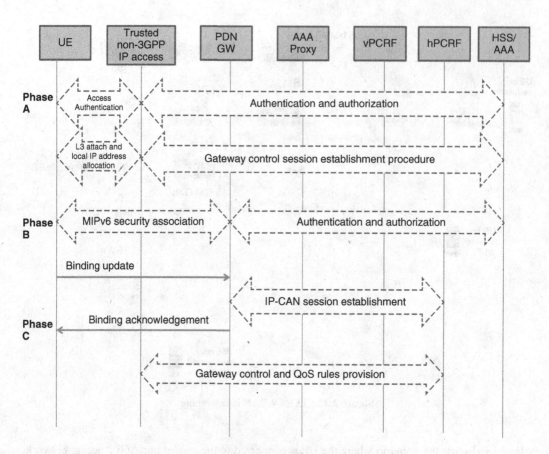

Figure 2.33   Client-assisted WLAN–LTE interworking

over the PMIPv6 tunnel set up between the ePDG (evolved Packet Data Gateway) and the PDN gateway. As the signal-to-noise ratio goes below some threshold, the mobile begins to search for an available 4G network and initiates an attach process through the MME and HSS. After successful authentication at the HSS over the S6a interface, a bearer path is created for the mobile with proper quality-of-service parameters. Depending upon the types of application supported by the mobile, bearer paths with different QoS parameters are created. After successful bearer path establishment and radio establishment, the data connectivity for the existing session changes. Data now traverses the PMIPv6/GTP tunnel set up between the SGW and the PDN gateway.

However, in the case of client-assisted S2c-based handover, the client uses the DSMIPv6 protocol to communicate with the PGW. When the client is on a non-3GPP network, the UE builds a DSMIPv6 tunnel to the appropriate P-GW and is assigned a care-of address. The UE uses the MIPv6 tunnel to communicate with the PGW. However, when the mobile switches back to the 3GPP network, it does not need to obtain a new address but instead uses the existing PMIPv6/GTP tunnel between the SGW and PGW.

In the second scenario, when the UE is connected to a nontrusted public Wi-Fi network, it accesses the PGW through an ePDG using the SWn and S2b interfaces (3GPP, 2009). In this case the UE has to

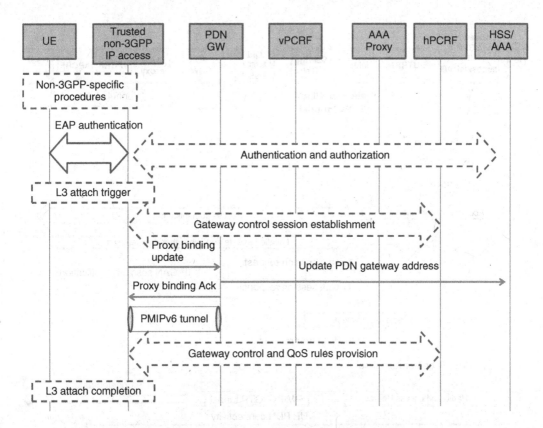

Figure 2.34   Network-assisted WLAN–LTE interworking

set up a secure tunnel with the ePDG through the untrusted WLAN network over the SWu interface. Similarly to the trusted-network scenario, the handover can be client-assisted or network-assisted. For network-assisted handover, the S2b interface, based on PMIPv6, offers handover capabilities similar to those of S2a. In this case the ePDG acts like a MAG (media access gateway), whereas the PGW acts like an LMA (local mobility agent). Figure 2.37 illustrates network-assisted handover from the WLAN (untrusted non-3GPP network) to the LTE network.

Figure 2.38 illustrates a client-assisted handover from the WLAN (untrusted non-3GPP network) to the LTE network. In this case, the S2c interface is used to provide client-assisted handover over untrusted access networks. However, the client uses the SWu interface to provide the security access to the packet core using IPSec.

2.9   Multicast Mobility

IP multicasting allows IP packets to be delivered from a single source to a group of receivers that are part of the same multicast group using a multicast routing protocol. Multicast packets are generally routed along a single shared tree or multiple source-based spanning trees for efficient content distribution. There are several proposed schemes to provide native IP multicast routing over a wide-area

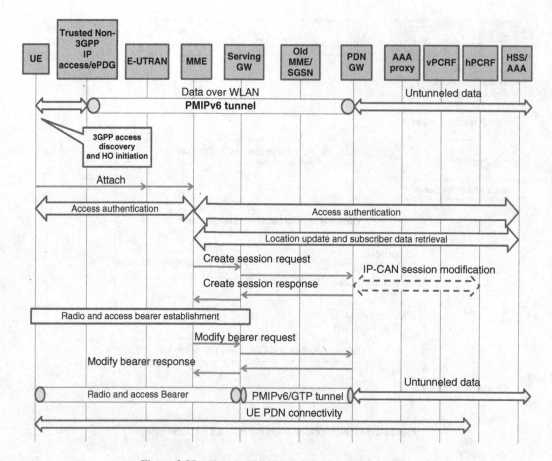

Figure 2.35 Flow for S2a-based WLAN–LTE handover

network, such as PIM (Deering et al., 1994), MOSPF (Moy, 1993), DVMRP (Waitzman et al., 1988), CBT (Ballardie et al., 1993), and BGMP (Thaler, 2004). Traditional multicasting techniques cannot handle a large number of distinct multicast groups and do not provide a means to handle multicasting when some routers are multicast-capable while others are. Recently, however, alternative techniques have been developed, such as Source Specific Multicast (SSM) (Holbrook and Cain, 2006), UMTP (UDP Multicast Tunneling Protocol) (Finlayson, 2003), and AMT (Thaler et al., 2002), that provide multicast support for nonmulticast-enabled networks while providing novel ways to support specific applications such as content distribution. Quinn and Almeroth (2001) explained many of the issues associated with multicast deployment over wide-area networks. Security, scalability, and interoperability are some of deployment issues associated with multicasting. However, we focus here on mobile-receiver issues. The multicast group "join" and "leave" latencies are some of the deployment issues related to multicast mobility. Joining and leaving multicast groups are usually handled through the Internet Group Management Protocol (IGMP) (Cain et al., 2002; Fenner, 1997), although we have developed some alternative application layer techniques, which are described in Chapter 9. In this section, we highlight only the mobility aspects of multicast traffic.

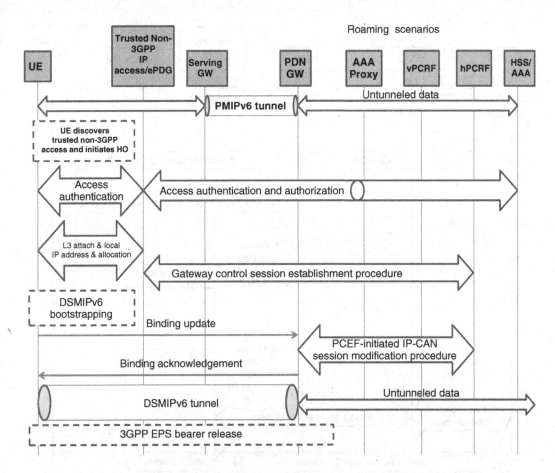

Figure 2.36 Flow for S2c-based LTE–WLAN handover

Unlike unicast traffic, multicast traffic distribution is receiver-driven, where the mobile receiver makes a request to its access router to join a specific multicast group and the router joins the upstream multicast tree. The mobile receiver periodically exchanges its group information with the router using IGMP (Cain et al., 2002) in an IPv4 network or multicast listener discovery (MLD) (Almesberger et al., 1997) in a IPv6 network. Mobile multicast introduces several challenges, namely general routing issues, mobile-receiver issues, mobile-source issues, and deployment issues. Movement of a sender and receiver can introduce encapsulation and decapsulation of packets and routing-state maintenance owing to dynamic creation of a multicast tree. The receiver issues include join latency, packet loss, packet duplication, packets out of order, and leave latency, which affect the performance of real-time traffic. Reverse path forwarding (RPF), packet loss, multicast scoping, and source discovery are some of the issues associated with multicast delivery when the source is mobile. We describe some of the work related to multicast mobility in this section.

Xylomenos and Polyzos (1997), Varshney and Chatterji (1999), and Acharya et al. (1995) described many of the architectural issues associated with mobility support for multicast traffic. Romdhani et al. (2004) categorized mobile multicasting primarily into four categories:

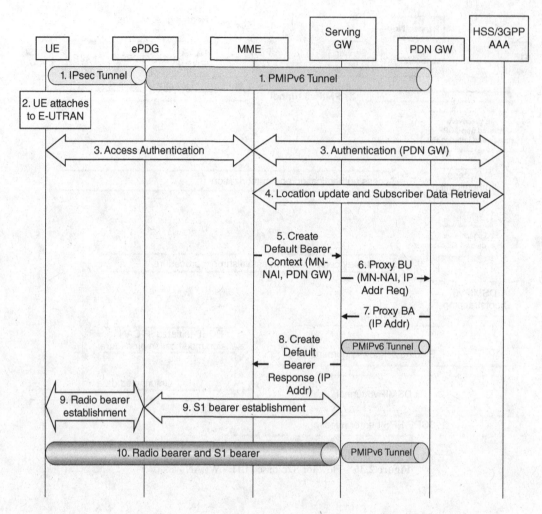

Figure 2.37   Flow for S2b-based WLAN (untrusted)–LTE handover

home-subscription-based, remote-subscription-based, hybrid, and non-IP multicast-based solutions. We describe two of these approaches and some of the mobile multicast protocols in each category.

Home-Subscription-Based Approach

The home-subscription-based or bidirectional-tunneling approach depends on the home network and associated Mobile IP entities such as the home agent and the foreign agent, and uses a multicast router located in the home network. It puts the burden of multicasting on the home agent by creating tunnels between the HA and the mobile or the foreign agent. However, tunneling multiple multicast packets to a foreign network is inefficient.

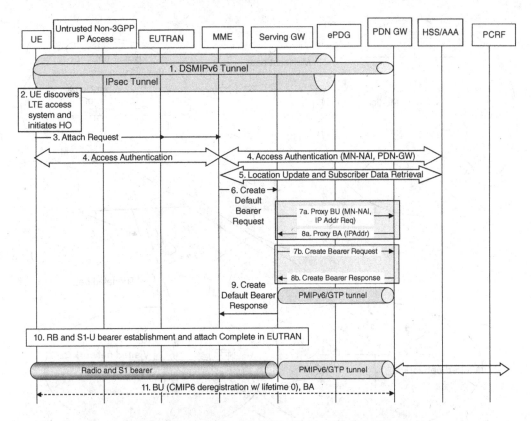

Figure 2.38   Flow for S2c-based WLAN (untrusted)–LTE handover

Figure 2.39 illustrates the home-subscription-based architecture. Figure 2.40 describes the associated flow. In this architecture, the HA is multicast-enabled and is responsible for periodically forwarding multicast group membership control messages to the mobile while it is away. To join a specific multicast group, the mobile node establishes a bidirectional tunnel with its HA and tunnels its membership message (e.g., IGMP) to the HA. It uses the same tunnel header used for routing unicast packets between the MN and the HA. When the HA receives the join request, it decapsulates and forwards it to the local multicast router on the home link. The local multicast router intercepts this membership message and sends the join message to the upstream router on the home network. Once the multicast branch is established, the HA forwards the multicast traffic destined for the mobile over the tunnel. Thus, the HA acts like a multicast proxy for the mobile. When the mobile moves to the new foreign subnet, it does not need to join the multicast tree again, as the HA has already joined the multicast tree on behalf of the mobile. Although this method has the advantage that the mobile does not need to rejoin the multicast tree when the mobile moves from one network to another, this scheme suffers from triangle routing and tunnel overhead, resulting in join latency. The HA needs to establish a per-MN tunnel to forward the multicast traffic. The HA is also the central point of failure in this scheme.

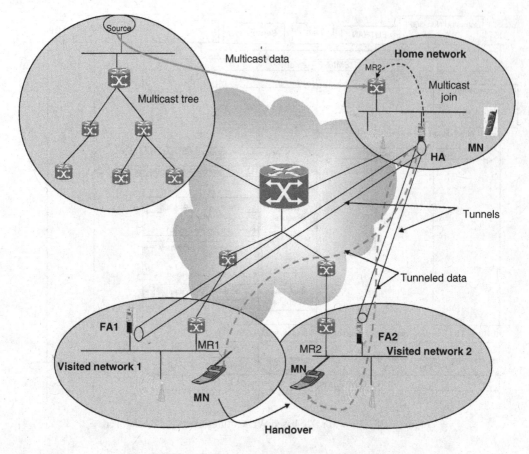

Figure 2.39   Multicast mobility: home-subscription-based

There are several mobile multicast protocols that belong to the home-subscription-based category and reduce some of the drawbacks associated with the basic home-subscription-based approaches. We highlight some of these multicast protocols here. Mobile Multicast (MOM) (Williamson et al., 1998) proposes to reduce the explosion problem in bidirectional tunneling by electing one designated HA called the designated multicast service provider (DMSP). The MOM protocol uses a single tunnel between the HA and the FA instead of using a per-MN tunnel. The FA, in turn, uses link-level multicast to complete the delivery. Range-based MOM (Lin and Wang, 2000) takes the MOM approach one step further and elects a multicast agent close to the FA to tunnel multicast packets to the foreign network. MPDSR (Multicast Protocol with Dynamic Service Range) (Yang and Park, 2001) enhances the RBMOM protocol in the Mobile IPv6 environment. The main goal of this approach is to determine an optimal service range and avoid service disruption. It does so by introducing two other network elements, namely the core source node (CSN) and boundary foreign agent (BFA). In order to eliminate possible disruption triggered by the handover procedure, the mobile receiver prejoins the multicast group prior to the handover.

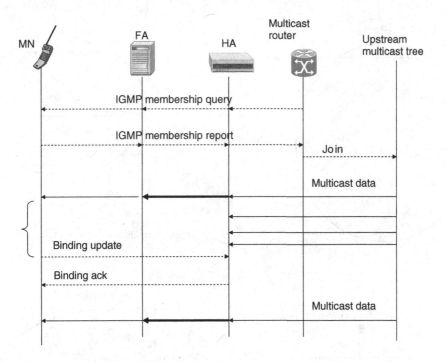

Figure 2.40   Flow for home-subscription-based architecture

Remote-Subscription-Based Approach

The remote-subscription-based approach takes the burden off the home agent, eliminates tunneling, and avoids the duplication of multicast packets being tunneled to foreign networks. However, this requires that after each handoff, the user must rejoin a multicast group. In addition, the multicast trees used to route multicast packets must be updated after every handoff to track the members of the multicast group.

As shown in Figure 2.41, unlike the case in the home-subscription-based approach, in the remote-subscription-based approach the mobile receiver joins the multicast group via a local multicast router (MR1) on the foreign network. The mobile sends its membership report message to the local multicast router (MR1) in the visited network, where the mobile currently resides. The mobile node and the multicast router take care of group management using IGMP for IPv4 and MLD for IPv6. The local multicast router intercepts this message and joins the desired multicast group by joining the upstream routers. Figure 2.42 shows the flow associated with the remote-subscription-based approach. Since this approach does not depend upon the home network or any of the home network elements, the mobile obtains a new care-of address (CoA) after the handover and sends a new membership report to the multicast router (MR2) in the new network. While MR2 is in the process of joining the upstream multicast tree, the router in the previous network (MR1) leaves the upstream multicast tree if there are no other receivers tuned to the multicast group. This is called *leave latency*, as defined in Appendix B. After the mobile node has moved to the new network, it does not have to wait to complete Mobile IP registration before it is able to

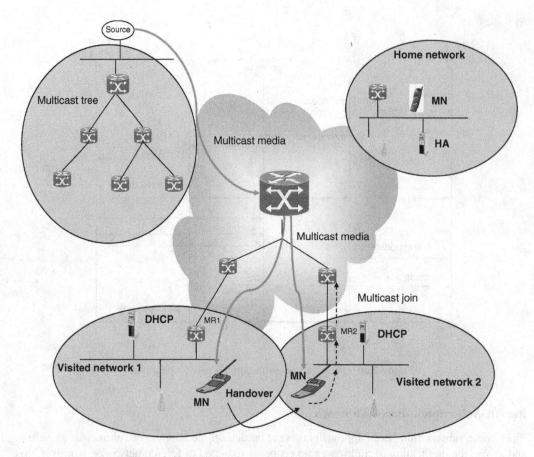

Figure 2.41   Multicast mobility: remote-subscription-based

join the multicast tree in the new network. Thus, the join latency in the remote-subscription-based approach is less than that in the home-subscription-based approach.

There are several remote-subscription-based mobile multicast protocols available that aim to reduce the join latency and tunnel convergence problem. These protocols include remote-subscription-based Mobile IP, Mobicast (Tan and Pink, 2000a), hierarchical SSM (Kim et al., 2001), remote subscription with a multicast agent, preregistration with a mobility support agent (MSA) (Wu, 1999), a multicast agent protocol (MMA) (Shin et al., 2000), and a timer-based mobile multicast protocol (TBMOM) (Park, 2002). The mobility support agent protocol offers a solution based on preregistration to reduce the join latency and associated packet loss. The multicast agent protocol MMA uses a forwarding technique between the foreign networks instead of the preregistration method. This reduces the packet loss due to join latency. The timer-based mobile multicast protocol TBMOM selects a foreign multicast agent (FMA) to store the information about the membership of the mobile members in the foreign network. The TBMOM protocol reduces multicast packet loss since unicast tunnels can be set up between the FMAs of different foreign networks. The hierarchical SSM-based (where "SSM" stands for "single-source multicast" here) and Mobicast-based approaches provide hierarchical mobility management by introducing

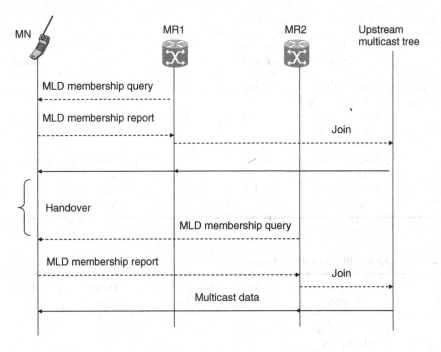

Figure 2.42   Flow for remote-subscription-based architecture

hierarchical mobility agents such as a border gateway router (BGR) and a domain foreign agent (DFA), respectively. Both of these approaches split the multicast path into a two-level hierarchy and thus limit the distribution of multicast traffic in the lower hierarchy whenever there is movement within a hierarchy.

We provide a summary of these two types of multicast protocols and qualitative comparisons of them with respect to some performance parameters, namely join latency, point of failure, tunnel convergence, and hierarchical mobility properties, in Table 2.4 and Table 2.5. The tunnel convergence problem arises when a number of mobiles belonging to different home networks are anchored at a specific foreign agent. If all of these mobiles are subscribed to a specific multicast group, then each HA responsible for the corresponding mobile tunnels the multicast packets to the FA. This will result in duplicate packets for the same multicast group. The number of duplicate packets will increase as the number of mobile hosts subscribed to the same multicast group increases. Thus, it is desirable to reduce the problem of tunnel convergence. We now briefly present the comparison of the join latency for the home-subscription- and remote-subscription-based protocols. For example, the average join latency in the home-subscription-based approach is independent of the multicast group size, as each mobile receives a multicast datagram from its home agent independently. The average join latency for MOM is moderate and is similar to that for the home-subscription-based approach. However, in the MOM protocol, disruption decreases with increasing group density owing to sharing of the DMSP by visiting members of the same group in the same foreign network. Compared with the home-subscription approach and MOM, the remote-subscription and multicast agent approaches provide less disruption since access to the multicast backbone is much closer. However, the multicast agent approach has lower disruption than the remote-subscription approach when the

Table 2.4 Multicast mobility – home subscription

Protocol	Join latency	Point of failure	Hierarchical mobility	Tunnel convergence	Key differentiator
Mobile IP home subscription	Remains same with group size	Home agent	No	One tunnel per MN	–
MOM	Reduces with increase in group size	DMSP	No	One tunnel for all MNs	Introduces DMSP
RBMOM	Smaller than MOM	DMSP	No	One tunnel for all MNs	Introduces service range
MPDSR	Least handoff delay	BFA	No	One tunnel for all MNs	Reduces tunnel length

Table 2.5 Multicast mobility – remote subscription

Mobility protocol	Join latency	Point of failure	Hierarchical mobility	Tunnel convergence	Differentiator
Remote subscription	Less than home subscription	FA or DHCP	No	N/A	
Multicast agent	Same as RS	MA	No	Tunnel between FA and MA	Uses coordinator MA between FAs
MSA	Less than RS (preregistration)	MSA	No	No tunnel	Preregistration
MMA	Less packet loss due to forwarding	MA and MF	No	One IP tunnel (MA and MF)	Forwarding technique reduces packet loss
TBMOM	Less than RS	DMSP	No	Does not use IP tunnel	Uses hybrid forwarding approach
Hierarchical SSM	Less delay than RS	BGR	Yes	IP tunnel between source and BGR	Reduces join latency for micromobility
Mobicast	Less packet loss buffering	DFA	Yes	Does not use IP tunnel	Buffering solves packet loss

multicast group size is small. As the group size increases, the remote-subscription method does better than the multicast agent approach. However, at a very high group density, both of these approaches offer the same performance, as the probability of finding other group members in a new IP subnet approaches 1. The mobility support agent approach reduces the join latency to a great extent because of its ability to provide preregistration before the handoff.

Multicast mobility protocols in their current form suffer from performance issues due to tunnel overhead and to join and leave latency. Thus, optimization techniques need to be applied to the existing multicast mobility protocols to reduce the join latency and leave latency. In Chapter 10, we present some of the optimization techniques that we have developed for multicast stream delivery that reduce the join latency in a hierarchical multicast environment.

2.10   Concluding Remarks

A careful survey and analysis of the handoff processes for cellular mobility protocols and several IP-based mobility protocols can be extrapolated to the common functions that are required to complete a handoff event. Since many of these functions, such as discovery and authentication, are access-dependent, one needs to take into account the access characteristics of the multilayer handoff operations for IP-based mobility when designing optimization techniques. Lessons learnt from the optimization techniques for cellular mobility can easily be applied to improve the optimization techniques for IP-based mobility. For example, the mobility-proxy-assisted forwarding technique for IP-based mobility benefits from GSM's MSC-anchored forwarding techniques to reduce packet loss by forwarding the media from the previous network. Soft-handoff techniques for CDMA-based cellular networks can be used as guidelines when designing IP-based fast-handoff mechanisms by applying bicasting or multicasting mechanisms. Thus, when one is designing any new mobility protocol for the next generation of networks or proposing a new optimization technique, it is very important to investigate the abstract handover primitives and study optimization techniques for cellular mobility.

3

Systems Analysis of Mobility Events

In this chapter, we analyze the mobility event resulting from the handoff of a mobile node. In particular, we analyze the discrete operations of the mobility event in order to design a formal mobility systems model. However, we limit the analysis of the model to infrastructure-based mobility only. The infrastructure-based mobility model assumes that the layer 2 and layer 3 network components of the core networks, namely access points, routers, and servers, are not mobile and only the last hop of the access network changes. Figure 3.1 shows several different functional components in an IP-based infrastructure-based mobility environment. This figure includes the layer 2 and layer 3 points of attachment, the configuration agent, the authentication agent, the authorization agent, and the signaling agent, which perform the primitive handoff operations. During a mobility event, these network components are engaged in distributed communication to take care of the handoff-related operations. Figure 3.1 also shows how the mobile node changes its point of attachment as it moves between the layer 2 point of attachment, the layer 3 point of attachment, different administrative domains, and various access technologies. The labels A, B, C, and D in Figure 3.1 show the locations of the mobile node as it connects to different points of attachment during its movement. The mobile starts from the location marked A and moves to the locations marked B, C, and D. During this process, the mobile is subject to layer 2, layer 3, and administrative-domain handoff (e.g., handoff from domain A and domain B). Each of these handoffs (e.g., L2 and L3 handoff) involves different types of handoff operations and different amounts of delay. For example, a layer 2 handoff contributes up to 900 ms delay in an 802.11 environment; a layer 3 handoff contributes up to 4 s of delay, inclusive of layer 2 delay; and an administrative-domain handoff involving heterogeneous access results in a handoff delay of up to 18 s (Dutta et al., 2005c). The networks N1 and N2 are shown as networks with different access technologies (e.g., 802.11 and CDMA). Figure 3.2 shows three different audio outputs from a mobile when it is subject to layer 2, layer 3, and heterogeneous handoff. These output waveforms represent the media disruption due to the delays caused by the handoff-related operations. Thus, it would be useful to investigate the delays related to each type of handoff operation. In this section, we analyze the delays due to the handoff-related operations in different layers.

Mobility Protocols and Handover Optimization: Design, Evaluation and Application, First Edition.
Ashutosh Dutta and Henning Schulzrinne.
© 2014 John Wiley & Sons, Ltd. Published 2014 by John Wiley & Sons, Ltd.

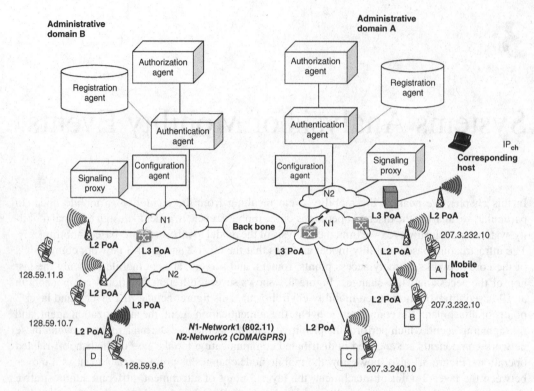

Figure 3.1   Functional components of infrastructure-based mobility

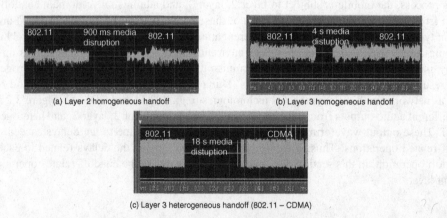

Figure 3.2   Media disruption due to handoff-related operations

3.1 Summary of Key Contributions and Indicative Results

Currently, there is no work that provides a systematic analysis of the handoff processes and the associated subprocesses that are executed in all layers during an IP-based handoff. Without such a systematic analysis of the handoff processes, it is difficult to determine the data dependency among the handoff processes, which could lead to the design of potential optimization techniques.

Based on the comparative analysis of mobility protocols presented in Chapter 2, we have systematically analyzed each of these handoff processes for an IP-based 4G network. This systematic analysis of the abstract functions highlights how these mobility functions are performed across several layers in an IP-based network. For example, a discovery function in layer 2 implies discovering a layer 2 point of attachment such as an access point, while discovering a layer 3 point of attachment involves discovering a default router. Similarly, authentication, security association, and configuration are other handoff components that take place across all layers. We have analyzed the basic primitive operations that are part of each of these handoff functions. Based on this handoff analysis, we have built several unoptimized handoff systems, such as systems based on MIP-based network layer mobility (MIPv4 and MIPv6) and SIP-based application layer mobility, and obtained experimental results for each of these unoptimized handoff components that show the delays associated with each of the mobility components (e.g., discovery, authentication, and configuration).

From this systematic analysis (or decomposition) of the handover operations and experimental results from some of the indicative unoptimized handoff systems that we built, it is possible to determine the delays associated with each of the primitive operations associated with the handoff event. Unlike other experimental analyses (Vogt, 2006) that were meant only for one type of mobility protocol (e.g., MIPv6), these experimental analyses were based on both network layer and application layer mobility protocols and worked across access techniques (e.g., CDMA and 802.11). Comparative analysis of our experimental results for different mobility protocols has verified the mapping between handover delays and common abstract functions independent of the type of mobility protocol and access mechanism.

In the rest of the chapter, we analyze the handover components and map these to different layers, present experimental results from several unoptimized mobility protocols that we experimented with, and map these delays to the respective handoff components.

3.2 Introduction

In order to determine the primitive operations that take place during a handoff event, we investigated and analyzed several cellular and IP-based mobility protocols in different layers. We have described the details of the handoff operations for a set of cellular and IP-based mobility protocols in Chapter 2. In Table 3.1, we summarize how the various cellular and IP-based mobility protocols perform similar sets of operations to support a handoff event. In particular, we illustrate how these mobility protocols carry out different handoff-related functions, namely network discovery, resource discovery, network detection, configuration, authentication, encryption, binding update, and media rerouting. The column headings in Table 3.1 represent the primitive operations that are part of a handoff event for the various types of mobility protocols represented in the row headings.

The operational aspects of these functions play an important role in the design of any specific mobility protocol. For example, the operational aspects of a discovery function could include many factors, such as the number of messages between the mobile and the server, and the discovery mechanism (e.g., network layer or application layer). The performance of any mobility protocol

Table 3.1 Mapping of basic operations of a mobility event

Mobility type	Access type	Network discovery	Resource discovery	Triggering technique	Detection technique	Configuration	Authentication	Encryption	Binding update	Media rerouting
GSM	TDMA	BCCH	FCCH	Channel strength	SCH	TMSI	SRES A3	DES	MSC control	Anchor
WCDMA	CDMA	Pilot	Sync channel	Channel strength	Frequency	TMSI	SRES/A3	AES	Network control	Anchor
IS-95	CDMA	Pilot	Sync channel	Channel strength	RTC	TMSI	Diffie–Hellman AES CAVE	AES	MSC	PDSN msC
802.11	CSMA/ CA	Beacon 11R	11R 802.21	SNR at mobile	Scanning channel number SSID	SSID channel number	802.1X EAPoL	WEP WPA 802.11i	Associate	IAPP
Cellular IP	Any	Gateway beacon	AP beacon	AP beacon ID	GW Beacon	MAC address	IKE	AES	Route update	Intermediate router
MIPv4	Any	ICMP RA, FA advertisement	ICMP	FA advertisement, L2 assisted	FA advertisement	FA-CoA Co-CoA	IKE PANA EAP	AES	MIP registration	FA RFA HA
MIPv6	Any	Stateless RA proactive	CARD 802.21 802.11R	Router adv.	Router prefix	CoA	IKE PANA EAP	AES	MIP update MIP-RO	CH MAP HA
SIP-based mobility	Any	Stateless RA ICMP router	802.21 802.11R	L3 router adv.	Router prefix ICMP	CoA AOR register	IKE EAP INVITE Exchange	AES SRTP	Re-INVITE INVITE	B2BUA CH RTPTrans

depends upon how efficiently these functions operate. In order to get a better understanding, we performed a comparative analysis of an application layer mobility protocol (SIP-based mobility) and a network layer mobility protocol (MIPv6), and have illustrated how each of these primitive operations is carried out for both of these mobility protocols in Dutta et al. (2006d). In the following, we briefly describe how some of the handoff operations are taken care of by each of these two mobility protocols.

3.2.1 Comparative Analysis of Mobility Protocols

In Chapter 2, we described mobility management techniques for both SIP-based mobility and Mobile IPv6 (MIPv6). In this section, we compare how some of the basic operations such as discovery, registration, binding update, configuration, location management, tunneling, and security operations are performed for each of these two mobility protocols.

In SIP-based mobility management, as part of the registration process, the mobile updates its IP address using the visited SIP proxy or home proxy. In the case of MIPv6, the mobile sends a binding update to the home agent and corresponding host. Thus the binding update to the home agent can be regarded as the registration process for MIPv6.

Both MIPv6 and SIP-based mobility require no foreign agent in the network, and thus use a co-located care-of address as the new identifier. SIP-based mobility for IPv4 networks depends mostly upon DHCP for IP address configuration. Both SIP-based mobility for IPv6 networks and MIPv6, however, use stateless autoconfiguration to configure a layer 3 identifier.

In the absence of route optimization, payload packets are tunneled between the mobile node and the home agent in both directions in Mobile IPv6. This specific tunneling mechanism uses IPv6 encapsulation as specified in RFC 2473 (Conta and Deering, 1998). In addition to the extra headers assigned to the original packet, additional time is spent owing to the processes of encapsulation and decapsulation. SIP-based mobility, on the other hand, does not make use of tunneling, as the media travel directly between the correspondent host and the mobile host. Thus, processing delay due to encapsulation and tunneling overhead is avoided when SIP-based mobility is used.

Binding update was defined in Chapter 2. In an IP-based environment, it is the process of notifying the correspondent node or another networking node such as a home agent about the new layer 3 identifier of the mobile so that data can be forwarded to the new address of the mobile after the handoff. In the case of Mobile IPv6, as the mobile obtains a new care-of address either via a stateful DHCP server or via stateless autoconfiguration, it notifies the correspondent host and the home agent. Since route optimization is an inbuilt mechanism for MIPv6, the new data obtained during mid-session mobility does not need to be rerouted via the home agent. On the other hand, SIP-based mobility sends the binding update as part of its re-INVITE signal to the correspondent host only, since there is no home agent for SIP-based mobility.

Mobile IPv6 does not by itself provide a fast-handoff mechanism. However, there are extensions to Mobile IPv6, namely FMIPv6 (Koodli, 2005) and HMIPv6 (Soliman et al., 2006), that provide fast handoff during mid-session mobility. FMIPv6 provides a fast-handoff mechanism by introducing several reactive and proactive mechanisms, whereas HMIPv6 introduces a mobility anchor point (MAP) to take care of intradomain mobility. Similarly, SIP-based mobility has been extended to provide fast handoff by introducing a network entity called a B2BUA (back-to-back user agent) into the hierarchy to limit the binding update or by forwarding transient traffic from the previous network.

Both of these mobility protocols provide security for the signaling and data. MIPv6 takes advantage of network layer security mechanisms such as IPSec to protect the signaling between the mobile and the home agent. It can also use an IPSec tunnel instead of an IP–IP tunnel between the mobile and the home agent to carry tunneled data. SIP-based mobility, on the other hand, can provide multilayer security. It can either choose IPSec to provide security in layer 3 for both signaling and media (RTP), or use S/MIME to secure the SIP signaling and use secure RTP (SRTP) to secure the media stream. However, SRTP can also be used to protect data in the case of MIPv6, but it needs a separate key distribution architecture, unlike the case for SIP-based mobility, where the key is distributed by an INVITE exchange method.

Mobile IPv6 provides return routability support by using CTI (care-of test init) and HTI (home test init) messages. This actually verifies the new care-of address of the mobile before the binding update is sent out. While it helps to avoid session hijacking and so on, it adds delay to the binding update procedure. SIP, on the other hand, does not support inherent return routability testing, but the new care-of address of the mobile can be verified by using cryptographic techniques such as SIP identity (Peterson and Jennigs, 2006).

3.3 Analysis of Handoff Components

Based on an analysis of the basic operations associated with both IP-based mobility protocols and cellular protocols, we can categorize the handoff process into six main phases, namely *network discovery and selection*, *network attachment*, *configuration*, *security association*, *binding update*, and *media rerouting*. Figure 3.3 decomposes the mobility event into several processes and subprocesses.

Each of these operations involves several network elements, namely the mobile node, access point, router, server, and correspondent node. Figure 3.4 shows how each of these suboperations involves

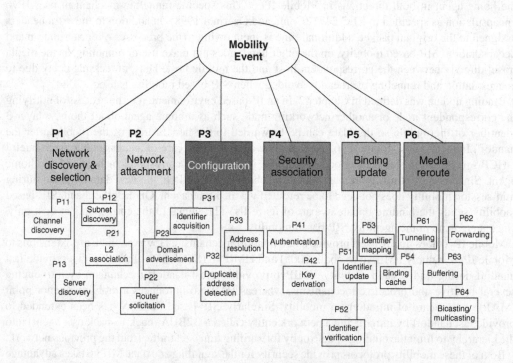

Figure 3.3   Systems decomposition of handoff

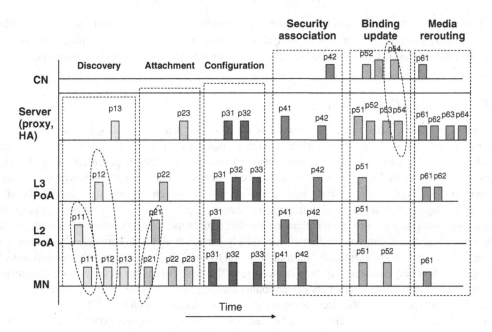

Figure 3.4   Handoff functions distributed across network elements

several components of the network in several layers. For example, layer 2 discovery involves inter-action between the mobile node and the access point, whereas layer 3 discovery involves the mobile node, the access point, and the router in the network. On the other hand, a binding update involves the core components of the network, such as the home agent, the server, and the core routers. We describe these handoff operations in detail.

Handover initiation, or handover preparation, precedes a handover process. The handover initiation process does not contribute to the handover delay. Depending upon who initiates the handover and who has primary control over it, a handoff process can be either mobile-initiated, network-initiated, mobile-controlled, or network-controlled Manner and Kojo 2004. During this phase, the mobile or the network node determines the need for a handoff based on measurements on the mobile, and initiates the preparation for the handoff operation. For example, during a mobile-initiated handoff, the mobile makes a decision about an impending handoff based on the signal-to-noise ratio (SNR) or the quality of service of the media traffic, and starts a network discovery process to determine the best available network to connect to. A handover process may result in reinitiation of layer 2 or layer 3 operations based on whether the mobile node loses its layer 2 or layer 3 connectivity, respectively. For example, if the mobile does not receive a layer 3 router advertisement, it initiates a layer 3 handoff process without necessarily disturbing the layer 2 association. On the other hand, a layer 2 unreachability event resulting from a low SNR or nonreceipt of a beacon will lead to a handover initiation process in layer 2, which may or may not involve handoff in layer 3.

The IEEE 802.21 working group (http://www.ieee802.org/21) has defined several event service primitives as part of its Media Independent Handover Function (MIHF). "Link Up," " Link Going Down," and "Link Down" are examples of such primitives, which can be used to provide the status of lower-layer events and expedite the handoff process. For example, since the layer 2 association takes place before any other upper-layer operations, it helps to send a layer 2 event notification, such as "Link Up," to execute an upper-layer mobility function such as layer 3 configuration. The SNR threshold may generate such an event notification. Thus, an event notification from a lower layer helps to establish a successful link to the new point of attachment in an expedited manner.

3.3.1 Network Discovery and Selection

The very first phase of the handoff process is the network discovery and selection phase. During this phase, the mobile or network discovers possible new points of attachment. This phase involves discovering both the neighboring networks and the resources required for handoff within the network. Once the target network is discovered, several resource parameters within the target network are retrieved, including the channel number, bandwidth, encryption algorithm, authentication server, registration server, and configuration server. The resource discovery process helps the mobile to associate with the proper channel number and proper authentication parameters in the new network so that it can communicate successfully. Arkko et al. (2008) provided an overview of network discovery and selection, and their application to handoff. Based on the type of access technology, the discovery process may be passive, where the node listens for network announcements, or active, where the node solicits network announcements. Different types of discovery process take place for each type of network, such as discovery of a new location area for GSM or discovery of a new routing area in the case of GPRS. For IP-based networks, the discovery process can span across all layers and include discovery of a layer 2 PoA (point of attachment), a layer 3 PoA, a subnet, or a domain. Here, a domain is defined as an administrative domain.

Each layer 2 access technology provides a different out-of-band mechanism for discovering networks and resources. For example, GSM uses a broadcast control channel (BCCH), CDMA uses a pilot channel, and 802.11 uses active and passive scanning to discover the new point of attachment. Based on the access characteristics of the target network, discovery of the appropriate network takes different amounts of time. For example, in an IEEE 802.11 network, without any optimization, an active-scanning operation takes about 500 ms (Mishra et al., 2003a), including the channel probe, authentication, and association delays. The timing of network discovery in GSM networks is determined by the frequency of the broadcast channel. In the case of IP-based mobility, some of the upper-layer detection mechanisms, such as foreign agent advertisement and router advertisement, can help discover the servers in the network.

Network selection is a process by which a mobile node or a network element analyzes the information discovered about its neighboring networks, and then selects a network to connect to. The selection may be based on criteria such as required QoS, cost, or user preferences. An appropriate selection mechanism helps the overall resource optimization process and increases the probability of successful handoff to the target network.

3.3.2 Network Attachment

After the mobile has selected the target network, it attempts to connect to the new network point of attachment. Network attachment can take place in several layers. A mobile attaches to an 802.11 access point by means of layer 2 association and connects to a router by means of layer 3 association. Event notifications from lower layers, such as the availability of a new point of attachment or the sudden lapse of an existing connection, is usually passed to the upper layers in order to initiate the subsequent handoff-related functions in an expedited manner. The IETF is currently working on standardizing a protocol for the detection of network attachment called DNA (Choi, 2005) that involves mechanisms in both layer 2 and layer 3 and can notify the upper layer about network detection.

3.3.3 Configuration

The configuration phase helps to prepare the reconnection path of the mobile. During this phase, the mobile obtains a temporary identifier in the network it is visiting, discovers servers such as the outbound SIP server and the default gateway, and maps its address to the appropriate network entity in the network. The configuration phase can be split into the following subphases: *identifier acquisition, duplicate address detection*, and *address resolution*. During the identifier acquisition process, the mobile acquires a new temporary L2 or L3 connection identifier at the new point of attachment in the network. For example, this could include a care-of address (CoA) in the case of IP-based mobility or a temporary mobile subscriber identity (TMSI) in the case of GSM. In an IPv4-based network, the mobile uses a DHCP server to obtain the care-of address, whereas in an IPv6-based network the mobile obtains the address either from a DHCP server or using a stateless autoconfiguration method.

The duplicate address detection process tests the uniqueness of the mobile's address in the network. Address uniqueness testing is performed differently in IPv4 and IPv6 networks. For example, when a mobile node (MN) uses stateless address autoconfiguration to assign an IPv6 address to its interface after getting a new router advertisement (RA), it sends a neighbor solicitation on the local link in order to verify whether any other node on the link has the same address. When a predetermined time has elapsed and the MN has not received any reply, the MN assumes that no other node on the link has this address and finally assigns that address to its interface. IPv4 nodes use an ARP broadcast to determine if any other node in the network has the same address.

Once the uniqueness of the address has been determined, the mobile and the default router keep the mapping between the IP addresses and MAC addresses of each other using an address resolution mechanism. This address resolution mechanism helps the mobile and the default router to communicate with each other in layer 2 in the same subnetwork.

In general, the completion of these processes requires a series of signaling messages between the mobile and servers in the network and thus contributes to the handoff delay.

3.3.4 Security Association

A security association is defined as a secure channel between two end points that applies a security policy and uses keys to protect information. Before a new communication path is established between the end points, the mobile node needs to authenticate itself and then establish a security association with other network nodes, such as routers, access points, or the correspondent host. Establishing a security association involves the procedures of *authentication and authorization* and a *key derivation*. The authentication and authorization processes allow the mobile to get access to network resources. The authentication process includes an exchange of messages between the mobile, the authenticator, and the authenticator server, for example an AAA server in the network. A successful authentication process generates a key that can be used to encrypt the data. The encryption process itself is considered to be separate from authentication and does not affect the handoff delay, unless the type of encryption algorithm changes after the handoff. However, the process of deriving the encryption key contributes to the handoff delay.

Security association may take place in several layers of the protocol stack. For example, during a layer 3 movement in an 802.11 access network, EAP messages need to be carried between the mobile node, the authenticator, and the authentication server to generate a shared master key, followed by a four-way handshake between the mobile node and the access point to generate a PTK (pairwise transient key) that can be used for encryption in layer 2. For an IP-based network, ISAKMP (Internet Security Association and Key Management Protocol) (Maughan et al., 1998) defines the mechanism for establishing a security association in layer 3 using IPSec. An IPSec security association is defined by an encryption algorithm, an authentication algorithm, and a shared session key.

Each mobility protocol also uses a different mechanism for authentication. As shown in Table 3.1, the mobile uses SRES (signed response) and the A3 algorithm for authentication in GSM. For 802.11 access networks, the mobile can use open system authentication, shared-key authentication by WEP, or IEEE 802.11i-based authentication in layer 2. Fathi et al. (2005) demonstrated how different authentication mechanisms affect the association and transport delay in layer 2. Similarly, layer 2 access-independent network layer authentication protocols such as PANA (Protocol for carrying Authentication to Network Access) (Jayaraman et al., 2008) add delay during layer 3 authentication. Georgides (2004) showed that it can take up to 4 s to complete the authentication and authorization process. These operations can use a combination of EAP (Extensible Authentication Protocol) (Aboba et al., 2004) over layer 2 for IEEE 802.1x-based authentication and EAP-TLS (Transport Layer Security) (Aboba and Simon, 1999) for layer 3 authentication. At the time of reconnection after handoff, the reauthentication process adds to the handoff delay. An authentication process is followed by an authorization process in some cases. For example, an 802.1x-based or PANA-based authenticator usually communicates with an AAA server to complete the authorization process. In the case of interdomain mobility, the AAA server changes. Thus, the authentication process is further delayed as the mobile needs to go through the authorization process using the new AAA server.

3.3.5   Binding Update

Binding update is the process by which a mobile can update its newly obtained network identifier so that the data after handoff can be rerouted to the new destination. A binding update process consists of three main phases, namely identifier update, identifier mapping, and binding cache update. This identifier update process associates the new network identifier with the permanent identifier of the mobile. As the mobile connects to a new point of attachment and obtains a new temporary network identifier (e.g., a TMSI in GSM, a CoA in MIPv6, or an FA-CoA in MIPv4) in the new network, it needs to update the correspondent host and home agent so that packets can be routed to the new destination.

Until the reassociation of the new identifier is complete and the binding cache in the correspondent node or home agent has been updated, the transient in-flight data continues to go to the old network and is lost in the absence of any optimization mechanism, such as buffering or packet forwarding. For application layer mobility, the binding update process also includes a registration process. By means of the registration process, the mobile establishes a mapping between the permanent locator, for example a URI (Uniform Resource Identifier), and the temporary identifier, for example a care-of address, for the proper functioning of location management. An optimized or hierarchical registration process expedites locating a user and provides faster delivery of the new data.

Binding update also needs to be authenticated. For example, MIPv6 introduces a return routability procedure and adds two additional messages, namely CTI (care-of test init) and HTI (home test init), to obtain a binding key that can authenticate the binding update message. However, this process contributes to additional delay in the binding update procedure after every handoff. Binding update can also be secured by way of IPSec. The security association for the binding update in an IP-based network can be uniquely identified by a tuple consisting of a security parameter index (SPI), an IP destination address, and a security protocol (AH or ESP) identifier.

3.3.6   Media Rerouting

The media rerouting phase is the last phase during the handover process and follows the binding update process. This phase involves rerouting of the media so that the delivery of data changes from the old path to the new path according to a predefined service guarantee.

Once the binding update is complete, the home agent or correspondent host updates its binding cache and the data from the correspondent node is routed to the mobile's new location. Media delivery can take place in several ways. In one of these, the media delivery can take place using the direct path between the correspondent node and the mobile node. In a second scenario, the media is delivered using an indirect path and uses a network entity called a home agent. However, until the time the new data is being sent directly to the mobile's new point of attachment (nPoA), there are packets in flight to the previous point of attachment. Media rerouting also takes care of rerouting in-flight packets. In both cases, the in-flight data can be captured and redirected to the new point of attachment. The media rerouting process may include several elementary operations, such as encapsulation, decapsulation, tunneling, buffering, and store-and-forward techniques. During the media rerouting process, transient data may be lost or delayed because of these operations. There is data overhead associated with the encapsulation, decapsulation, and tunneling operations. Thus, there is a need to optimize these operations to ensure that the media delivery delay is reduced after a handoff. Optimization techniques for media delivery are often referred to as route optimization methodologies. Route optimization techniques are defined in Chapter 6. As an example, operations in the network such as buffering or forwarding mechanisms can help to reduce packet loss but add delay to the traversal of packets. Thus, the buffering period needs to be adjusted to achieve a compromise between packet loss and one-way packet delay.

3.4   Effect of Handoff across Layers

In this section, we illustrate how these basic handoff operations affect multiple layers in an IP-based network. Based on the type of mobility, a subset of these operations is executed in each layer during a mobility event, and the overall handoff delay includes contributions from delays in all layers. For example, during an intradomain movement, the mobile is not subject to any delay due to the authorization process, unlike the case in interdomain mobility. Similarly, a layer 2 handoff does not involve delays due to layer 3 operations, namely layer 3 identifier acquisition and duplicate address detection. The following is a description of how optimization techniques can be useful for each of these basic operations in every layer.

3.4.1 Layer 2 Delay

In the 802.11 environment, channel scanning, probing, authentication, and association are the basic functions that contribute to the delay before a mobile completes network attachment in layer 2. Scanning is considered to be a discovery process in layer 2. Encryption and user authentication using WPA (Wi-Fi Protected Access) in conjunction with 802.1X (IEEE, 2006b) and EAP (Extensible Authentication Protocol) contribute additional delays because of the associated four-way handshake between the mobile and the access point. Mishra et al. (2003a) provided a comprehensive analysis of the delays due to the basic operations associated with a layer 2 handoff. Shin et al. (2004b) also discussed the probe, authentication, and association delays during layer 2 handoff in 802.11 networks. Both of these studies demonstrated that the probe delay constituted 90% of the total layer 2 handoff delay.

We have made measurements in our experimental test bed to study the layer 2 handoff-related operations for two different operating systems, namely Linux and Windows, using four different layer 2 drivers, namely Aironet, Orinoco, DLink, and Centrino. From a thorough analysis of the layer 2 handoff event, we found that the scanning and probing operations made the largest contributions to the delay. For example, in our experimental setup, using active scanning with the Orinoco driver in a Linux environment, it takes almost 100 ms for the probing action to be completed. This is followed by layer 2 open authentication and association, which take about 2 and 20 ms, respectively. We also experimented with layer 2 authentication, such as that in IEEE 802.11i, and found the average EAP and four-way handshake delay to be 79 and 616 ms for the nonroaming and roaming cases, respectively. The roaming scenario involves interaction with the AAA server during the mobile's handoff, whereas authentication is limited to the local authentication agent for the nonroaming scenario.

3.4.2 Layer 3 Delay

In an IP-based environment, network association in layer 3 involves several basic operations, such as discovery of a layer 3 point of attachment (e.g., default gateway discovery), IP address acquisition, duplicate address detection, neighbor reachability detection, local access authentication, and authorization. Each of these operations involves a number of message exchanges between the mobile and other network entities such as a router, a DHCP server, and an authentication server. From a periodic router advertisement, the mobile can discover the new layer 3 point of attachment. There are several protocols, such as DHCP (Droms 1997a), DHCPv6 (Droms et al., 2003), PPP (Simpson, 1994), and stateless autoconfiguration (Thomson and Narten, 1998), that can help the mobile to acquire an IP address. The time taken by each of these IP address configuration methods was described by Dutta et al. (2006b). Table 3.2 shows the IP address acquisition delays due to different types of configuration protocols, including zeroconf (Cheshire et al., 2005), a static (manual configuration) technique, and the proactive IP address acquisition technique. The times taken to configure an IP

Table 3.2 IP address acquisition delay

Layer 3 configuration method	DHCP with ARP	DHCP w/o ARP	IPv6 stateless	DHCPv6	PPP	FA CoA	Zeroconf	Manual	Proactive
Address acquisition delay	4 s	400 ms	160 ms	500 ms	8 s	2 s	5 s	100 ms	4 ms

address using these protocols differ owing to various factors, such as variation in the number of signaling messages needed to acquire the IP address and the duplicate address detection technique. In a typical interdomain mobility scenario, there are additional operations, such as reauthentication, reauthorization, and profile verification by the AAA servers, during the handoff process. In some cases, the AAA-related messages traverse all the way to the home AAA server before network service is granted to the mobile in the new network. Thus, it is desirable to have an optimized method of interaction between the mobile and the AAA server during an interdomain handoff. After the layer 3 identifier has been reconfigured, other layer 3 functions, such as the binding update from the mobile and media redirection at the correspondent host, contribute an additional delay.

3.4.3   Application Layer Delay

A mobile node suffers from application layer delay during a handoff operation because some of the handoff-related operations are performed in the application layer. The application layer delay is mostly attributed to operations such as binding update in the application layer, processing at the end hosts, registration, and upper-layer encryption such as TLS (Transport Layer Security) and SRTP (Secured RTP). SIP-based mobility is a typical example where the binding update is done in the application layer. SIP-based binding update (e.g., re-INVITE) contributes to the delay because of a three-way-exchange between the mobile and the correspondent host during the identifier rebinding process.

3.4.4   Handoff Operations across Layers

Figure 3.5 illustrates a sample protocol flow where the mobile is subject to subnet handoff and uses MIPv6 as the mobility protocol. In this specific example, the mobile moves from the old layer 2 point of attachment (oPoA) to a new layer 2 point of attachment (nPoA), and in the process it disconnects from the previous access router (pAR) and connects to the new access router (nAR), resulting in a change in the layer 3 point of attachment. Figure 3.5 shows the message exchanges between different components in the network that are used to take care of handoff operations, namely layer 2 discovery, layer 3 discovery, authentication, layer 2 security association, address acquisition, duplicate address detection, and binding update. Each of these operations needs a different number of signaling messages and carries a varying amount of data payload. Thus, the consumption of systems and network resources (e.g., CPU cycles, battery power, and bandwidth) and the amount of time taken to complete each of these operations will vary.

In order to study the effect of the different components on handoff delay, we experimented with secured interdomain mobility using both network layer (e.g., Mobile IP) and application layer (e.g., SIP-based) mobility techniques (Dutta et al., 2004a,b). Interdomain mobility introduces additional operations, such as local authentication and interaction with an authorization server, resulting in additional delay. These mobility prototypes include a combination of network detection, registration, configuration, authentication, security, location management functions, and roaming support for wireless Internet (Dutta et al., 2001). We used DRCP (a variant of DHCP for configuration) and PANA (Jayaraman et al., 2008) for secure access control; Diameter (Calhoun et al., 2003a) for authorization; and ESP (Encapsulating Security Payload) within IPSec (Kent and Atkinson, 1998a) and SRTP (Baugher et al., 2004) for encryption.

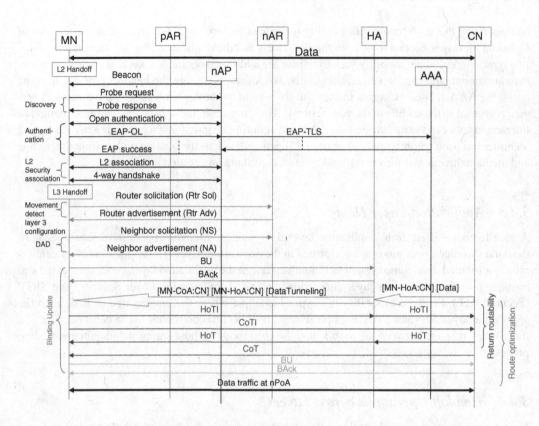

Figure 3.5 Protocol flow for MIPv6-based operations

Figure 3.6 and Figure 3.7 show the call flows for Mobile IP-based and SIP-based interdomain handoff, respectively. ERC1 and ERC2 are two edge router controllers that also act as a DRCP server, PANA server, and IPSec terminating point. These flows demonstrate the protocol interaction among several functional components, such as the SIP user agent or Mobile IP client, the DRCP server, the PANA server, the home agent, the IPSec server, and the AAA server, when the mobile moves from its home domain to a visited domain and return to the home domain. During this movement, the mobile changes its layer 2 and layer 3 points of attachment, and then reestablishes the layer 3 authentication, authorization, and security association using IPSec. During this interdomain handoff, the mobile needs to communicate with the visited AAA server for reauthorization.

Table 3.3 shows the timing delay for some of the functional components during interdomain handoff when SIP-based mobility and MIP are used as the mobility protocols. The end system processing time is not shown in this table. These experimental results demonstrate delays due to several handoff components, namely the layer 2 beacon interval, subnet and domain discovery, IP address acquisition, local authentication, and authorization, and also the delay due to binding update, such as SIP re-INVITE and MIP registration. "Prehandoff media" represents the time when the media are received prior to handoff in the old network, and "posthandoff media" represents the time the media are received at the new point of attachment after the handoff. In both cases, IKE (Internet Key Exchange) took almost 5 s for successful establishment of the IPSec-based security association.

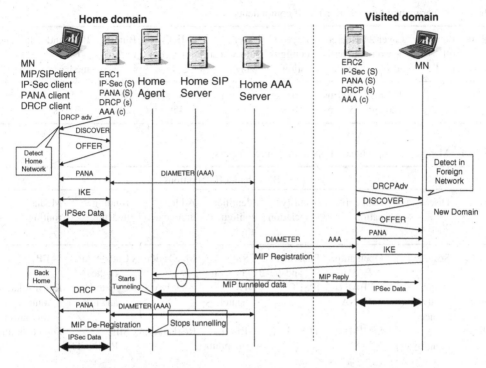

Figure 3.6 Call flow for Mobile IP-based interdomain handoff

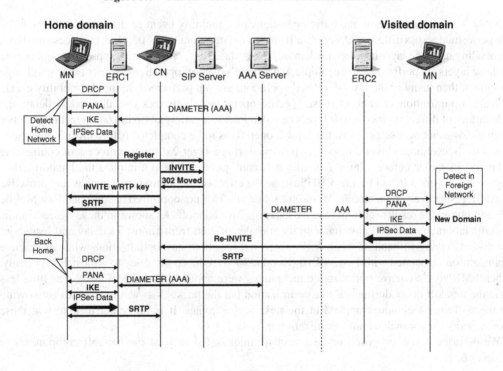

Figure 3.7 Call flow for SIP-based interdomain handoff

Table 3.3 Experimental timings for handoff components

Mobility type	Prehandoff media	Layer 2 beacon adv.	Router advertisement	Layer 3 configuration	Layer 3 authentication delay	IKE process	Binding update	Posthandoff media	In-handoff delay
SIP	51.7 s	120 ms	500 ms	80 ms	10 ms	5130 ms	240 ms	53 s	1.3 s
MIP	23.8 s	120 ms	500 ms	80 ms	10 ms	4580 ms	20 ms	31.1 s	7.3 s

Table 3.4 Mapping of basic handoff operations across layers

Layer	Basic handoff operations						
	Discovery	Authentication	Security association	Identifier configuration	Address uniqueness	Binding update	Media routing
Layer 2	Scanning	Open auth, EAPoL	Four-way handshake	ESSID, beacon	MAC address	Update ARP cache	IAPP
Layer 3	Router advertisement	L3 EAP IKE PANA	IPSEC	DHCP stateless	ARP DAD	Update HA and CN	Encapsulation, tunneling, forwarding
Application layer	AAA discovery	S/MIME	TLS SRTP	URI–IP mapping	Registration	SIP Re-INVITE	Direct routing

Table 3.4 summarizes how the basic operations of a mobility event as described in Section 3.1 are performed across different layers in an IP-based environment with 802.11 as the access medium. Depending upon the layer in which the handoff takes place (e.g., layer 2 or 3), appropriate operations in those layers are performed. For example, if a mobile's handoff operation does not involve a change in subnet, then many of the layer-3-related operations are not performed during the mobility event.

While optimization of each of these specific operations minimizes overall handoff delay, the scheduling of different tasks across layers can also lead to a variety of optimization techniques. For example, it is not necessary for all the layer 2 operations to be completed before layer 3 operations can start. Scheduling a layer 3 discovery process during a layer 2 discovery process or configuring a layer 3 identifier before a layer 2 identifier is configured will help minimize the handoff delay. Figure 3.8, Figure 3.9, and Figure 3.10 illustrate the effects of nonoptimized, reactive, and proactive handoff, respectively, on a mobile. We used Mobile IPv6 for nonoptimized handoff and Fast Mobile IPv6 to obtain the results for both reactive and proactive handoff. As shown in these figures, in this specific laboratory test bed environment, the mobile suffered from almost 7 s delay and lost about 70 packets owing to handoff when MIPv6 was used without any optimization, whereas reactive optimization associated with Fast MIPv6 (Koodli, 2005) reduced the delay to almost 1.5 s. Finally, when FMIPv6's proactive optimization techniques were applied, the handoff delay was little less than the handoff delay during reactive optimization but the packet loss was reduced to zero owing to the buffering techniques applied at the next access router. It is important to note that these experiments did not include any optimization in layer 2.

We describe a few proposed optimization techniques for each of the handoff components in Chapter 6.

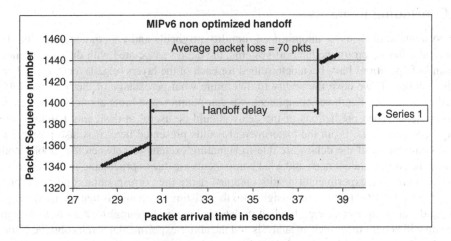

Figure 3.8 Results for FMIPv6 nonoptimized handoff

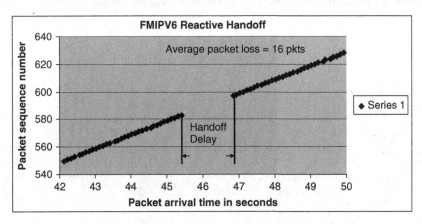

Figure 3.9 Results for FMIPv6 reactive handoff

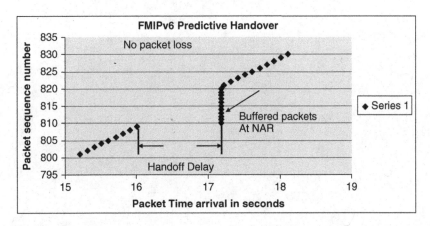

Figure 3.10 Results for FMIPv6 proactive handoff

3.5 Concluding Remarks

From a systematic analysis of unoptimized mobility protocols and experimental results obtained from several different prototype handoff systems, the delays associated with different components of the handoff operations have been determined for each of the layers (e.g., layer 2, layer 3, and the application layer). If we have the ability to determine what percentage of the total handoff delay is contributed by each handover component, we can obtain some hints about how to design new mobility systems. Such predictions of the delay should be useful in designing handoff optimizations for different layers. From the experimental results presented here, it is also possible to find out what components of the delays are due to signaling exchanges between the mobile node, the server, and the correspondent node and what components are due to processing in the network nodes and end systems. The experimental results obtained using the corresponding optimized protocols (e.g., MIPv6 vs. FMIPv6) give some insight into the existing mechanisms that are used to optimize certain handover components (e.g., binding update and media forwarding). Any new optimization mechanism will benefit from such an analysis and the above experimental results obtained from both optimized and unoptimized systems, but it will need to take into account the amount of resources available, such as processing power and battery power, to achieve a balance between handoff delay and systems resources.

4

Modeling Mobility

In this chapter, we develop a system model for the mobility event by incorporating the state transitions associated with the basic operations that take place during handoff. This system model decomposes the mobility event into various tasks and subtasks and analyzes the primitive operations.

4.1 Summary of Key Contributions and Indicative Results

In the absence of any formal mechanism to analyze the dynamics of handoff systems, it is difficult to predict or verify the systems performance of unoptimized handover or any specific handoff optimization technique. Without a model for a mobility system, it is difficult to design new mobility protocols or new optimization techniques for existing mobility protocols in a systematic way.

We have modeled the handoff-related processes as a discrete-event dynamic system (Cao and Ho, 1990) and used deterministic timed-transition Petri nets to build various unoptimized mobility models and devise associated optimization techniques. We have performed data dependency analysis and resource analysis of handoff-related operations to determine the possible sequences of operations and investigated behavioral properties such as deadlocks and liveness associated with handoff operations using Petri nets.

The proposed framework for the mobility model has the following key features:

- It allows us to analyze the data dependency among the handover components and illustrates how the handoff operations are distributed over the network components in the different layers.
- It allows us to investigate the resource dependency of various handoff operations using an experimental test bed.
- It allows us to design a model of a mobility system using timed-transition Petri nets based on an analysis of the data dependency and resource dependency.

The key benefits of the mobility model are as follows:

- The model can predict the systems performance for optimized handoff operations.
- The model can be used to design an optimal path for the sequence of execution of events based on the expected performance and the resource constraints.

Mobility Protocols and Handover Optimization: Design, Evaluation and Application, First Edition.
Ashutosh Dutta and Henning Schulzrinne.
© 2014 John Wiley & Sons, Ltd. Published 2014 by John Wiley & Sons, Ltd.

- The model can verify systems behavior (e.g., deadlocks) during handover.
- The model can be used to design various Petri-net-based approaches (e.g., using the Floyd algorithm, the resource–time product, or matrix-based solutions) to evaluate mobility models for handoff components.
- The system model can be used to investigate parallelism and opportunities for optimization during a handoff operation. Using the model, one can predict or verify the systems performance of an unoptimized handover and of any specific handoff optimization technique.
- The model can predict the performance of any mobility protocol in any specific deployment scenario, such as intradomain, interdomain, or heterogeneous handoff.
- The model can also be used to analyze the trade-off between performance metrics and resources when a mobility event includes parallel, optimistic, or speculative operations.

In the rest of the chapter, we discuss related work, the dynamics of the proposed Petri-net-based mobility model, give details of mechanisms that can be used to evaluate the systems performance, and investigate the opportunities for optimization using the model-based approach.

4.2   Introduction

The behavioral properties of a system are those properties that characterize the interactions among different components of the system. These properties define how the system operates for various working procedures. In order to understand the dynamic behavioral properties of a handoff system, study the interactions among the primitive operations of handoff, and investigate the trade-off between performance metrics such as handoff delays and resource utilization during handoff operations, it is important to design a formal model of a mobility system. For example, some questions related to the systems dynamics might be: Will the system ever reach a conflict state during a given sequence of handoff operations, where one operation cannot proceed because another operation has not yet started? Under what conditions will there be a conflict due to resource sharing? How does the system behave in the absence of conflicts? Can we obtain the required performance measures for the system under conditions of resource constraints? Such a model should also be useful for defining the control aspects of the system, so that it can provide guidance about how to rectify anomalies and unwanted behavior such as deadlocks. Nonavailability of data and resources due to sharing could possibly lead to deadlock situations, which need to be avoided by scheduling the handoff operations properly. The model should be able to analyze the important behavioral properties of the mobility event, such as the possibility of deadlocks during a handoff operation, to verify the correctness of operation of the system and predict possible sequences of transitions during handoff that will meet certain performance criteria.

4.3   Related Work

There have been several related efforts to model certain aspects of mobility management. Marsan et al. (2001) used a Petri-net-based model to analyze the performance characteristics of wireless Internet access for GSM and GPRS systems. Amadio and Prasad (1998) modeled IP mobility using a process calculus and applied this to a specific protocol, Mobile IPv6 (Johnson et al., 2004). This process-calculus-based method used a software agent approach to model the mobility event. Tutsch and Sokol (2001) provided a Petri-net-based performance evaluation of bandwidth partitioning, meant for multimedia sessions in wireless access. Molina-Ramirez et al. (1994) and Jaimes-Romero et al. (1997) used a Petri-net-based approach to model and analyze channel allocation schemes in cellular systems. Mostafa and Cicak (2006) developed a Petri-net-based model to calculate the end-to-end delay between communicating hosts in a Mobile IP environment. However, none of these papers attempted to model the interaction of different functional components of a handoff system or study the behavioral characteristics such as liveness and deadlocks of the system. None of these models was intended to systematically analyze the elementary operations involved in a mobility event or serve as a basis for optimizing the operations for a mobility event.

Based on our systems analysis of the mobility event, we have developed a formal model to study the design of an IP-based handoff system. This model can be applied to analyze any type of mobility protocol, study the functional components, and predict the systems performance during handoff.

4.4 Modeling Mobility as a Discrete-Event Dynamic System

A mobility event can be viewed as a perturbation to the steady state of a communicating node that may affect the various layers in the protocol stack. As the communicating node is subjected to hand-off, it goes through a series of sequential discrete states before it attains a steady state by returning to the communicating state again. Each of the basic operations described in Chapter 3 can be modeled as a transition event where the mobile moves from one state to another. As a result of the series of state transitions during the mobility event, the mobile's communication is interrupted because of the delay associated with each transition. A discrete-event dynamic system (DEDS) (Cao and Ho, 1990) is a type of system where the state space is discrete and state changes are driven by external or internal events. A mobility event can thus be modeled as a DEDS, since many of its components exhibit either concurrent, sequential, or competitive activities (i.e., two operations competing for the same resources) among many of the handoff operations that are part of the mobility event.

We considered several tools as possible candidates to model and analyze mobility events using the DEDS-based approach. These included software languages such as Esterel (Boussinot et al., 1991), Markov chains, queuing theory, minimax algebra (Cuninghame-Green, 1991; Patsidou and Kantor, 1991), Gantt charts (Hormozi and Dube, 1999), and Petri nets (Murata, 1989). Esterel is a synchronous programming language tailored to the development of reactive applications that show both concurrency and determinism. However, it cannot analyze behavioral aspects of a system such as precedence relations among events, mutual exclusion, deadlocks, and liveness. The existing algebra-based techniques such as minimax algebra cannot be used to analyze the dynamics of a handoff system, as these techniques can only be applied to the subclass of DEDSs that are described by a max-linear time-invariant system, whereas handoff systems are time-variant. Markov-chain-based solutions are more suitable for systems that exhibit nondeterministic or stochastic behavior, unlike the behavior of handoff systems, which is deterministic in nature. Although queueing theory can be used for performance evaluation, it involves a stochastic frame-work and is meant for average long-term evaluation, and is therefore unsuitable for handoff systems, which are deterministic and show transient behavior. Gantt charts can demonstrate the dependencies among events but cannot verify the correctness of a model.

Finally, we chose Petri nets as the tool to model the mobility event for the following reasons. Petri nets can be used to model properties such as process synchronization, asynchronous events, sequential operations, concurrent operations, and conflicts or resource sharing. Timed Petri nets (Zuberek, 1991) provide a uniform environment for modeling, design, and performance analysis of discrete-event systems. In a timed Petri net, times are assigned to the transitions and/or places of the Petri net. Timed Petri nets were originally developed by Ramchandani (1974), where the firing times were associated with the transitions of the net, and tokens were removed from the transitions' input places at the beginning of a firing. The graphical representation makes Petri nets intuitively a very appealing way to represent temporal events (i.e., events related to time). As a mathematical tool, a Petri net model can be described by a set of linear algebraic equations or by other matrix-based mathematical models reflecting the behavior of the system. Several software tools are available (e.g., TimeNet (Zimmermann et al., 1999a), a MATLAB®-based Petri net tool (Matcovschi et al., 2003), SPNP (Hirel et al., 2000), UltraSAN (Sanders, 1995), and TOMSPIN (Klas and Lepold, 1992)) to automate the analysis of a Petri net model and obtain a performance evaluation. Timed-Petri-net models can also be used with other existing techniques, such as simulated annealing (Zimmermann et al., 1999b), to perform a trade-off analysis between different performance metrics, such as handoff delay, handoff probability, and systems resources.

The modeling of mobility events can be viewed as analogous to the modeling of a flexible manufacturing system, as both exhibit a series of sequential operations. Zuberek and Kubiak (1999) modeled and analyzed simple schedules for manufacturing cells. Similar techniques can be applied to conduct performance analyses of models of mobility systems. We used deterministic timed Petri nets (Murata, 1989) to model the mobility event and derive the relevant optimized models by applying the appropriate concurrent and proactive mechanisms.

4.5 Petri Net Primitives

In this section, we define some fundamentals of Petri nets and describe some of the primitives of Petri nets that are essential for mobility analysis.

Formally, a Petri net can be defined as follows:

$$PN = (P, T, I, O, M_0),$$

where

1. $P = \{P_1, P_2, \ldots, P_m\}$ is a finite set of places.
2. $T = \{t_1, t_2, \ldots, t_m\}$ is a finite set of transitions, $P \cup T \neq \emptyset, P \cap T = \emptyset$.
3. $I : (P \times T) \rightarrow N$ is an input function that defines directed arcs from places to transitions, where N is a set of nonnegative integers.
4. $O : (P \times T) \rightarrow N$ is an output function that defines directed arcs from transitions to places.
5. $M_0 : P \rightarrow N$ is the initial marking.

A *marking* is an assignment of *tokens* to the places of a Petri net. Tokens are assigned to and can be thought of as residing in the places of a Petri net. The number and position of tokens may change during the execution of a Petri net. The tokens are used to define the execution of a Petri net.

If $I(p, t) = k(O(p, t) = k)$, then there exist k directed (parallel) arcs connecting place P to transition t (and transition t to place P). If $I(p, t) = 0(O(p, t) = 0)$, then there exist no directed arcs connecting p to t (or t to p).

In Petri net modeling, a transition is enabled after a token is made available from the places representing shared resources. Once an operation is completed, these tokens are returned to the places that represent the shared resources, thus making those resources available for other operations. The following are the firing rules for any transition in a Petri net:

1. A transition t is said to be enabled if each input place P of t contains at least a number of tokens equal to the weight of the directed arc connecting p to t.
2. An enabled transition t may or may not fire, depending on any additional dependency.
3. The firing of an enabled transition t removes from each input place p a number of tokens equal to the weight of the directed arc connecting p to t. It also deposits in each output place P a number of tokens equal to the weight of the directed arc connecting t to P.

Zhou and Robbi (1994) discussed the basic operations of Petri nets, and their precedence, concurrency, and conflict relationships. Figure 4.1 shows some of the fundamental operations of a Petri net that can be applied in scheduling a set of handoff operations. We give a brief description of these Petri-net-based operations below.

1. *Sequential operations.* If one operation follows another, then the places and transitions representing these should form a cascade or sequential relation in a Petri net. Figure 4.1 illustrates the situation where transition t_2 can fire only after the firing of t_1. This imposes the constraint "t_2 after t_1." Such precedence constraints are typical of the execution of the parts of a DEDS. Also, this Petri net construct models the causal relationships among the activities.
2. *Concurrent operations.* If two operations are initiated by an event, they form a parallel structure starting with a transition where two places are two outputs of the same transition. In Figure 4.1(c),

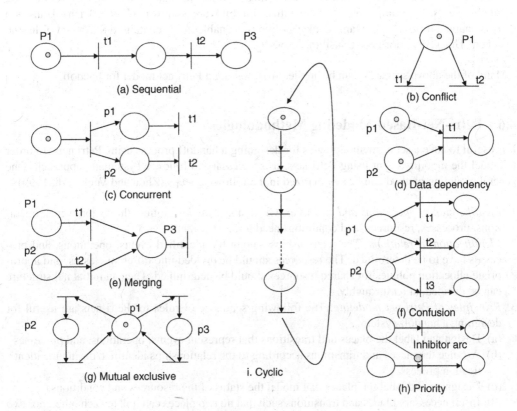

Figure 4.1   Petri net primitive operations

the transitions t_1 and t_2 are concurrent. A necessary condition for transitions to be concurrent is the existing of a forking transition that deposits a token in two or more output places.

3. *Conflicting operations*. If either of two operations can follow an operation, then the two transitions form two outputs from the same place. In Figure 4.1(b), t_1 and t_2 are in conflict. Both are enabled, but the firing of either of these transitions leads to the disabling of the other transition. The resulting conflict may be resolved in a purely nondeterministic way or in a probabilistic way by assigning appropriate probabilities to the conflicting transitions.

4. *Cyclic operations*. If a sequence of operations follow one after another and completion of the last operation initiates the first one, then a cyclic structure is formed among these operations. Figure 4.1(i) shows a cyclic structure in a Petri net.

5. *Synchronization*. Figure 4.1(d) shows an example of data dependency or synchronization among events. In this example, t_1 is enabled only when both p_1 and p_2 receive a token. The arrival of tokens in each of the two places could be the result of availability of data, resulting from some other complex operations elsewhere.

6. *Mutually exclusive operations*. Two processes are mutually exclusive if they cannot be performed at the same time owing to constraints on the usage of a shared resource. Two types of mutual exclusiveness, namely parallel and sequential mutual exclusion, were discussed by Zhou and DiCesare (1991). Figure 4.1(g) shows an example of how a resource may be shared by two processes and thus those processes cannot be completed at the same time.

7. *Inhibitor arcs or priority transitions*. Modeling of priorities can be achieved by introducing inhibitor arcs. An example of a Petri net with an inhibitor arc is shown in Figure 4.1(h). Transition t_1 is enabled if place p_1 contains a token, while t_2 is enabled if p_2 contains a token and p_1 has no token. This gives priority to transition t_1 over t_2.

Many of the above scenarios can be applied to designing a Petri net model for handoff.

4.6 Petri-Net-Based Modeling Methodologies

It is useful to have a set of methodologies for designing a handoff process using Petri nets. In order to model the mobility event using Petri nets, it is necessary to follow a systematic approach. One general modeling method can be summarized in the following steps (Zhou and Venkatesh, 1999):

1. *Identification of operations and resources*. Given a system description, the major events, operations, processes, resources, and conditions need to be identified.

2. *Identification of relations*. The relationships among the identified events, operations, and processes have to be determined. The resources should be divided into different classes, and appropriate allocation policies for shared resources should be determined. Then an initial net structure can be decided, at least roughly.

3. *Principles of Petri net modeling*. The following sequence of rules for Petri nets are useful for designing a handoff system.
   (a) Design and label the places and transitions that represent events, operations, and processes.
   (b) Arrange the places and transitions according to the relationships identified during the identification process.
   (c) Designate and label the places that model the status of the resources and conditions.
   (d) Insert necessary places and transitions such that no two places can link to each other, nor two transitions.

(e) For each transition, draw an input arc to it from a place if enabling it requires the resource(s), the truth of the condition, or the completion of the operation(s) represented in the place. Draw an output arc from the transition to a place if firing it releases resource(s), changes the condition(s), or signals the initiation of the operation for the place.

(f) The number of input arcs from a place to a transition should equal the number of tokens (often implying resources) required in that place to enable the transition. The number of output arcs from a transition to a place should equal the number of tokens for the place to be produced when the transition fires.

(g) Determine the initial number of tokens over all places according to the system's initial state.

(h) Associate other characteristics, for example timing, with places, transitions, and arcs, if needed. For example, for a deterministic timed Petri net, timing information should be added to the transitions.

In the following sections, we describe how we have applied some of the above methodologies in formulating a Petri net model for handoff.

4.7 Resource Utilization during Handoff

In Petri net terminology, a resource can be represented as a structural implicit place. "Structural implicit" means that if we have an arbitrarily large amount of resources (i.e., the number of tokens in the places representing the resources is arbitrarily large), then the marking of these places does not limit concurrent processing. However, in most cases most operations take place under some resource constraints.

Taking into account the constraints in any handoff system, the resources can be classified as those dedicated to specific tasks, and shared resources. A dedicated resource can be represented as a place with a single input and a single output arc only, and a shared resource can be represented as a place with multiple input and multiple output arcs. Resources of the same kind may be represented by a place with a number of tokens corresponding to the amount of resources. Initiation of an operation often requires several kinds of conditions and available resources; this is modeled as a transition with several input places. Completion of an operation may release some resources and change the status of the conditions; this is modeled as a transition with several output places. When a set of processes are executed concurrently, they can share a set of common resources. These resources can be represented as places (e.g., $R_1, R_2, R_3, R_4, R_5, \ldots, R_n$) to model the availability of resources.

Figure 4.2 shows an example where a shared resource place, R, is shared by two processes, A and B. Assuming that process A starts first, process B cannot start until the sequence of operations for process A is complete and the resource token has been released back to resource place R after transition t_4 has fired.

Several types of resources are utilized to complete the handoff-related operations during a mobility event. For the purpose of analysis, we categorize these resources into four different types, namely battery power (P_T), memory (P_M), bandwidth (P_B), and CPU processing cycles (P_C). These resources can be consumed by different parts of the network, namely the mobile node, access point, access router, network server, and correspondent host, during each of the handoff operations. Here, we describe in detail three of these resources that we have modeled.

Battery power is consumed in several operations, namely transmit, receive, and display operations, and reading and writing operations on the disk of a system. Based on the type of system, the energy consumption will vary between the different types of operation a mobile system has to perform

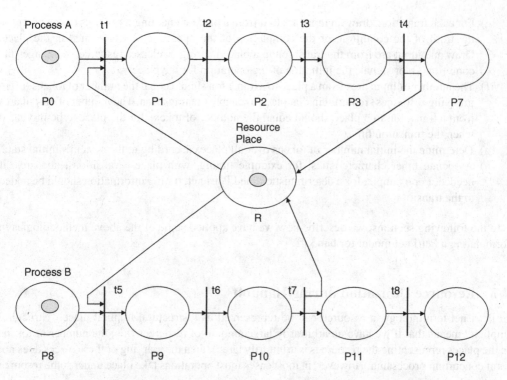

Figure 4.2 Sequence of operations in a Petri net with a shared resource

(Jones et al., 2001). For example, a laptop will consume a lot more battery power for display on the monitor than for other operations. However, since we are interested in the handoff operations that involve signaling only, we consider only the signaling-related operations related to communication. Shnayder et al. (2004) classified battery power usage for different types of operations in a wireless sensor system (Hempstead et al., 2005). The latter paper demonstrated that a wireless sensor system consumes most power during transmit and receive operations. Pering et al. (2006) demonstrated that wireless interfaces consume approximately 70% of the total power. Thus, in our analysis, for resource-sharing purposes, we considered the energy expended in the transmit and receive operations for these signaling messages. Caching consumes memory and processing power in the mobile node and correspondent node but does not use bandwidth.

The per-user effective bandwidth during a communication depends on several multiplexing points in the network, the type of access network (e.g., 802.11), the routers in the core of the network, the associated air access, and system-defined allocations of bandwidth (via priority mechanisms, class-based queuing, and weighted fair queuing) that provide segregation of signaling traffic from bearers. The handoff operations involve a set of round trip signaling messages between network components (e.g., between the mobile and the server). Each of these signaling messages shares transmission resources in the mobile's network interface, channel resources in the access network, and bandwidth associated with the router in the core of the network.

The access to channel resources will vary between different access techniques. For example, in an IEEE 802.11-based network, when one mobile has access to channel resources and is transmitting data, other mobile nodes will have deferred access to those channel resources. Thus, channel resources in the access network are shared by different mobile nodes. However, if there is only one mobile node in the network, per-user bandwidth is shared by different applications (e.g., the Web and ftp). Kravets and Krishnan (2000) discussed application-driven power management for mobile communication. That paper discusses how the scheduling of transmission of applications can help save energy. Handoff-related signaling could be considered as one type of application, and it is important to investigate what fraction of per-user effective bandwidth can be utilized to take care of handoff-related signaling. In modern computer networks, although bandwidth may not be an issue, handoff-related signaling may be affected if a large number of mobiles attempt to hand off to a target cell at the same time or a mobile attempts to hand off to a cell that is heavily overloaded, resulting in MAC layer retransmission.

CPU cycles are consumed in processing the signaling messages at the mobile, router, and correspondent host and in performing operations such as encryption, tunneling, encapsulation, and forwarding. Potlapally et al. (2003) analyzed the energy consumption characteristics of different cryptographic algorithms for various types of security-related operations such as key setup, encryption, and decryption. That paper also showed the overhead associated with authentication.

The number of signaling messages for each of the handoff operations varies, each signaling message consists of different numbers of bytes, and these signaling messages span different components in the network. Thus, the amounts of bandwidth required for each of these operations are different and depend upon the number of signaling messages and of bytes per message. For example, since the discovery process involves scanning neighboring channels, it involves additional signaling messages over the air compared with the authentication process, which involves messages in the core of the network. On the other hand, the authentication process involves more processing power and CPU cycles for key derivation. Similarly, the resource usage for each of the other handoff components varies accordingly. Depending upon the sequence of handoff operations, the resources may need to be shared when several handoff operations occur concurrently.

In Section 4.8, we illustrate the number of bytes exchanged during several different handoff-related signaling messages and the amount of resources utilized to support various types of handoff-related transitions. We model the handoff operations using three different types of resources (e.g., battery power, effective user bandwidth, and user processing power) and a set of data places, along with a set of transitions between these places.

4.8 Data Dependency Analysis of the Handoff Process

Analysis of the data dependencies and task dependencies among the operations required in a handoff event can determine the extent of parallelism among the operations and the types of proactive operations that may be possible. In this section, we introduce the concept of data and task dependency, which can be applied to model handoff operations using Petri nets. By applying these task dependencies, data dependencies, and resource availability in a Petri net model, it is possible to determine the schedule or sequence of events during handoff operations.

4.8.1 Petri-Net-Based Data Dependency

Cruz Filho et al. (2000) provided an example of how Petri nets can be used to analyze data dependency. Maciel et al. (2001) proposed a method for estimating the number of functional units taking timing constraints into account, and considered the data dependency as an input to the estimation

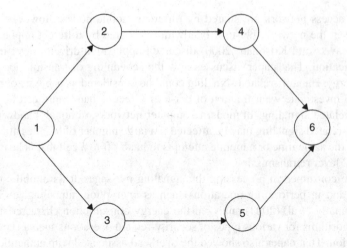

Figure 4.3   Data dependency relationships

process. Belhe and Kusiak (1993) discussed how a timed Petri net can be used for performance evaluation in a design process and illustrated how the task dependency could be used to generate a Petri net. Before we introduce the data dependency analysis of the handoff event, we shall give some examples of how a data dependency between events maps to a Petri net model and the associated resource constraints.

Figure 4.3 shows an example of the data dependency relationships among the activities of a specific phase of a design process. At each stage of the design process, decisions are made based on the data or information available at that stage, which is provided by the preceding stages. Thus, a network of data dependency relationships among design activities can be established. An arrow can be drawn from design activity i to activity j, indicating that activity j depends upon activity i. In Figure 4.3, activities 2 and 3 depend on the data from activity 1, activity 4 depends on the data from activity 2, activity 5 depends on the data from activity 5, and activity 6 depends on the data from both activities 4 and 5. Figure 4.4 shows how an equivalent Petri net can be formed based on the data dependency graph shown in Figure 4.3.

Figure 4.5 adds resources to the preceding Petri net model, which was based on the dependency graph. Places p_7, p_8, p_9, and p_{10} represent the resource places. Each of the resource places is equipped with a certain amount of resources, shown as tokens. The resource requirements for different activities are expressed by capacities on the arcs. The availability of a specific data item from previous states and the availability of the required resources enable a specific task to go forward. Similar task dependency methods can be applied to the handoff process.

4.8.2   Analysis of Data Dependency during Handoff Process

This section describes the dependency among several components of the handoff process. It is helpful to analyze the sequence of tasks for these handoff components and investigate the data that a specific task may depend upon. Completion of one task may result in data that is needed for the execution of another task. Thus, one task may depend upon another task.

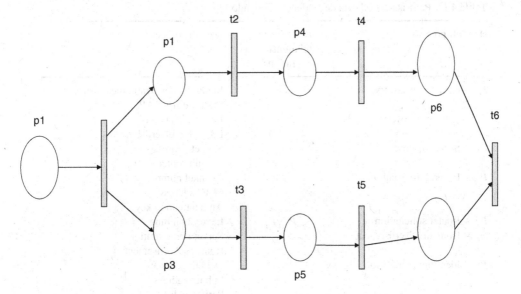

Figure 4.4   Petri net model for handoff based on data dependency

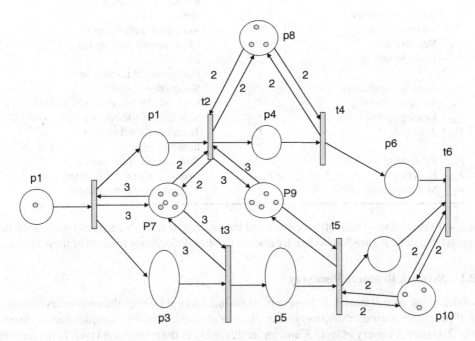

Figure 4.5   Timed-Petri-net model with resource constraints

Table 4.1 Petri-net-based data dependency for handoff

Handoff process	Precedence relationship of operations	Data the process depends on
P_{11} – Channel discovery	P_{00}	Signal-to-noise ratio measurement
		Distance from AP
P_{12} – Subnet discovery	P_{21}, P_{22}	Layer 2 beacon ID
		L3 router advertisement
P_{13} – Server discovery	P_{12}	Subnet address
		Default router address
P_{21} – Layer 2 association	P_{11}	Channel number
		MAC address
		Authentication key
P_{22} – Router solicitation	P_{21}, P_{12}	Layer 2 binding
P_{23} – Domain advertisement	P_{13}	Server configuration
		Router advertisement
P_{31} – Identifier acquisition	P_{23}, P_{12}	Default gateway
		Subnet address
		Server address
P_{32} – Duplicate address detection	P_{31}	ARP
		Router advertisement
P_{33} – Address resolution	P_{32}, P_{31}	New identifier
P_{41} – Authentication	P_{13}	Address of authenticator
P_{42} – Key derivation	P_{41}	PMK (pairwise master key)
P_{51} – Identifier update	P_{31}, P_{52}	L3 address
		Uniqueness of L3 address
P_{52} – Identifier verification	P_{31}	Completion of CoTI
P_{53} – Identifier mapping	P_{51}	Updated MN address at CN and HA
P_{54} – Binding cache	P_{53}	New care-of address mapping
P_{61} – Tunneling	P_{51}	Tunnel end-point address
		Identifier address
P_{62} – Forwarding	P_{51}, P_{53}	New address of the mobile
P_{63} – Buffering	P_{62}, P_{51}	Acquisition of new identifier
P_{64} – Multicasting/bicasting	P_{51}	Acquisition of new identifier

Table 4.1 shows several handoff processes and the associated data dependencies among the set of operations, based on Figure 3.3. We briefly discuss the data dependency for each of these operations.

4.8.2.1 Network Resource Discovery

Network and resource discovery is the very first task that takes place once the mobile has initiated the handoff procedure. Resource discovery takes place in several layers, for example channel discovery in layer 2, subnet discovery in layer 3, and server discovery in the application layer. In the following, we describe the dependencies among the subprocesses that are part of the network discovery process according to Figure 3.3.

1. *Channel discovery*. Channel discovery by the mobile node depends upon successful receipt of beacons or pilot signals from the neighboring access points or base stations. For example, in the

normal handover scenario for an 802.11 network, the mobile needs to be in the coverage area of the target network to be able to receive the SSID from the broadcast channel. Thus, the ability to receive beacons over a broadcast address is a precedence operation for the channel discovery operation to start. However, this precedence condition may not be necessary if the mobile uses a proactive discovery technique such as the 802.21-based information service (Dutta et al., 2005b), where the mobile can receive channel information from the neighboring networks based on the current position of the mobile. Thus, the channel discovery process can stay out of the critical path of the handoff operation if the target channel is discovered prior to the mobile's handoff.

2. *Subnet discovery*. Since subnet discovery is a layer 3 process, the mobile needs to complete the layer 2 association process P_{21} before it can receive router advertisements to discover that it has indeed moved to a new subnet. The mobile depends upon the prefix from the router advertisement before it can discover a new subnet. Although router advertisement takes place at a predefined rate based on the router configuration, the mobile can always send a router solicitation message to trigger a router advertisement. This helps it to discover a layer 3 point of attachment in the absence of or delay in the router advertisement. In most cases, lower-layer events such as layer 2 association and network attachment event notifications such as "Link up" trigger a router solicitation from the mobile, which in turn triggers the router advertisement.

3. *Server discovery*. The server discovery process can take place only after subnet discovery is complete. The server discovery process usually includes discovery of a DNS server and an outbound SIP server that provide additional services such as domain lookup and routing of signaling messages, respectively. The server discovery process P_{13} depends upon the completion of the subnet discovery process P_{12} and the identifier acquisition process P_{31}. The mobile needs to have acquired the default router address and subnet prefix before it can complete the server discovery process.

4.8.2.2 Network Attachment

The next phase of the handoff operation is network attachment. The network attachment process involves association with the new point of attachment in different layers. Layer 2 association, router solicitation, and domain advertisement are three different tasks that are part of this process.

1. *Layer 2 association*. In order to complete a successful layer 2 association, the mobile needs to have access to the channel number, the MAC address of the access point, and security keys for authentication. While the channel number can be obtained during the discovery or scanning process, the mobile node needs to have access to a WEP-based security key for shared-key authentication or to acquire a preshared key to use the EAP authentication method in the case of 802.11i-based authentication. In the case of open system authentication, WEP keys can be used to encrypt the data.

2. *Router solicitation*. The router solicitation message from the mobile triggers the layer 3 attachment process by soliciting a router advertisement from the router. However, this operation takes place only after the mobile has established a layer 2 binding with the access point, so that the mobile can communicate with the access point using the MAC address.

3. *Domain advertisement*. Similarly, the domain advertisement process helps with network attachment in the application layer. This process helps the mobile to determine the administrative domain it belongs to. The mobile needs to have completed server configuration or router advertisement. The mobile depends upon data from the server configuration or router advertisement process before it can proceed with the domain advertisement process.

4.8.2.3   Mobile Configuration

The next phase of the handoff process is configuration, which includes several tasks such as identifier acquisition, duplicate address detection, and address resolution. The following is a description of the dependence among the subprocesses of the configuration process.

1. *Identifier acquisition*. The mobile node depends upon data such as the default gateway and the subnet prefix in order to complete the identifier acquisition task. These data are obtained as a result of the completion of tasks such as domain advertisement (P_{23}) and subnet prefix discovery (P_{12}).
2. *Duplicate address detection*. The mobile node depends upon the completion of the acquisition of the layer 3 identifier in the new network before it can determine the uniqueness of the identifier. Thus, the duplicate address detection process cannot start until a new identifier has been obtained in the new network. Thus, it is not possible to start process P_{32} before process P_{31} is complete.
3. *Address resolution mapping*. Address resolution, or mapping of the IP address with the MAC address, can be completed only after a new address has been obtained. However, duplicate address detection and address resolution could take place in parallel, reducing the extent of sequential operation.

4.8.2.4   Security Association

The following paragraphs describe the instances of data dependency among the subprocesses that constitute the security association process.

1. *Authentication*. A successful authentication operation depends upon successful discovery of the authenticator. After the discovery of the authenticator, the mobile can communicate with the authenticator, which in turn communicates with the authentication server (e.g., an AAA server) to complete the authentication and authorization process. This process involves an exchange of signaling messages between the mobile node and the authenticator. Thus, the data from the operation P_{13} in Table 4.1 that provides the information about the authenticator acts as precedence data for the authentication subprocess P_{41}.
2. *Key derivation*. The derivation of an encryption key (e.g., a pairwise transient key (PTK)) that helps to establish a security association between the mobile and the authentication server by encrypting the data is possible only after a successful authentication process that generates a pairwise master key (PMK) is complete. Thus, there is a precedence relationship between the derivation of the PMK by way of EAP-based authentication and the generation of the PTK. However, the authentication process can always take place before the mobile moves to the target network (proactive authentication), thereby leaving only the generation of the PTK to take place after the handover to the new network. This is one way of optimizing the authentication process.

4.8.2.5   Identifier Update

Updating the new identifier at the home agent is possible only after the mobile has obtained a new layer 3 address. Thus, data such as the new layer 3 address and the test of its uniqueness are prerequisite data before the mobile can start the binding update process. In some cases, such as in MIPv6, this identifier update needs to be verified against the possibility of an unlawful binding update before the binding update is complete. This verification is usually done by means of a return routability process, whereby the mobile node and correspondent node exchange a pair of keys to authenticate each other. Similarly, identifier mapping and the establishment of a binding cache entry depend upon successful completion of the processes of identifier update and identifier verification.

4.8.2.6 Data Forwarding

Tunneling, forwarding, and buffering are the remaining handoff-related operations that have to take place before the handoff is complete and data is received at the mobile. A tunnel can only be set up after the mobile has learnt the address of the home agent and the home agent has learnt the new care-of address of the mobile. The tunneling operation can be completed only after binding update is over.

Similarly, forwarding of data can only take place after the mobile has obtained a new care-of address and has sent a binding update to the local forwarding agent (e.g., a foreign agent in the previous network) and the home agent. Depending upon where the buffering takes place (i.e., at the edge of the network or at the home agent), forwarding and buffering operations can take place in parallel. Multicasting and bicasting are two forms of techniques for forwarding transient data during a handoff. We describe some of these techniques in Chapter 6.

4.9 Petri Net Model for Handoff

Figure 4.6 shows a high-level view of how a set of discrete states associated with a mobility event can be represented in a formal framework by using a timed-Petri-net approach (Murata, 1989). This high-level Petri net model shows a handoff operation where a mobile goes from a *connected* state to a *disconnected* state and then returns to a connected state.[1] Each place P_i represents various stages of the mobility event, and the transitions t_i represent the time taken to complete various sets of operations between the stages. Each of these stages can be considered as multiple subsystems of the system representing the mobility event.

Table 4.2 describes the places that represent various stages of the handoff event, the shared resources, and the transitions associated with Figure 4.6. The transitions represent the respective operations, and delays are associated with each of these handoff operations. Although a deterministic delay is assumed for the transitions here, this framework can also be applied if the transitions follow other types of delay distribution, such as exponential, uniform, or finite discrete. The transition time will vary depending upon the processing speed and shared resources available. Shared resources are, ideally, represented by tokens that become available before a transition is fired, leading to the next stage in the handoff process.

In Figure 4.6, each of the places P_0–P_7 represents various stages of the mobility event as described in Chapter 3, and the places P_B, P_M, and P_P represent shared resources, namely effective user bandwidth, battery, and processing power, respectively, that are used during a mobility event. The effective user bandwidth is a shared resource that is determined by the available network bandwidth in the access and core networks, whereas battery power and CPU cycles are shared resources used by the same mobile node for multiple operations within the mobile.

The path shown by the dotted lines in Figure 4.6 representing place P_7 illustrates an example of parallel operation where a hierarchical binding update (t_6) occurs in parallel with a regular binding

---

[1] "Connected state" and "disconnected state" are defined in Appendix B.

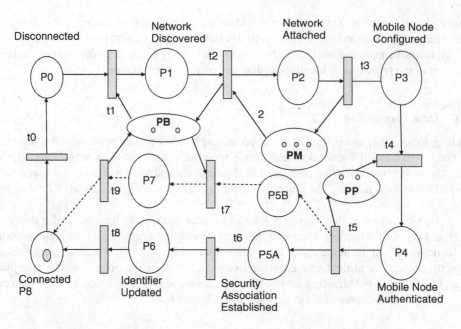

Figure 4.6   A generalized high-level timed-Petri-net model for handoff

update (t_7). After a security-related operation (t_5) is over, both the hierarchical binding update (t_6) and the regular binding update (t_7) can take place in parallel. P_{5A} and P_{5B} are the intermediary places in this model after transition t_5 is fired.

Each of these transition processes may consist of several subprocesses. Table 4.3 shows the sub-transitions for each of the transitions. Transition process t_1 (network resource discovery) consists of several subprocesses that are represented by several subtransitions, namely t_{11}, t_{12}, and t_{13}. These subtransitions define layer 2 channel discovery, layer 3 subnet discovery, and server discovery, respectively.

The hierarchical property of Petri nets (Zuberek, 2000) can be applied to each of these transition processes to study the interaction at the subprocess level. For example, as shown in Figure 4.7, transition process t_1 (network resource discovery) consists of several subprocesses (places), namely channel discovery (p_{11}), subnet discovery (p_{12}), and server discovery (p_{13}), that are connected by several subtransitions, namely t_{11}, t_{12}, and t_{13}. Similarly, Figure 4.7 also shows how the hierarchical nature of a Petri net can be applied to the network attachment (t_2) and configuration (t_3) processes. Although only three processes have been highlighted in Figure 4.7, the hierarchical property of Petri nets can be applied to other operations as well, as shown in Table 4.3.

Figure 4.8 shows a subnet-level view of a mobility system model as represented by the modular nature of the Petri net. It includes the state transitions in each layer and illustrates the component-level interactions in each layer and across layers. Several subprocesses in each of the layers, such as discovery, layer 2 authentication, layer 3 address acquisition, layer 3 authentication, security associ-ation, and binding update, are illustrated here. Reduction in the number of transitions, minimization of the time for each of the state transitions, parallelization of many of the state transitions within each layer and across layers, and reduction in number of states will contribute to the overall optimization of the handoff process.

Table 4.2 Description of places and transitions for handoff

Place	Description
P_0	Mobile is in disconnected state
P_1	Network and resources discovered
P_2	Target network selected
P_3	Mobile node is configured and registered
P_4	Mobile node is authenticated
P_{5A}, P_{5B}	Security association is established
P_6	Binding update is complete
P_7	Intradomain binding update is complete
P_8	Mobile is in connected state
P_B	Bandwidth resources
P_M	Battery resources
P_P	CPU resources

Transition	Description
t_0	Mobile node gets disconnect trigger
t_1	Mobile node discovers network and resources at the new PoA
t_2	Mobile node selects the network
t_3	Mobile node goes through configuration and registration
t_4	Mobile node goes through authentication process
t_5	Mobile node goes through key derivation and security association process
t_6	Mobile goes through binding update process
t_7	Mobile node goes through hierarchical binding update
t_8	Data is redirected to the mobile node
t_9	Network buffering during handoff

Table 4.4 illustrates how three different types of resources (bytes transferred, CPU cycles, and battery power) are used in various handoff operations as described in Chapter 3. These values are based on the experimental results from our mobility test bed presented in Chapter 6. "Bytes exchanged" refers to the total number of bytes exchanged during the transmission and receipt of signaling messages at the mobile during a specific handoff operation. The CPU cycle samples were taken from an Oprofile (Levon and Elie, 2005) analysis of the mobile for the various operations during the handoff. The battery power consumption is based on the energy expended, which is based on the numbers of bytes received and transmitted by the mobile. Feeney and Nilsson (2001) modeled the per-packet energy consumption as a fixed component due to channel access and a variable component proportional to the size of the packets transmitted and received. That paper also showed how different amounts of energy were needed for transmit and receive operations. Thus, the expended energy can be described using a linear equation as follows:

$$\text{Energy} = m \times \text{size} + b, \tag{4.1}$$

where m is the amount of energy spent per byte in a message and b is the fixed amount of energy per channel access. It is important to note that the value of m will vary for both transmit and receive operations, and the energy cost per channel access will also vary for those operations. Khalaf-Bitar and Rubin (2009) cited examples of the amount of energy spent in transmitting and receiving a bit

Table 4.3 Atomic operations during handoff

Transition	Handoff operation	Subtransitions	Suboperations
t_0	Disconnect trigger	t_{00}	Layer 2 unreachability test
		t_{01}	Layer 3 unreachability
t_1	Network discovery	t_{11}	Discover layer 2 channel
		t_{12}	Discover layer 3 subnet
		t_{13}	Discover server
t_2	Network attachment	t_{21}	Layer 2 association
		t_{22}	Router solicitation
		t_{23}	Domain advertisement
t_3	Mobile configuration	t_{31}	Identifier acquisition
		t_{32}	Duplicate address detection
		t_{33}	Address resolution
t_4	Authentication	t_{41}	Layer 2 open authentication
		t_{42}	Layer 2 EAP
t_5	Security association	t_{51}	Master key derivation
		t_{52}	Session key derivation
t_6	Binding update	t_{61}	Identifier update
		t_{62}	Identifier verification
		t_{63}	Identifier mapping
		t_{64}	Binding cache
t_7	Hierarchical binding update	t_{71}	Fast binding update
		t_{72}	Local caching
t_8	Media redirection	t_{81}	Tunneling
		t_{82}	Forwarding
		t_{83}	Buffering
t_9	Local data redirection	t_{91}	Local id mapping
		t_{92}	Multicasting/bicasting

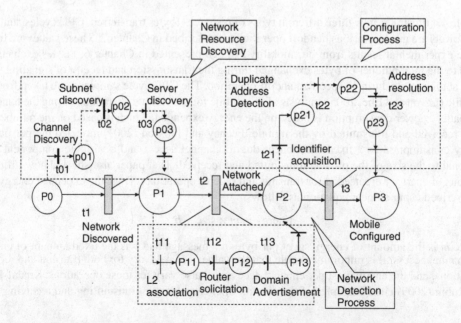

Figure 4.7   Hierarchical decomposition of Petri-net-based handoff model

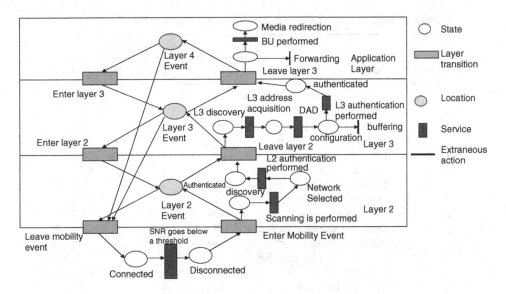

Figure 4.8   Layered modeling with a Petri net

in an 802.11-based network. According to that paper, at 50 mW transmission power and 11 Mb/s transmission rate, it takes about 1.3 μJ of energy to transmit a byte of data, and about 1 μJ to receive a byte at that rate. If the radio interface works at a slower rate, it takes more energy to transmit a byte. For example, the mobile will consume about 7 μJ per byte when the radio operates at 2 Mb/s. According to the Sierra wireless card's CDMA 1X EvDo interface, the energy consumption rate at 200 mW transmission power is about 1.4 W. Thus, it takes about 6 μJ to transmit one byte at 1.8 Mb/s uplink speed using a CDMA 1X-EvDO interface.

In order to determine what percentage of energy is used for handoff-related operations, it is important to find out the total amount of data and voice that a mobile node transmits and receives on a daily basis and the average number of handoffs that a mobile is subjected to. From an experimental calculation, we found that a mobile node transmitted and received about 7 Kb of data when MIPv6 was used as the mobility protocol. This amount will vary if a different mobility protocol (e.g., SIP or ProxyMIP) is used instead. For example, a SIP-based mobility protocol exchanges additional data for the binding update, resulting in about 10 Kb of data, but Proxy MIPv6 requires less data to carry out the binding update operation and this results in about 6 Kb of data exchange. Thus, the total amount of data transfer and the number of times the mobile needs to access a channel will also vary. A recent online survey showed that a PDA user was subjected to 100 Mb of data over a period of a month, and FCC data (www.fcc.gov) showed that a mobile was subjected to about 5 Mb of voice traffic over the same period. Thus, a mobile is subjected to about 3.5 Mb of voice and data transfer per day. The number of times a PDA is subjected to handover depends upon the mobility pattern of the type of user (e.g., a salesperson or a researcher) and the characteristics of the access method (e.g., CDMA or 802.11). An experimental study (Thajchayapong and Peha, 2006) showed that a PDA was subjected to an average of 10 handoffs a day in a microcellular environment, amounting to about 70 Kb of data transfer due to handoff, when FMIPv6 was used as the mobility protocol in an IEEE 802.11-based environment. Thus, handoff contributes about 2% of the total energy spent in a microcellular environment. The handoff rate in a cellular environment will be smaller than in a microcellular environment. The approximate handoff rate can be estimated for the example of a

Table 4.4 Resource assignment for each of the suboperations

Transition	Operation	Resource consumption		
		Bytes	CPU cycles	Battery (nJ)
t_{00}	Layer 2 unreachability	43	5	51600
t_{01}	Layer 3 unreachability	86	3	103200
t_{11}	Discover layer 2 channel	109	3	130800
t_{12}	Discover layer 3 subnet	110	4	132000
t_{13}	Discover server	450	5	540000
t_{21}	Layer 2 association	99	2	118800
t_{22}	Router solicitation	70	4	84000
t_{23}	Domain advertisement	226	4	271200
t_{31}	Identifier acquisition	1426	5	1711200
t_{32}	Duplicate address detection	164	6	196800
t_{33}	Address resolution	60	3	72000
t_{41}	Layer 2 open authentication	94	3	112800
t_{42}	Layer 2 EAP	2842	6	3410400
t_{43}	Four-way handshake	504	4	604800
t_{51}	Master key derivation (PMK)	0	10	0
t_{52}	Session key derivation (PTK)	0	6	0
t_{61}	Identifier update	352	4	422400
t_{62}	Identifier verification	148	6	177600
t_{63}	Identifier mapping	0	8	0
t_{64}	Binding cache	0	3	0
t_{71}	Fast binding update	110	3	132000
t_{72}	Local caching	0	6	0
t_{81}	Tunneling	60	2	72000
t_{82}	Forwarding	100	2	120000
t_{83}	Buffering	120	3	144000
t_{91}	Local id mapping	40	4	48000
t_{92}	Multicasting/bicasting	192	2	230400

daily commuter in the USA as follows. The average cell size in a cellular network varies from 5 to 25 miles and, on average, the one-way commute distance for a mobile subscriber in the USA is about 20 miles a day. Thus, a commuting mobile subscriber will be subjected to an average of 5–6 handoffs per day. However, this value does not include the scenario where a mobile subscriber with a dual-mode radio can switch over to hotspot access (by IEEE 802.11) or where there are many overlapping cells in a highly populated, dense urban area.

It is assumed that the tokens in the resource places represent different amounts of handoff resources (e.g., 1 bandwidth token = 100 Kbits, 1 battery power token = 50 mJ, and 1 CPU token = 2 cycles). We now give examples of Petri net models for various handoff operations.

Figure 4.9 shows the relative resource usage during a handoff operation based on the values in Table 4.4. Based on the resource requirements for each of these suboperations, we model four specific operations, namely the *discovery*, *attachment*, *authentication*, and *configuration* processes, along with the shared resources needed for these operations.

Figure 4.10 models the discovery process by having a limited number of tokens available in the resource places. The resource places are p_3, p_4, and p_5, which represent bandwidth (network capacity), CPU cycles, and processing power, respectively.

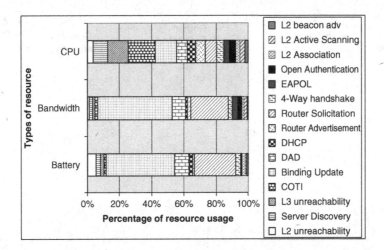

Figure 4.9   Relative resource usage during handoff

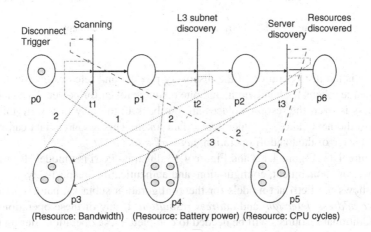

Figure 4.10   Petri net model for discovery

As soon as the mobile gets a disconnect trigger (e.g., the SNR goes below a certain threshold), the mobile is ready to perform a scanning operation. In order to complete the scanning operation, it needs to get a disconnect trigger indication from place p_0, which is considered as a data token; two tokens from resource place p_3; one token from resource place p_4; and two tokens from resource place p_5. Thus, in order for the scanning operation to be completed, the resource places p_3, p_4, and p_5 must have enough tokens available, leading to the firing of transition t_1. Once the transition t_1 has fired, the resources are released for subsequent operations. As the mobile moves to the next operation (e.g., layer 3 subnet discovery), it may need additional resources of a specific type and release other types of resources. For example, in this specific case, layer 3 subnet discovery needs additional battery power, since the mobile needs to perform additional transmission and receipt of messages. After the layer 3 subnet operation is over, it releases one token back to place p_3, the bandwidth place.

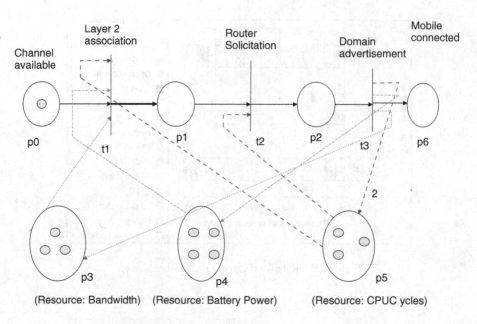

Figure 4.11   Petri net model for network attachment

Server discovery is the last operation of the discovery process and releases one resource token for CPU, three resource tokens for bandwidth, and one resource token for power. Thus, after the server discovery process is over, the resource places are put back to the way they were initially and the mobile moves to the next state, p_6, which has a data token. This resource data can be used as an input to the next stage of the handoff operation, *network attachment*.

Similarly, Figure 4.11, Figure 4.12, and Figure 4.13 illustrate Petri net models for the subsequent operations of network attachment, configuration, and authentication, respectively.

Figure 4.14 shows the Petri net models for the configuration subtasks, namely *identifier acquisition*, *duplicate address detection*, and *address resolution*. If any of these operations are done in parallel, then additional resources will be needed to complete these operations during that period.

Figure 4.15 shows the model when the four operations of discovery, attachment, configuration, and authentication work together to complete part of the handoff operation. While running operations in parallel may reduce the amount of time needed to complete the handoff, resource sharing and non-availability of data may lead to a deadlock situation where a specific operation cannot be completed because of a lack of resources.

Figure 4.16 shows the Petri net model based on the data dependency relationships shown in Table 4.1 when there are no concurrent handoff operations. Figure 4.17 and Figure 4.18 illustrate the dependency graphs of two different sequences of operations that involve some level of concurrency. Figure 4.17 illustrates the dependency graph when security association takes place during scanning, and Figure 4.18 shows an example when both security association and subnet discovery take place during scanning. Figure 4.19 and Figure 4.20 show the corresponding Petri net models in terms of places and transitions. These models are shown without the resources in the data dependency graphs. They show clearly how specific data dependencies might affect the sequence of execution

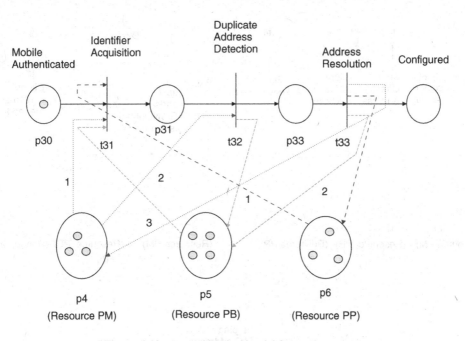

Figure 4.12   Petri-net-based model for configuration

of the handoff operations. Petri-net-based models can be used to analyze various types of mobility events, namely intrasubnet, intratechnology, intersubnet, and intertechnology handovers, involving various mobility protocols. This analysis can be carried out with many existing Petri-net-based software tools, as mentioned in Section 4.3.

Out of the several tools available, we used two specific tools to analyze Petri net models of handoff process, namely TimeNet and a MATLAB®-based tool. Figure 4.21 shows how we modeled MIPv6 (Johnson et al., 2004) using TimeNet. This takes into account the hierarchical nature of Petri nets (Zuberek, 2000) and shows the interaction between different states during a mobility event at a granular level. The hierarchical nature of Petri nets provides an ability to decompose a specific system event into smaller tasks. However, we used a MATLAB®-based Petri net tool (Matcovschi et al., 2003) to model the various handoff components and associated optimization techniques and to analyze the associated behavioral properties. This toolbox is equipped with a user-friendly graphical interface. Its functions cover the key topics of analysis such as coverability trees, structural properties (including invariants), time-dependent performance indices, and max-plus state-space representations. Some of the MATLAB®-based results for various handoff optimization techniques are presented in Chapter 11.

4.10   Petri-Net-Based Analysis of Handoff Event

Many of the behavioral properties of a handoff system, such as reachability and deadlocks, which lead to the allowable sequences of transitions during a handoff event, can be verified using Petri net analysis methods. There are two types of Petri net analysis methods: *reachability tree analysis*

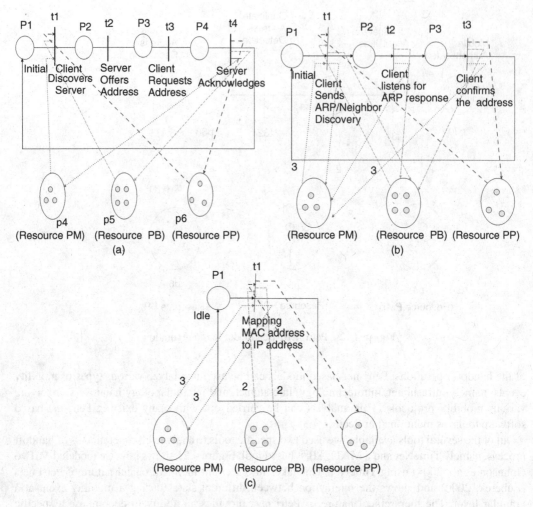

Figure 4.13 Petri net model for authentication

and *matrix equations*. In this section, we analyze deadlocks and the effectiveness of reachability analysis and matrix equations for detecting deadlocks, and determine the sequences of transitions that are possible during a handoff event.

4.10.1 Analysis of Deadlocks in Handoff

Deadlock is a highly unfavorable situation in which a set of tasks request resources held by other tasks at the same time. A deadlock situation involving one set of tasks can easily propagate to other tasks, eventually crippling the whole system so that it remains blocked indefinitely, and so deadlock can bring unnecessary costs to a system such as long unavailability of the system and low use of critical and expensive resources. The concept of liveness in Petri nets is closely related to that of

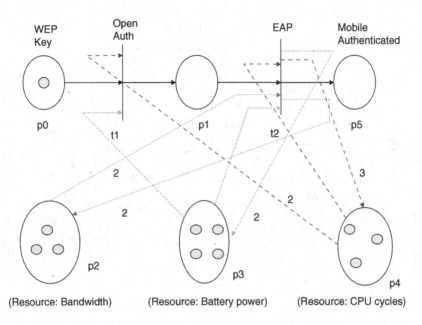

Figure 4.14 Petri net models for configuration subtasks

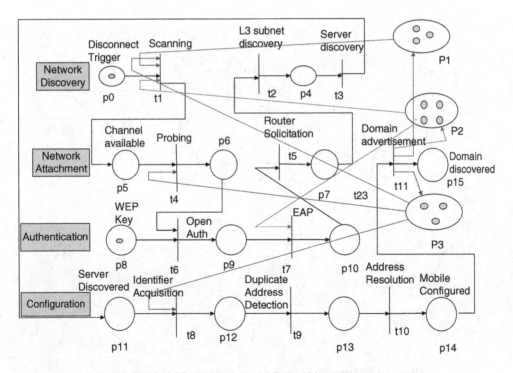

Figure 4.15 Petri net model for combined operations

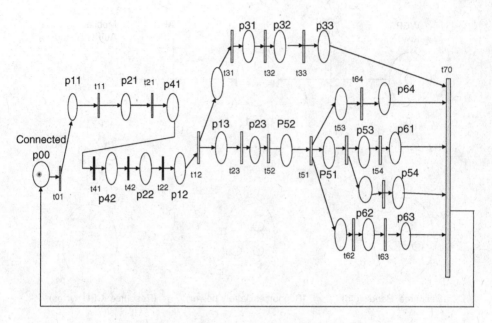

Figure 4.16   Petri net model based on data dependency

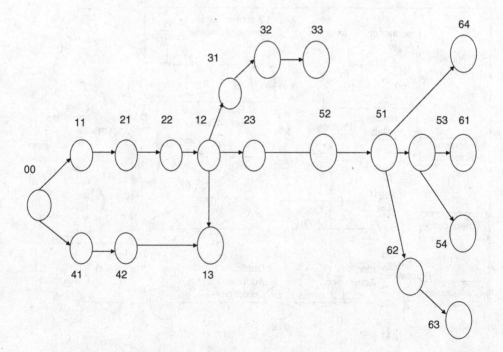

Figure 4.17   Dependency graph: security association during scanning

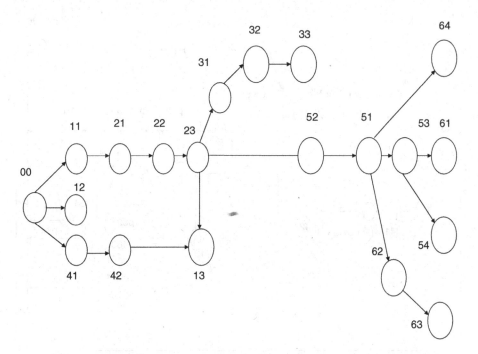

Figure 4.18   Dependency graph: security association and discovery during scanning

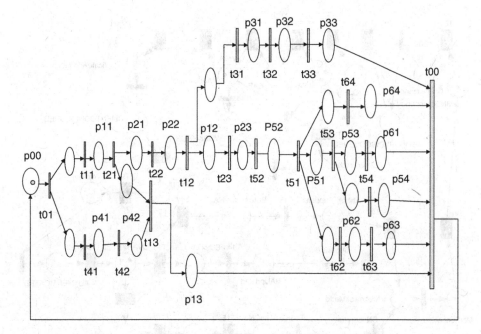

Figure 4.19   Petri net model without resources: security association during scanning

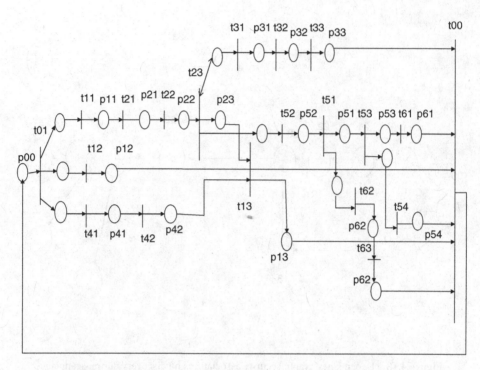

Figure 4.20 Petri net model: security association and subnet discovery during scanning

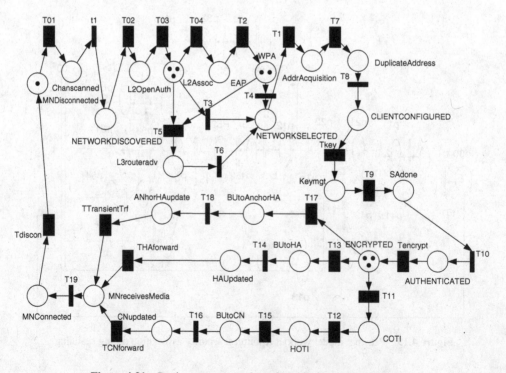

Figure 4.21 Petri net representation of Mobile IPv6 using TimeNet

deadlock, which has been studied extensively in the context of operating systems. Coffman et al. (1971) discussed the following four necessary conditions that must hold for a deadlock to occur:

1. *Mutual exclusion.* A resource is either available or allocated to a process that has exclusive access to that resource.
2. *Hold and wait.* A process is allowed to hold resources while requesting more resources.
3. *No preemption.* A resource allocated to a process cannot be removed from that process until it is released by the process itself.
4. *Circular wait.* Two or more processes are arranged in a chain in which each process waits for resources held by the process next in the chain.

A deadlock in a Petri net is a transition (or set of transitions) that cannot fire. A transition is live if it is not deadlocked. A transition t_j of a Petri net P is potentially fireable in a marking M_i if there exists a marking $M_j \in R(P, M_i)$ and t_j is enabled in M_j. A transition is live in a marking M_i if it is potentially fireable in every marking in $R(P, M_i)$.

We show a simple example to illustrate the deadlock problem in Figure 4.22. Consider a system with two different resources q and r and two processes a and b. If both of the processes need both of the resources, it will be necessary for the resources to be shared. To accomplish this, we require each process to request a resource and later release it. Figure 4.22 illustrates these two processes and the resource allocation with a Petri net. The initial marking indicates that the resources q (p_4) and r (p_5)

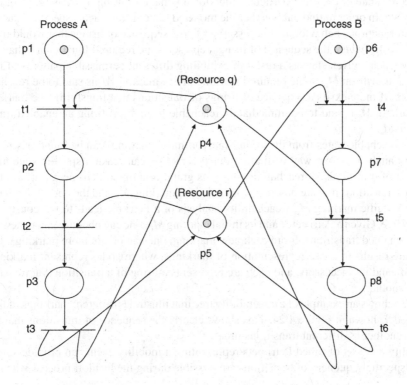

Figure 4.22   Example of deadlock arising from resource sharing

are available and processes *a* and *b* are ready. By inspecting the Petri net, we can see that several sequences of transitions are possible.

If we do a reachability analysis, it is evident that the execution sequences $t_1 t_2 t_3 t_4 t_5 t_6$ and $t_4 t_5 t_6 t_1 t_2 t_3$ do not result in deadlock. However, if a specific execution sequence consists of $t_1 t_4$, where process *a* has *q* and wants *r*, and process *b* has *r* and wants *q*, then the system is deadlocked and neither process can proceed.

These conditions for deadlock can also be applied to a system model representing a handoff event. In the case of a Petri net model representing a handoff event, deadlock may happen because of the absence of data from a preceding event that may be needed to successfully enable a transition. Lack of resources such as power, bandwidth, CPU cycles, or memory due to resource sharing can also result in deadlock. Thus, either the absence of data from a preceding event or a lack of resources may prevent a transition from firing and result in deadlock. A specific deadlock scenario can be analyzed by performing a reachability analysis. By investigating the reachable markings, one can determine which specific sequences of operations may lead to a deadlock situation.

Chapter 11 illustrates some examples of specific sequences of schedules that may lead to a blocking event and how these can be detected by using reachability analysis.

4.10.2   *Reachability Analysis*

An important issue in designing a model of a handoff system is to determine whether the system can reach a specific state or exhibit a particular functional behavior during its transition from one state to another state. In order to find out whether the modeled system can reach a specific state as a result of a required functional behavior, it is necessary to find a sequence of firings that would transform a marking M_0 to M_i, where the sequence of firings represents the required functional behavior. A real system may reach a given state as a result of exhibiting different permissible patterns of functional behavior that transform M_0 to the required marking M_i. A marking M_i is said to be reachable from a marking M_0 if there exists a sequence of firings of transitions that transforms the marking M_0 to M_i. The marking M_i is said to be immediately reachable from M_0 if firing an enabled transition in M_0 results in M_i.

The list of reachable states from the initial state (or initial marking) can be found in various ways. Reachability analysis is one way of finding out if a mobile can reach a specific state after going through a set of transitions. A reachability analysis graph of a timed Petri net clearly indicates the sequences of transitions that are possible from an initial marking M_0 and the ones that are not reachable from a specific marking M_i. Reachability analysis of a Petri net leads to the construction of a coverability tree. Given a Petri net N and its initial marking M_0, one can obtain as many new markings as there are enabled transitions. From each new marking, one can obtain more markings. Repeating the procedure results in a tree representation of markings, where nodes represent markings generated from M_0 and its successors, and each arc represents a firing of a transition that transforms one marking to another.

Figure 4.23 shows an example of a coverability tree that illustrates different markings of the sample Petri net model shown in Figure 4.24. This shows clearly the sequence of transitions that lead from one state to another as different transitions fire.

Reachability analysis of timed Petri nets representing a mobility event can provide some insight into which specific sequences of transitions are possible during the handoff process when a mobile

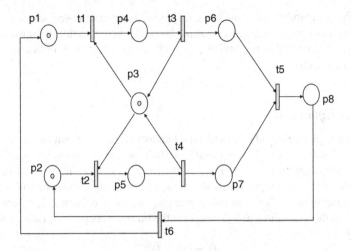

Figure 4.23   Coverability tree

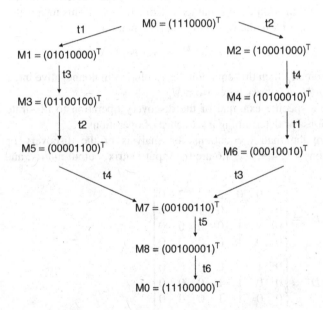

Figure 4.24   A Petri net with shared resources

goes from a connected to a disconnected state and then back to the connected state. It can also determine if a specific target transition is attainable from a given transition. For example, it is possible to find out whether a mobile that is in the configuration stage can proceed to the binding update stage if it follows a specific sequence of operations.

By looking at the coverability tree, one can easily determine specific sequences of executions that should be avoided. For example, a specific marking is deadlocked if no further transitions are possible from that marking. It is desirable to prevent deadlocks in a system by avoiding the transitions that give rise to these deadlocks.

4.10.3   Matrix Equations

A second approach to analyzing a Petri-net-based handoff system is based on a matrix view of Petri nets. Matrix-based analysis can answer questions such as whether or not a specific sequence of transitions is attainable from a given marking. A Petri net (P, T, I, O) has two matrices D^- and D^+, which represent the input and output functions, respectively. Each of these matrices has m rows and n columns, where the m rows represent the transitions and the n columns represent the places. D^- defines the inputs to the transitions and D^+ defines the outputs. As per Peterson (1981), a matrix D can be defined by

$$D = D^+ - D^-. \tag{4.2}$$

The development of matrix-based analysis for Petri net theory provides a useful tool for solving the reachability problem. If we assume that a marking M_j is reachable from a marking M_i, then there exists a sequence σ of transition firings that leads from M_i to M_j. This means that $f(\sigma)$ is a solution, in nonnegative integers, for x in the following matrix equation:

$$M_j = M_i + x.D. \tag{4.3}$$

If M_j is reachable from M_i, then this equation has a solution in nonnegative integers; if the equation has no solution, then M_j is not reachable from M_i.

We now describe a specific example of the discovery operation to illustrate the usefulness of matrix-based equations for determining a sequence of operations.

The following are the matrix equations for analysis of the discovery operation shown in Figure 4.10. Equations 4.4, 4.5, and 4.6 are the input matrix, output matrix, and incidence matrix, respectively.

$$D^- = \begin{bmatrix} 1 & 0 & 0 & 2 & 2 & 2 & 0 \\ 0 & 1 & 0 & 0 & 1 & 0 & 0 \\ 0 & 0 & 1 & 0 & 0 & 0 & 0 \\ 0 & 0 & 1 & 0 & 0 & 0 & 0 \end{bmatrix}, \tag{4.4}$$

$$D^+ = \begin{bmatrix} 0 & 1 & 0 & 0 & 0 & 0 & 0 \\ 0 & 0 & 1 & 1 & 0 & 0 & 0 \\ 0 & 0 & 1 & 1 & 0 & 0 & 0 \end{bmatrix}, \tag{4.5}$$

$$D = D^+ - D^- = \begin{bmatrix} -1 & 1 & 0 & -2 & -2 & -2 & 0 \\ 0 & -1 & 1 & 1 & -1 & 0 & 0 \\ 0 & 0 & 0 & -1 & 3 & 2 & 1 \end{bmatrix}. \tag{4.6}$$

Given a firing sequence $\sigma = t_1 t_2 t_3$, this translates into a firing vector $f(\sigma) = (1, 1, 1)$. By applying Equation 4.3, it is possible to find the final marking for the firing vector $f(\sigma) = (1, 1, 1)$. $M_j = (1, 0, 0, 3, 4, 3, 0) + (1, 1, 1) \cdot D$ produces the final marking $M_j = (0, 0, 0, 3, 4, 3, 1)$.

Thus, it is evident that M_j is reachable from the initial marking M_i with a sequence of transitions $t_1 t_2 t_3$ that corresponds to the firing vector $(1,1,1)$. The firing vector determines the number of times a specific transition is fired. A similar matrix analysis can be done for other handoff-related operations such as configuration, authentication, and attachment to determine reachability.

4.11 Evaluation of Systems Performance Using Petri Nets

The systems performance of a timed-Petri-net model representing a mobility event can be verified in several ways, as described by Holliday and Vernon (1987), Ramamoorthy and Ho (1980), Zuberek (1980), and Murata (1985). Here, we briefly describe the analysis by Ramamoorthy and Ho (1980) for DTTPN (deterministic timed-transition Petri net) models. We then present three approaches to studying the systems performance of a mobility model using Petri nets. The first approach illustrates that the minimum cycle time can be obtained from the Petri net model using the number of directed circuits, the number of transitions, and the timing associated with each transition. The minimum cycle time is an indicator of the maximum performance. The second approach involves computing the shortest distance by using the Floyd algorithm (Floyd, 1962). The third approach is based on a scalar measure called the resource–time product (RTP) for timed-Petri-net models. The resource–time product is defined as the product of the number of tokens (resources) and the length of the time these tokens are reserved for firing an enabled transition. Using the concept of the resource–time product, one can derive a lower bound on the minimum interval at which each transition can initiate its firing.

Here are some related definitions, as described by Wang (1998).

Definition 4.1 In a Petri net, *a directed path* is a sequence of places and transitions $p_1 t_1 p_2 t_2 \ldots p_k$ if transition t_i is both an output transition of place p_i and an input transition of place p_{i+1} for $1 \le i \le k-1$.

Definition 4.2 In a Petri net, *a directed circuit* is a sequence of places and transitions $p_1 t_1 p_2 t_2 \ldots p_k$ if $p_1 t_1 p_2 \ldots p_k$ is a directed path and p_1 equals p_k.

Definition 4.3 A Petri net is *consistent* if and only if there exists an assignment of nonzero integers to its transitions such that at every place, the sum of the integers assigned to its input transitions equals the sum of the integers assigned to its output transitions.

Definition 4.4 A Petri net is *decision-free* if and only if, for each place in the net, there is exactly one input arc and exactly one output arc.

4.11.1   Cycle-Time-Based Approach

The Petri net model representing the general mobility system is actually a decision-free Petri net, where the minimum cycle time is an indicator of the maximum performance. The cycle time is represented as $C = \max\ T_k/N_k : k = 1, 2, 3, \ldots, q$, where T_k is the sum of the execution times of the transmissions in circuit k, N_k is the total number of tokens in the places in circuit k, and q is the number of circuits in the net. In the case of a system model representing a mobility event, these values can vary depending upon the number of transitions and the sequence of transitions involved in a cycle.

Table 4.5 shows how the overall cycle time obtained using the directed-circuits approach is affected when some optimization techniques, namely hierarchical binding update, proactive

Table 4.5　Cycle time for Petri net model with mobility optimizations

Type of optimization	Loops in state transition path	T_i	N_i	T_i/N_i
No optimization	$P_0 t_1 P_1 t_2 P_2 t_3 P_3 t_4 P_4 t_5 P_5 t_6 P_6 t_7 P_{10}$	24 t	1	24 t
Hierarchical binding update	$P_0 t_1 P_1 t_2 P_2 t_3 P_3 t_4 P_4 t_5 P_5 t_8 P_7 t_9 P_{10}$	19 t	1	19 t
Proactive	$P_0 t_9 P_{10}$	2 t	1	2 t
Maintaining security binding	$P_0 t_1 P_1 t_2 P_2 t_3 P_5 t_6 P_6 t_7 P_{10}$	19 t	1	19 t

discovery and configuration, and anchor-based security association, are applied to the generalized Petri net system mobility model shown in Figure 4.6. These optimization techniques are presented in detail in Chapter 6. In this table, the duration of each transition is represented in terms of a unit of time "t.".

4.11.2　Floyd-Algorithm-Based Approach

In order to verify the systems performance of a mobility event in which a specific optimization methodology has been applied and to determine whether the system satisfies the desired threshold for the cycle time, one can generate a place matrix P and a transition matrix Q. The entry (A, B) in the matrix P equals x if there are x tokens in place A and place A is connected directly to place B by a transition. The entry (A, B) in the transition matrix Q equals t_i if A is an input place of transition t_i and B is one of its output places. The entry (A, B) contains a symbol w if A and B are not connected directly. Given a threshold value C of the cycle time, one can generate a distance matrix $CP - Q$. Then, using the Floyd algorithm (Floyd, 1962), one can determine a matrix S. By inspecting the values of the diagonal elements of S, it is possible to determine if the system provides the desired performance. Sample pseudocode for the Floyd algorithm is given in Figure 4.25.

Using the Floyd algorithm, there are three possible cases that give information about the system performance:

1. If all diagonal entries of the matrix S are positive (i.e., $CN_k - T_k > 0$ for all circuits), the system performance is higher than the given requirement.
2. If some diagonal entries of the matrix S are zeros and the rest are positive (i.e., $CN_k - T_k = 0$ for some circuits and $CN_k - T_k > 0$ for the others), the system performance just meets the given requirement.
3. If some diagonal entries of the matrix S are negative (i.e., $CN_k - T_k < 0$ for some circuits), the system performance is lower than the given requirement.

A set of matrices that represent a sample Petri net model consisting of four places with a given value of $C = 3$ are shown in Equations 4.7 and 4.8. In this specific example, all the diagonal entries of the matrix S are nonnegative. Thus, the system performance just meets the threshold requirement of $C =$

Floyd algorithm

```
/* Assume a function edgeCost(i,j) which returns the cost
   of the edge from i to j (infinity if there is none).
   Also assume that n is the number of vertices and
   edgeCost(i,i)=∅
*/ \\
int path[][];\\

/* A 2-dimensional matrix. At each step in the algorithm,
   path[i][j] is the shortest path from i to j using
   intermediate vertices (1..k-1). Each path[i][j] is initialized
   to edgeCost(i,j) or infinity if there is no edge between
   i and j. */

procedure FloydWarshall ()\\
  for k: = 1 to n \\
    for each (i,j) in (1..n)\\
      path[i][j] = min ( path[i][j], path[i][k]+path[k][j]);\\
```

Figure 4.25   Pseudocode for the Floyd algorithm

3. By means of such an analysis, one can easily determine if a specific Petri net model representing a handoff model can meet the desired system performance requirements.

$$P = \begin{bmatrix} 0 & 1 & 0 & 0 \\ 0 & 0 & 1 & 0 \\ 0 & 0 & 0 & 1 \\ 1 & 0 & 0 & 0 \end{bmatrix}, \quad Q = \begin{bmatrix} w & 2 & w & w \\ w & w & 3 & w \\ w & w & w & 4 \\ 3 & w & w & w \end{bmatrix}, \tag{4.7}$$

$$CP - Q = \begin{bmatrix} \infty & 1 & \infty & \infty \\ \infty & \infty & 0 & \infty \\ \infty & \infty & \infty & 1 \\ 0 & \infty & \infty & \infty \end{bmatrix}, \quad S = \begin{bmatrix} 0 & 1 & 1 & 1 \\ 0 & 0 & 0 & 1 \\ 0 & 1 & 0 & 1 \\ 0 & 1 & 1 & 0 \end{bmatrix}. \tag{4.8}$$

In Chapter 11, we compare various optimization techniques using the cycle-time-based and Floyd-algorithm-based approaches.

4.11.3   Resource–Time Product Approach

The resource–time product is defined as the product of the number of tokens (resources) and the length of time these tokens are reserved for firing an enabled transition. It is assumed that each transition in the timed-Petri-net model has a given time delay for completing its firing. Using the concept of the resource–time product, one can derive a lower bound on the minimum interval at which each transition can initiate firing. Since the resource–time product is based on an S-invariant and a T-invariant, we shall describe details of these two invariants, as presented originally by Murata (1985).

4.11.3.1 Invariant Analysis

Here, we provide details of S- and T-invariants and describe how these are calculated.

An integer solution x of the homogeneous equation

$$A^T x = 0 \qquad (4.9)$$

is called a T-invariant. A is the incidence matrix. The nonzero entries in a T-invariant represent the firing counts of the corresponding transitions, which belong to a firing sequence that transforms a marking M_0 back to M_0.

Similarly, an integer solution y of the transported homogeneous equation

$$Ay = 0 \qquad (4.10)$$

is called an S-invariant. The nonzero entries in an S-invariant represent weights associated with the corresponding places so that the weighted sum of the numbers of tokens on these places is constant for all markings reachable from an initial marking.

It is assumed that the timed Petri net N under consideration functions periodically with a period of τ units of time, during which time the initial marking is reproduced and transition i fires z_i times, where $z_i > 0$ for each i. Thus, we have $Z_i = cX_i$ or, in terms of n-vectors $z = [z_i]$ and $x = [x_i]$,

$$z = cx \geq x, \qquad (4.11)$$

where c is a positive integer and x is a minimal T-invariant of N such that $x_i > 0$ for each i. τ_m is called the minimum cycle time for the timed Petri net, where the time delay of each firing of a transition is given.

When transition i is enabled, a_{ij}^- tokens will be reserved in place j for at least d_i units of time. Thus, the resource–time product due to these a_{ij}^- tokens being reserved is $a^- * d_i$. Since transition i fires z_i times during a period of τ units of time, the resource–time product for firing transition i is $a^i * d_i * z_i$. By applying this to all of the transitions in N, one obtains the total resource–time product for place j as follows:

$$\sum_{i=1}^{n} a_{ij}^- d_{ij} z_i.$$

Thus, the resource–time products for all m places are given by a column vector that can be expressed as the matrix product $(A^-)^T Dz$, where D is the diagonal matrix of the delays d_i.

In the above resource–time products, we have considered only tokens that are reserved for firing. As per Murata (1985), on further analysis it is found that

$$\tau \geq y^T (A^-)^T Dx / y^T M(t_0). \qquad (4.12)$$

Therefore, a lower bound on the minimum period may be given by

$$\tau_{min} = \max(y T(A^-)^T Dx / y^T M(t_0)), \qquad (4.13)$$

where the maximum is taken over all independent minimal-support S-invariants, $y_k \geq 0$. For a timed marked graph, each directed circuit C_k yields a minimal-support S-invariant y_k. Thus, Equation 4.9 can be reduced to $\tau_{min} = \max(\text{Total delay in } C_k / M_0(c_k))$, which is equivalent to the circuit analysis approach.

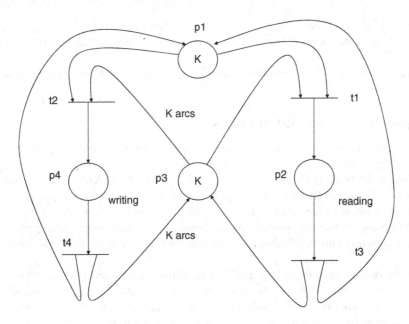

Figure 4.26   Petri net for resource–time product calculation

We now describe an example of how a cycle time may be derived using the RTP approach. Figure 4.26 shows a Petri net for a reader/writer system, where k tokens in place p_1 represent k processes, which may read and write in a shared memory represented by p_3. As per the Petri net model, up to k processes may be reading concurrently, but when one process is writing, no other processes can be reading or writing. Equation 4.14 shows the incidence matrix A and the forward incidence matrix A^-:

$$A = \begin{bmatrix} -1 & 1 & -1 & 0 \\ -1 & 0 & -k & 1 \\ 1 & -1 & 1 & 0 \\ 1 & 0 & k & -1 \end{bmatrix}, \quad A^- = \begin{bmatrix} -1 & 1 & -1 & 0 \\ -1 & 0 & -k & 1 \\ 1 & -1 & 1 & 0 \\ 1 & 0 & k & -1 \end{bmatrix}. \tag{4.14}$$

The system performance or cycle time for a handoff event can be measured using any of the above approaches. Although determining the cycle time using the directed-circuit method is a reasonable approach as it gives a visual solution, it has shortcomings, since one needs to enumerate a number of circuits for a system containing a large number of places and transitions. The Floyd algorithm-based approach or the RTP-based approach is preferred when the number of places and transitions is large. In order to determine the cycle time, one needs to determine the S- and T-invariants.

In the specific example shown in Figure 4.26, there are two independent S-invariants, $y_1 = (1101)^T$ and $y_2 = (011k)^T$, and two independent T-invariants, $x_1 = (1010)^T$ and $x_2 = (0101)^T$. The firing count vector of a firing sequence leading back to the initial marking after each transition has fired at least once is given by $x = x_1 + x_2 = (1111)^T$. Let the time delay of transition i be d_i, $i = 1, 2, 3, 4$. Since the initial marking is given by $M(t_0) = (k0k0)^T$, by evaluating Equation 4.11 for the two S-invariants y_1 and y_2 we obtain

$$\tau_m = \text{Max}(d_1 + d_2 + d_3 + d_4)/k, \quad (d_1 + kd_2 + d_3 + kd_4)/k = d_2 + d_4 + (d_1 + d_3)/k.$$

This means that the transitions t_1, t_2, t_3, and t_4 can be initiated at most once every $[d_2 + d_4 + (d_1 + d_3)/k]$ units of time, since $x = (1111)^T$ is the minimal T-invariant such that $x_i > 0$ for each i in this timed Petri net.

In Chapter 11, we show how some of these approaches can be used to model handoff systems.

4.12   Opportunities for Optimization

In a Petri net model representing a handover process, the cycle time is representative of the systems performance. The order of execution of the handoff processes and the resources in the system play an important role in determining the cycle time. Figure 4.27 illustrates a Petri net representation of the possible sequences of execution for a pair of processes such as *Pa* and *Pb* that are part of a mobility event. Thus, the sequence of execution of subtasks during a mobility event affects the overall cycle time, and the amount of system resources expended may vary because of parallel execution of events. The amount of resources utilized during a parallel operation increases the peak amount of resources utilized.

Thus, several optimization techniques can then be applied to the generalized mobility model to obtain corresponding optimized models. These techniques can also be applied to the subprocesses within the overall system or in a hierarchical manner for each of these subprocesses. Performance evaluation can then be performed for each of these optimized models, and these models can be compared with the nonoptimized models. Most of the optimization techniques that we will describe in Chapter 6 can be mapped into sequential, concurrent, and proactive modes of operation using the Petri net model. In Figure 4.27, the proactive mode of operation can be equated to a variation of the case where operation *Pa* happens during operation *Pb*. This implies that some parts of operation *Pb*

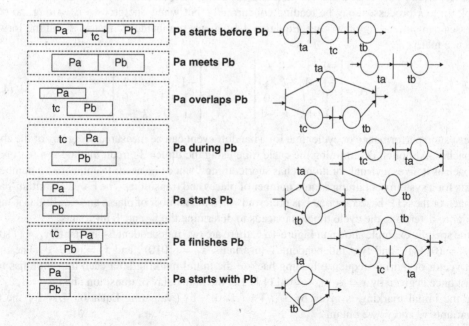

Figure 4.27   Petri net mapping for sequence of events

start before operation *Pa*. In the following section, we describe a few ways of optimizing handoff events using different sequences of operations.

4.12.1 Analysis of Parallelism in Handoff Operations

If there are two operations that are not mutually exclusive, then these operations can be done in parallel. Two operations can be done in parallel only if there is no resource constraint and if the starting of one operation does not depend upon data from the other operation. However, in some resource-constrained environments, these operations may need to share common resources. In such a situation, there may be deadlocks due to lack of resources.

Analysis of the sequence of events and the precedence relationships among these operations can help us to analyze the opportunities for parallelism. In some situations, when the execution of a specific event does not depend upon data from another event, these two events do not need to occur in sequence and there is an opportunity for parallelism. Here are some examples of parallelism that are possible during handoff operations.

- *Example 1*. A layer 3 point of attachment can be discovered during layer 2 association if some of the router advertisement information (e.g., the prefix of the router advertisement or the default gateway) can be passed as part of the layer 2 beacon. One way this can be achieved is by a beacon-stuffing approach (Chandra et al., 2007), whereby layer 3 information can be passed as part of the layer 2 beacon advertisement. This approach results in parallel operation, leading to simultaneous discovery of layer 2 and layer 3 points of attachment. Thus, it reduces the time taken compared with the situation if those operations had been done in sequence.
- *Example 2*. In normal situations, duplicate address detection is performed only after an identifier has been obtained. However, a proactive lookup operation to test the uniqueness of an address before it is assigned to the mobile can help speed up the identifier configuration process. Alternatively, address uniqueness can be tested during the process of address acquisition.
- *Example 3*. Another possible opportunity for parallelism is execution of the authentication and key derivation processes in parallel. Authentication is usually performed using EAP over L2 or EAP over L3, which generates a pairwise master key. Generation of a pairwise transient key is followed by a regular authentication process. If these two operations can be done in parallel, the handoff delay due to authentication will be reduced.
- *Example 4*. Forwarding, buffering, and multicasting can all be done in parallel to reduce the loss of data at the cost of additional bandwidth resources.

4.12.2 Opportunities for Proactive Operation

In some cases, where parallel operations among the handoff processes are not possible, many of the operations can be completed proactively. Proactive operations mean that some of the handoff-related operations can be done in the current network before the mobile moves to the new network. Network discovery, IP address acquisition processes, and authentication processes are some of the handoff-related operations that can be performed ahead of time in the serving network before the mobile moves to the new network. However, additional resources are needed to support these proactive operations. We will describe many such proactive mechanisms for handoff in Chapter 6.

4.13 Concluding Remarks

Based on an analysis of the data dependency and resources for the various handoff operations, it is possible to build a mobility model using timed-transition Petri nets for both optimized and unoptimized handoff systems. Using a hierarchical approach based on Petri nets, each of the handoff components can be modeled independently, based on the primitive operations associated with that handoff component. For example, layer 3 configuration is based on several layer 3 primitive operations, namely IP address acquisition, duplicate address detection, and address resolution. Petri net models for each of these components (e.g., discovery and configuration) can be synthesized to build a complete handoff system. Depending upon the type of handoff and whether one wishes to investigate the systems performance (e.g., cycle time) or check for anomalies of the handoff system (e.g., deadlocks), one can apply a specific Petri-net-based methodology. For example, in order to verify the systems performance, one can use the Floyd algorithm approach; in order to determine the exact cycle time, one can use a circuit-time-based approach.

On the other hand, if one needs to determine system anomalies such as deadlocks, one needs to use reachability analysis. By investigating a Petri net model for an unoptimized handoff system, one can determine the opportunities for parallelism where some of the handoff-related operations could run in parallel, but these parallel operations may need some modifications to the behavior of current systems. For example, to have the ability to discover a layer 3 point of attachment (default router) during the discovery of a layer 2 point of attachment (access point), one needs to modify the layer 2 beacon so that it can carry the router's address or the subnet prefix. Thus, a mobility model can help in analyzing the performance of existing handoff systems and also offers the ability to predict the behavior of new handoff systems that can be designed based on the available resources and expected systems performance.

5

Layer 2 Optimization

With the growth of IEEE 802.11-based wireless LANs, Voice over IP (VoIP) and similar applications are now commonly used over wireless networks. A mobile station performs a handoff whenever it moves out of the range of one access point and tries to connect to a different one. This takes a few hundred milliseconds, causing interruptions in VoIP sessions. This chapter introduces a new handoff procedure that reduces the MAC layer handoff latency, in most cases to a level where VoIP communication becomes seamless. This new handoff procedure reduces the discovery phase using a selective scanning algorithm and a caching mechanism.

5.1 Introduction

IEEE 802.11-based wireless LANs have seen very fast growth in the last few years and VoIP is one of the most promising services to be used in mobile devices over wireless networks. One of the main problems in VoIP communication is the handoff latency introduced when moving from one access point to another. As we will show below, the amount of time needed for a handoff in an 802.11 environment is too large for seamless VoIP communication. We have been able to reduce the handoff latency using a modified handoff procedure, where the modifications are limited to the mobile device and are compatible with standard 802.11 behavior.

In the following, we first briefly discuss the work done in this particular area, followed by a brief description of how we tackled the problem. Then, in Section 5.3, we describe the IEEE 802.11 architecture, focusing on the management frames and the handoff process; then we briefly describe the HostAP driver and how we modified it in order to implement our new algorithm. In Section 5.5, we illustrate how we were able to reduce the total handoff latency to an average value of 129 ms by using a selective scanning procedure and to an average of 3 ms by using a caching mechanism. In Sections 5.7 and 5.8, we present the environment of the experiments and the measurement results.

5.2 Related Work

A lot of work has been done to reduce the handoff latency when roaming between different subnets, and many new schemes for Mobile IP and route optimization have been proposed. In this chapter, we focus on reducing handoff latency in the MAC layer. As we will describe in Section 5.4, the MAC

Mobility Protocols and Handover Optimization: Design, Evaluation and Application, First Edition.
Ashutosh Dutta and Henning Schulzrinne.
© 2014 John Wiley & Sons, Ltd. Published 2014 by John Wiley & Sons, Ltd.

layer handoff latency can be divided into three components: probe delay, authentication delay, and association delay.

Shin et al. (2004a) focused on reducing the reassociation delay. The reassociation delay was reduced by using a caching mechanism on the access point (AP) side. This caching mechanism was based on the IAPP protocol for exchanging client context information between neighboring APs. The cache in the AP was built using the information contained in an IAPP Move-Notify message or in the reassociation request sent to the AP by the client. By exchanging the client context information with the old AP, the new AP does not require the client to send its context information in order to reassociate, hence reducing the reassociation delay.

Pack and Choi (2002) and Pack and Choi (2003) focused on the IEEE 802.1x authentication process. This process is performed after the station (STA) has already associated with a new AP. The IEEE 802.1x authentication delay was reduced by using the Frequent Handoff Region (FHR) selection algorithm.

Mishra et al. (2003a) showed that the discovery phase (i.e., the scanning time) was the most time-consuming part of the handoff process, accounting for over 90% of the total handoff delay, while the (re)association time contributed only a few milliseconds. Our work follows a novel approach and reduces the total handoff latency by reducing the scanning time. This is achieved by using a selective scanning algorithm and a caching mechanism. This caching data structure is maintained on the client side and no changes are required in the existing network infrastructure or the IEEE 802.11 standard, unlike the case in Shin et al. (2004a). All the required changes are done on the client-side wireless card driver.

Park et al. (2004) also proposed a selective scanning algorithm. However, their proposition relies on the use of neighbor graphs. This approach requires changes in the network infrastructure and use of IAPP. The scanning delay was defined as the duration from the first probe request message to the last probe response message. This definition does not take into consideration the time needed by the client to process the received probe responses. This processing time represents a significant part of the scanning delay and increases significantly with the number of probe responses received. In the current work and in Mishra et al. (2003a), the time required for processing the probe responses received by the client is taken into consideration.

5.3   IEEE 802.11 Standards

There are currently four IEEE 802.11 standards: 802.11a, 802.11b, 802.11g, and 802.11n. The 802.11a standard operates in the 5 GHz ISM band and uses a total of 32 channels, of which only eight do not overlap. The 802.11b and 802.11g standards both operate in the 2.4 GHz ISM band and use 11 of the 14 possible channels. Of these 11 channels, only three do not overlap. While 802.11b can operate up to a maximum rate of 11 Mbit/s, the 802.11g and 802.11a standards can operate up to a maximum rate of 54 Mbit/s. The 802.11g standard is backwards-compatible with the 802.11b standard, while the 802.11a standard, because of the different ISM band, is not compatible with the other two 802.11n uses multiple antennas to increase the data rates over 802.11a and 802.11g with a significant increase in net data rate from 54 Mbps to 600 Mbps.

We will focus our attention on the IEEE 802.11b standard, although most of the concepts and notions described here are still valid for 802.11a and 802.11g. As we said earlier, 802.11b operates in the 2.4 GHz ISM band. Its 14 channels are distributed over the range from 2.402 to 2.483 GHz (see Figure 5.1), each channel being 22 MHz wide. In the US, only the first 11 channels are used. Of these 11 channels, only channels 1, 6, and 11 do not overlap. So, in a well-configured wireless network, all or most of the APs will operate on channels 1, 6, and 11. Also, to avoid cochannel interference, two adjacent APs should never be on the same channel.

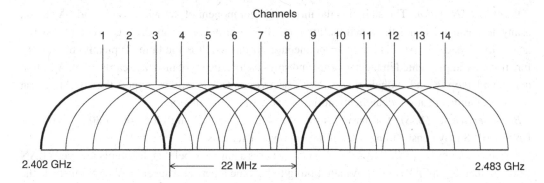

Figure 5.1 Channel frequency distribution in IEEE 802.11b. (Source: Schulzrinne 2004. Reproduced with permission of Henning Schulzrinne.)

5.3.1 The IEEE 802.11 Wireless LAN Architecture

The 802.11 architecture (Figure 5.2) is composed of several components and services that interact to provide station mobility to the higher layers of the network stack. We outline the components as follows, as described in Jain (2003).

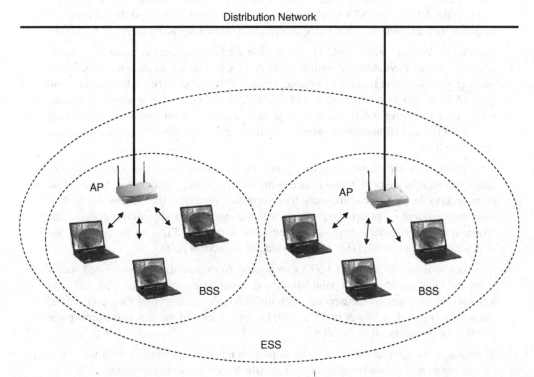

Figure 5.2 IEEE 802.11 architecture. (Source: Schulzrinne 2004. Reproduced with permission of Henning Schulzrinne.)

Wireless LAN station. The station is the most basic component of the wireless network. A station is any device that contains the functionality of the 802.11 protocol: the media access control (MAC) layer, the physical layer (PHY), and a connection to the wireless medium. Typically, the 802.11 functions are implemented in the hardware and software of a network interface card (NIC). A station may be a laptop PC, a handheld device, or an access point. All stations support the 802.11 station services of authentication, deauthentication, privacy, and data delivery.

Basic service set (BSS). The basic service set is the basic building block of an 802.11 wireless LAN. A BSS may consist of a group of any number of stations.

Service set identifier (SSID). A service set identifier is a unique label that distinguishes one WLAN from another. So, all APs and STAs attempting to become a part of a specific WLAN must use the same SSID. Wireless STAs use the SSID to establish and maintain connectivity with APs.

5.3.2   IEEE 802.11 Management Frames

The IEEE 802.11 management frames enable stations to establish and maintain communications. The following are common IEEE 802.11 management frame subtypes; the descriptions are quoted from Geier (2002).

Authentication frame: The 802.11 authentication is a process whereby the access point either accepts or rejects the identity of a STA. The STA begins the process by sending an authentication frame containing its identity to the access point. With open system authentication (the default), the STA sends only one authentication frame, and the access point responds with an authentication frame as a response indicating acceptance (or rejection).

Association request frame: 802.11 association enables the access point to allocate resources for and synchronize with a STA. A STA begins the association process by sending an association request to an access point. This frame carries information about the STA (e.g., supported data rates) and the SSID of the network it wishes to associate with. After receiving the association request, the access point considers associating with the STA, and (if accepted) reserves memory space and establishes an association ID for the STA.

Association response frame: An access point sends an association response frame containing an acceptance or rejection notice to the STA requesting association. If the access point accepts the STA, the frame includes information regarding the association, such as association ID and supported data rates. If the outcome of the association is positive, the STA can utilize the access point to communicate with other STAs on the network and systems on the distribution (i.e. Ethernet) side of the access point.

Reassociation request frame: If a STA roams away from the currently associated access point and finds another access point having a stronger beacon signal, the STA will send a reassociation frame to the new access point. The new access point then coordinates the forwarding of data frames that may still be in the buffer of the previous access point waiting for transmission to the STA.

Reassociation response frame: An access point sends a re-association response frame containing an acceptance or rejection notice to the STA requesting reassociation. Similar to the association process, the frame includes information regarding the association, such as association ID and supported data rates.

Disassociation frame: A station sends a disassociation frame to another station if it wishes to terminate the association. For example, a STA that is shutdown gracefully can send a disassociation frame to alert the access point that the STA is powering off. The access point can then relinquish memory allocations and remove the STA from the association table.

Beacon frame: The access point periodically sends a beacon frame to announce its presence and relay information, such as timestamp, SSID and other parameters regarding the access point, to STAs that are within range.

Probe request frame: A station sends a probe request frame when it needs to obtain information from another station. For example, a STA would send a probe request to determine which access points are within range.

Probe response frame: A station will respond with a probe response frame, containing capability information, supported data rates, etc., after it receives a probe request frame.

5.4   Handoff Procedure with Active Scanning

Handoff is a procedure executed when a mobile node moves from the coverage area of one AP to the coverage area of another AP. The handoff process involves a sequence of messages being exchanged between the mobile node and the participating APs. This sequence of messages can be divided into three types, namely probe, authentication, and association, which will be described later in detail. The transfer from the old AP to the new AP results in state information being transferred from the former to the latter, consisting of authentication, authorization, and accounting information. This can be achieved by the Inter Access Point Protocol (IAPP) that is currently under draft in IEEE 802.11f, or by a proprietary protocol specific to a vendor. We have briefly described IEEE 802.11f and the associated Inter Access Point Protocol in Chapter 2.

5.4.1   Steps during Handoff

The handoff process can be divided into two logical steps: discovery and reauthentication (Mishra et al., 2003a).

5.4.1.1   Discovery

The discovery process involves the handoff initiation phase and the scanning phase.

When the STA is moving away from the AP it is currently associated with, the signal strength and the signal-to-noise ratio of the signal from the AP decrease. This causes the STA to initiate a handoff. The STA now needs to find other APs that it can connect to. This is done by the MAC layer scanning function.

Scanning can be accomplished in either passive or active mode. In passive scan mode, the STA listens to the wireless medium for beacon frames. Beacon frames provide a combination of timing and advertising information to the STAs. Using the information obtained from the beacon frames,

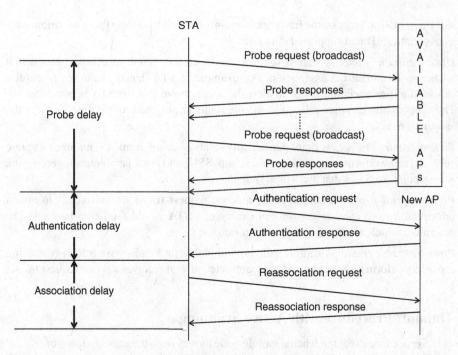

Figure 5.3 Handoff process using active scanning. (Source: Schulzrinne 2004. Reproduced with permission of Henning Schulzrinne.)

the STA can elect to join an AP. During this scanning mode, the STA listens to each channel of the physical medium to try to locate an AP.

Active scanning (Figure 5.3) involves transmission of probe request frames by the STA over the wireless medium and processing of the probe responses received from the APs. The basic procedure for the active scan mode includes the following steps (Committee et al., 1997):

1. Using the normal channel access procedure, namely carrier sense multiple access with collision avoidance (CSMA/CA), gain control of the wireless medium.
2. Transmit a probe request frame which contains the broadcast address as its destination.
3. Start a probe timer.
4. Listen for probe responses.
5. If no response has been received by minChannelTime, scan the next channel.
6. If one or more responses are received by minChannelTime, stop accepting probe responses at maxChannelTime and process all received responses.
7. Move to the next channel and repeat the above steps. After all channels have been scanned, all information received from probe responses is processed so that the STA can select which AP to join next.

5.4.1.2 Reauthentication

The reauthentication process involves authentication and reassociation with the new AP, as well as transfer of the STA's credentials from the old AP to the new AP. Authentication is a process by which the AP either accepts or rejects the identity of the STA. The STA begins the process by sending an authentication frame and authentication request, informing the AP of its identity; the AP responds with an authentication response, indicating acceptance or rejection. After successful authentication, the STA sends a reassociation request to the new AP, which then sends a reassociation response back to the STA, containing an acceptance or rejection notice.

The steps are as follows:

1. *Probe messages.* Once the STA decides to look for other APs, the probing process starts. The STA starts sending out probe requests and then processes the received probe responses, based on the active scanning algorithm described above. The time involved in this probing process is called the probe delay.
2. *Authentication messages.* Once the STA has decided to join an AP, authentication messages are exchanged between the STA and the selected AP. The time consumed by this process is called the authentication delay.
3. *Reassociation messages.* After a successful authentication, the STA sends a reassociation request and expects a reassociation response back from the AP. These messages are responsible for the reassociation delay.

5.5 Fast-Handoff Algorithm

As described by Shin et al. (2004a) and confirmed by experiments, the probe delay constitutes the biggest part (over 90%) of the handoff latency (Figure 5.4); for this reason, we focus on minimizing this delay.

In order to reduce the probe delay, we focused our attention on two different aspects of the problem. First, we had to reduce the probe delay by improving the scanning procedure, using a selective scanning algorithm; second, we had to minimize the number of times the scanning procedure was needed. This second point was achieved by the use of a caching mechanism. We will now describe the two algorithms that we used to achieve these aims.

5.5.1 Selective Scanning

As we described in Section 5.3, of the 14 possible channels that can be used according to the IEEE 802.11b standard, only 11 are used in the USA and, of these 11 channels, only three do do not overlap. These channels are 1, 6, and 11.

The selective scanning algorithm (Figure 5.5) is based on this idea. In selective scanning, when an STA scans APs, a channel mask is built. In the next handoff, this channel mask is used during the

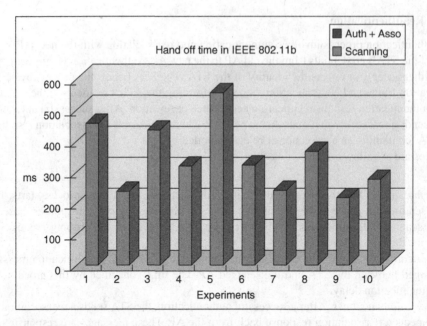

Figure 5.4 Handoff time in IEEE 802.11b. (Source: Schulzrinne 2004. Reproduced with permission of Henning Schulzrinne.)

scanning process. When this is done, only a well-selected subset of channels is scanned, reducing the probe delay. In the following, we describe the selective scanning algorithm in detail:

1. When the driver is first loaded, it performs a full scan (i.e., it sends out a probe request on all channels, and listens to the responses from the APs).
2. A channel mask is set by turning on the bits for all of the channels in which a probe response was heard as a result of step 1. In addition, the bits for channels 1, 6, and 11 are also set, as these channels are more likely to be used by APs.
3. The best AP, that is, the one with the strongest signal strength, is selected from among the scanned APs, and the STA connects to that AP.
4. The channel that the STA connects to is removed from the channel mask by resetting the corresponding bit, as the likelihood of there being an adjacent AP on the same channel as the current AP is very small. So, the final formula for computing the new channel mask is "scanned channels (from step 2) + 1 + 6 + 11 − the current channel."
5. If no APs are discovered with the current channel mask, the channel mask is inverted and a new scan is done. If still no APs are discovered, a full scan on all channels is performed.

5.5.2 Caching

The selective scanning procedure described above reduced the handoff latency in our experiments (Section 5.7) to 30–60% of the values obtained in the original handoff (see Figure 5.6), bringing the average value of the total handoff latency to 130 ms, to be compared with an original handoff latency

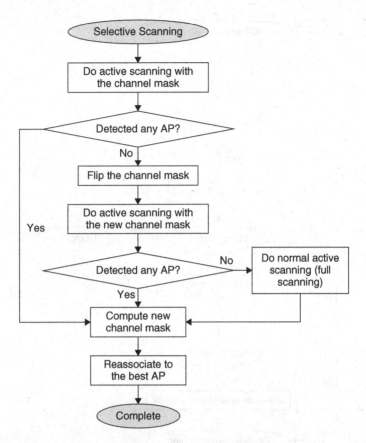

Figure 5.5 Selective scanning procedure. (Source: Schulzrinne 2004. Reproduced with permission of Henning Schulzrinne.)

Table 5.1 Handoff delay (ms) for 802.11b in link layer

	Experiment										
	1	2	3	4	5	6	7	8	9	10	Avg.
Original handoff	457	236	434	317	566	321	241	364	216	274	343
Selective scanning	140	101	141	141	141	139	143	94	142	101	129
Caching	2	2	4	3	4	2	2	2	2	2	3

of 343 ms (see Table 5.1). For seamless VoIP, it is recommended that the overall latency does not exceed 50 ms (Time, 2000). Further improvement was achieved by using an AP cache (Figure 5.7). The AP cache consisted of a table (Table 5.2) that used the MAC address of the current AP as the key. Corresponding to each key entry in the cache was a list of MAC addresses of APs adjacent to the current one which were discovered during scanning. This list was automatically created while roaming.

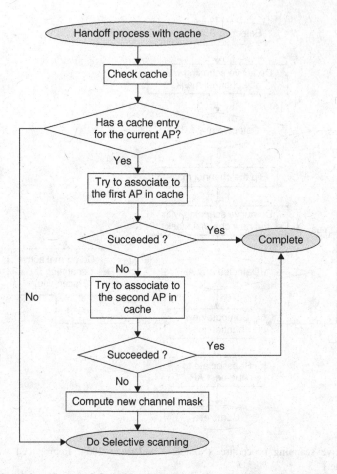

Figure 5.6 Bridging delay. (Source: Schulzrinne 2004. Reproduced with permission of Henning Schulzrinne.)

Table 5.2   Cache structure

	Key	Best AP	Second best AP
1			
2	MACl (Ch1)	MAC2 (Ch2)	MAC3 (Ch3)
⋮			
10			

The cache in the experiment had a size of 10, meaning that it could store up to 10 keys, and a width of 2, meaning that for each key, it could store up to two adjacent APs in the list. In the following, we describe how the caching algorithm works:

1. When an STA associates with an AP, the AP is entered in the cache as a key. At this point, the list of AP entries corresponding to this key is empty.
2. When a handoff is needed, we first check the entries in the cache corresponding to the current key.

3. If no entry is found (cache miss), the STA performs a scan using the selective scanning algorithm described in Section 5.5.1. The best two results based on signal strength are then entered in the cache with the old AP as the key.
4. If an entry is found (cache hit), we issue a command to associate with this new AP. On success, the handoff procedure is complete.
5. If the STA fails to connect to the first entry in the cache, the second entry is tried and if association with the second entry fails as well, our selective scanning algorithm is used.

From the above algorithm, we can see that scanning is required only if a cache miss occurs; every time we have a cache hit, no scanning is required. Usually, using the cache, it took less than 5 ms to associate with the AP. But, when the STA failed to associate with a new AP, the firmware waited for a long time, up to 15 ms. These values may vary from chipset to chipset. To reduce this "time to failure," we used a timer. The timer expires after 6 ms, and the STA then tries to associate with the next entry in the cache. In selective scanning, when the timer expires, the STA performs a new selective scan using the new channel mask.

A cache miss does not significantly affect the handoff latency. As mentioned above, when a cache miss occurred, the time to failure was only 6 ms. If the first cache entry missed and the second one hit, the additional handoff delay was only 6 ms. When both cache entries missed, the total handoff delay was 12 ms plus the selective-scanning time, all of this still resulting in a significant improvement compared with the original handoff time.

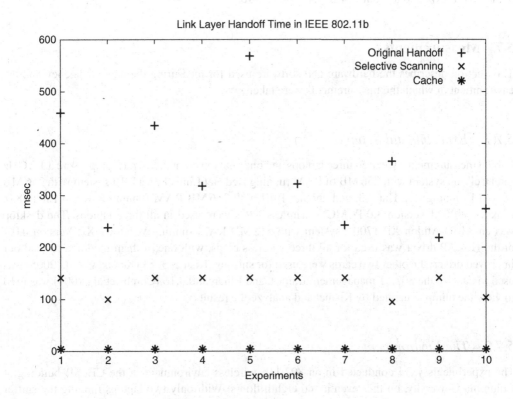

Figure 5.7   Caching procedure

5.6   Implementation

To implement our new algorithm, we had to modify the handoff procedure. Usually, the handoff procedure is handled by firmware; using the HostAP driver (Malinen, 2004), we were able to emulate the whole handoff process in the driver.

5.6.1   The HostAP Driver

The HostAP driver is a Linux driver for wireless LAN cards based on Intersil's Prism2/2.5/3 802.11 chipset (Malinen, 2004). Wireless cards using this chipset include the Linksys WPC11 PCMCIA card, the Linksys WMP11 PCI card, the ZoomAir 4105 PCMCIA card, and the D-Link DWL-650 PCMCIA card.

This driver supports a so-called host AP mode, that is, it takes care of the IEEE 802.11 management functions in the host computer and acts as an access point. This does not require any special firmware for the wireless LAN card. In addition to this, it has support for normal station operations in the BSS and possibly also in an independent basic service set (IBSS).

The HostAP driver supports a command for scanning APs, can handle the scanning results, and supports a command for joining to a specific AP. It is also possible to disable the firmware handoff by switching to manual mode. By enabling this mode, we were able to control the card functionalities at the driver level and use our fast-handoff procedure.

5.7   Measurements

This section describes the hardware and software used for measuring the handoff latency and the environment in which the measurements were taken.

5.7.1   Experimental Setup

For the measurements, we used three laptops and one desktop computer. The laptops were a 1.2 GHz Intel Celeron system with 256 MB of RAM running Red Hat Linux 8.0, a P-III system with 256 MB of RAM running Red Hat 7.3, and another P-III with 256 MB RAM running Red Hat Linux 8.0. Linksys WPC11 version 3.0 PCMCIA wireless NICs were used in all three laptops. The desktop was an AMD Athlon XP 1700+ system with 512 MB RAM running Windows XP. Version 0.0.4 of the HostAP driver was used for all three wireless cards, with one of them modified to load our improved driver; the other two cards were used for sniffing. Kismet 3.0.1 (Kershaw et al., 2005) was used to capture the 802.11 management frames, and Ethereal 0.9.16 (Combs et al., 2004) was used to view the dump generated by Kismet and analyze the result.

5.7.2   The Environment

The experiments were conducted in an 802.11b wireless environment in the CEPSR building at Columbia University, on the seventh and eighth floors. With only two laptops running the sniffer (Kismet), many initial runs were first conducted to get to know the wireless environment, specifically the channel numbers that the APs were operating on, the handoff starting points, and the next AP and channel that the STA would connect to.

The environment in which the packet loss measurements were taken differed from the one above. The measurements were again taken in the CEPSR building at Columbia University, on the seventh and eighth floors, but some rogue APs had been removed. This change in the environment caused a reduction in the original handoff time and, consequently, a drastic reduction in the packet loss. This will be shown in detail in Section 5.8.2.

5.7.3   Experiments

After gathering sufficient information about the environment, we started taking the actual handoff measurements. One sniffer was set to always sniff on channel 1 (as the first probe request is always sent out on channel 1 in normal active scanning), and the other sniffer was set to sniff on the next channel the STA was expected to associate with. For the measurement, the system clocks of the three laptops were synchronized using the Network Time Protocol (NTP). Also, to avoid multipath delays, the wireless cards were kept as close as physically possible during the measurements.

To measure the packet loss, in addition to the three laptops, the desktop was used as a sender and receiver. A UDP packet generator was used to send and receive data packets. Each UDP packet contained the packet sequence number in the data field to improve accuracy by linking the packet sequence number to its timestamp on all three laptops.

5.8   Measurement Results

In the following sections, we present the results for the total handoff time and packet loss.

5.8.1   Handoff Time

Figure 5.8 and Table 5.1 present the results that we obtained. As can be seen, with selective scanning, the handoff latency improved considerably, with an average reduction of 40%. But even this reduced time is not good enough for seamless VoIP. However, using the cache, the handoff latency time dropped to a few milliseconds, making it possible to have seamless VoIP. This huge reduction was possible because scanning, which took more than 90% of the total handoff time, was eliminated by using the cache.

5.8.2   Packet Loss

Table 5.3 presents the results we obtained when the STA performing the handoff was the receiver. Table 5.4 presents the results we obtained when the STA performing the handoff was the sender. Table 5.5 provides a summary of the results for the delay and packet loss for both the mobile sender and the mobile receiver in the original handoff scenario, and in the scenarios with selective scanning and with caching.

To measure the packet loss, UDP packets were transmitted to and from the STA, to simulate voice traffic during the handoff. Transmitting data packets adds to the handoff delay. This delay is caused by the fact that data packets are transmitted during the handoff process, in particular between the last

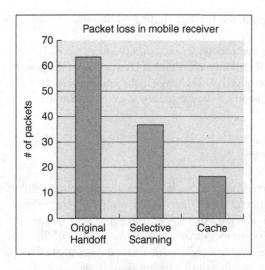 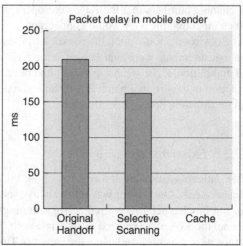

Figure 5.8   Link layer handoff time in 802.11b

Table 5.3   Packet loss during handoff in mobile (number of packets; 20 ms interval)

					Experiment						
	1	2	3	4	5	6	7	8	9	10	Avg.
Original handoff	36	55	32	79	37	122	134	32	69	36	63
Selective scanning	88	24	26	19	31	28	46	26	64	18	37
Caching	16	15	14	14	16	15	23	21	15	14	16

Table 5.4   Delay during handoff in mobile sender (ms)

					Experiment						
	1	2	3	4	5	6	7	8	9	10	Avg.
Original handoff	281	229	230	210	209	227	185	174	189	168	210
Selective scanning	185	132	147	131	204	182	164	133	151	184	161
Caching	0	0	0	0	0	0	0	0	0	0	0

probe response and the authentication request. This behavior is only noticed when the STA performing the handoff is the sender (see Table 5.5). When the STA performing the handoff is the receiver, no such delay is introduced. However, a new delay is introduced. This new delay, the bridging delay (Mishra et al., 2003a), is caused by the time needed to update the MAC addresses to the Ethernet switches forming the distribution system. In particular, when handoff happens and the STA associates with the new AP, the switch continues to send packets addressed to the STA through the old AP, which, after many retries, discards them. This behavior persists for about 140 ms (see Figure 5.6),

Table 5.5 Summary of results

	Handoff time in mobile receiver (ms)	Packet loss in mobile receiver (no. of packets)	Handoff time in mobile sender (ms)	Packet delay in mobile sender (no. of packets)
Original handoff	182.5	63.2	201.5	210.70
Selective scanning	102.1	37.0	141.1	161.7
Caching	4.5	16.3	3.9	0

after which the MAC addresses have been updated and the switch starts forwarding the data packets addressed to the STA through the new AP. This results in additional packet loss.

As can be seen in Figure 5.9, when the receiver performs the handoff, the packet loss drops to about 60% of the value obtained with the original handoff when selective scanning is used and to 40% when caching is used. When caching is used, the effect of the bridging delay is particularly prominent. Table 5.5 shows how, even though the handoff time is only a few milliseconds when caching is used, the packet loss is still considerable.

Figure 5.9 also shows the average packet delay introduced by the handoff procedure when the sender performs the handoff. For VoIP sessions, packets with a delay exceeding 100 ms can be considered lost. Table 5.4 shows the packet delay when the original handoff scheme, selective scanning,

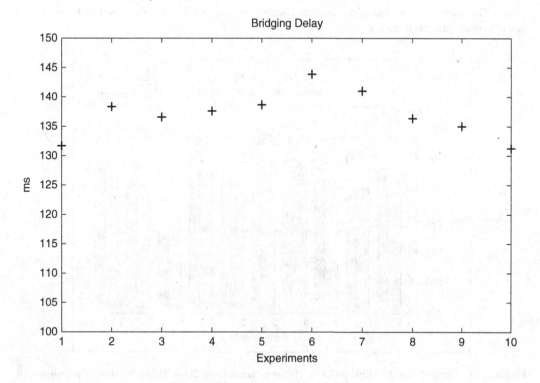

Figure 5.9 Packet loss and packet delay

and caching were used. Even though selective scanning reduces this delay by about 25%, the caching mechanism is necessary in order to achieve seamless VoIP communication.

As a note, the behavior of the selective scanning algorithm is not dependent on the environment, whereas the performance of the original handoff scheme is very much affected by it. Table 5.1 and Figure 5.8 refer to an environment in which rogue APs were present. Table 5.5 refers to a clean environment, without the presence of rogue APs. As we can see, the behavior with selective scanning is very consistent, whereas the performance of the original handoff scheme deteriorates with the environment.

5.9   Conclusions and Future Work

In this chapter, we have described the handoff procedure and demonstrated how the handoff latency can be substantially reduced with the use of a caching mechanism combined with a selective scanning algorithm. We have demonstrated how, in the best case, we were able to reduce the handoff latency to an average value of 129 ms using only the selective scanning algorithm and to an average value of 3.0 ms using the caching mechanism (Table 5.1). This reduction in handoff latency also considerably decreased packet loss and packet delay (Table 5.5). By using a dynamic channel mask (see steps 4 and 5 of the selective scanning algorithm in Section 5.5.1), scanning a subset of channels can be used as a generic solution.

Another important result of our work is that when selective scanning and caching are used, the probing process, which is the most power-consuming phase in active scanning, is reduced to a minimum. This makes it possible to use an active scanning procedure in devices such as PDAs where power consumption is a critical issue.

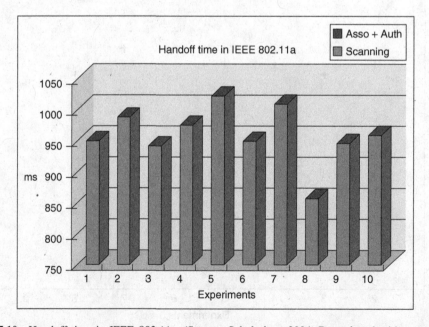

Figure 5.10   Handoff time in IEEE 802.11a. (Source: Schulzrinne 2004. Reproduced with permission of Henning Schulzrinne.)

With the new IEEE 802.11g standard out, similar testing can be done with this standard using the current algorithm. The extension of the current algorithm to the new 802.11g standard will require only minor adjustments, if any.

Figure 5.10 shows the original handoff time in an IEEE 802.11a network. As can be seen, the discovery phase is still the most time-consuming phase of the handoff process. Future testing will be done in an IEEE 802.11a environment. The extension of the proposed algorithm to this standard will require minor adjustments such as modification of the channel mask (selective scanning) and improved cache dimensioning and management.

The channel with the best signal is not necessarily the best channel to connect to, because it could be much more congested than a channel with a lower signal strength. Because of this, a heuristic which considers bit rate information together with signal strength should achieve optimal performance.

A procedure in which an AP somehow knew its neighboring APs and could provide that information to the STA could be used to fill the cache. This could also be combined with a positioning algorithm such as GPS or some other Wi-Fi positioning algorithm, allowing real-time filling and refreshing of the cache according to the actual position of the STA, always resulting in a cache hit.

Two very critical issues are the cache size and the cache management policy. A good cache policy, together with the use of AP neighboring and other heuristics, can achieve seamless VoIP sessions.

6

Mobility Optimization Techniques

In this chapter, we describe the key techniques that we have developed to optimize several basic operations of a mobility event in different layers, and highlight the associated key principles for those optimization techniques. These techniques are based on a few fundamental principles, such as reduction of the number of signaling messages during the basic operations, minimizing the traversal distance of the data, reduction of data and signaling overhead, minimization of lookup cost, caching, parallelization of sequential handoff operations, proactive operations, cross-layer triggers, and localization of binding updates. We present the application of these techniques to the optimization of the various handoff components described in Chapter 3 and describe experimental results obtained with some of these optimization techniques.

In Chapter 11, we describe the application of these optimization techniques to a Petri-net-based mobility model and then present experiments with a few of the proactive techniques to evaluate the overall systems performance.

6.1 Summary of Key Contributions and Indicative Results

We summarize in the following the key contributions made by some of the proposed optimization mechanisms for the various handoff components that we analyzed in Chapter 3. In this section, we highlight the key technical problems the proposed mechanisms address, briefly describe the proposed mechanisms, and highlight some of the key advantages.

We describe the detailed approach, related work, and experimental results for each of these mechanisms in Sections 6.3–6.11.

6.1.1 Discovery

Discovery of the network elements and resources during handover in a heterogeneous network depends upon the respective layer 2 discovery mechanism. These mechanisms introduce delays that are dependent on the underlying access characteristics. Currently, there is no existing mechanism that can discover the network elements in neighboring networks in an access-independent manner prior to the handover of a mobile.

Mobility Protocols and Handover Optimization: Design, Evaluation and Application, First Edition.
Ashutosh Dutta and Henning Schulzrinne.
© 2014 John Wiley & Sons, Ltd. Published 2014 by John Wiley & Sons, Ltd.

We have designed an application layer discovery mechanism that enables the discovery of network parameters and resources in the target network in an access-independent manner prior to the mobile's handover to the target network. The mobile then caches the addresses of the discovered network elements, namely the access point's MAC address, the channel number, the router's IP address, and the IP address of the authentication server, so that it can perform many of the handoff-related operations, namely authentication, security association, and configuration, ahead of time.

The following are some of the key benefits of the proposed application layer discovery mechanism:

- Using the proposed mechanism, a mobile can discover the network elements in all layers of the target network (e.g., access points, routers, and AAA servers) without depending upon the underlying access mechanism. While lower-layer discovery mechanisms such as IEEE 802.11u (Gast, 2005) allow the mobile to discover the higher-layer network parameters, 802.11u mechanisms are limited to 802.11 networks only. Gloserv (Arabshian and Schulzrinne, 2004) is a prior application layer discovery mechanism that discovers different types of services, but it does not discover the network elements. The proposed mechanism can discover the network parameters in an access-independent manner (e.g., regardless of whether the access is by 802.11 or by CDMA) using application layer protocols.
- The proposed mechanism reduces network discovery delays by proactively discovering the network elements and caching these locally in the client. While there are access-dependent optimization techniques (Montavont et al., 2005; Shin et al., 2004b; Velayos and Karlsson, 2004) to reduce the discovery latency of 802.11 networks during a mobile's handover, the proposed mechanism is access-independent and optimizes the network discovery operation for both homogeneous and heterogeneous handover.

We describe related work, details of the proposed mechanisms, and experimental results in Section 6.3.

6.1.2 Authentication

In general, layer 2 authentication is dependent upon the respective access mechanism (e.g., CDMA or 802.11) and is performed after the mobile hands over to the new network. Although some of the existing layer 2 access mechanisms (e.g., 802.11) support a preauthentication mechanism (e.g., 802.11i), where part of the authentication process (e.g., derivation of preshared keys) is performed before the mobile moves to the target network, this preauthentication support is limited to intrasubnet handoff only. Thus, a mobile will still be subject to authentication delay during handover between subnets.

The proposed network-layer-assisted layer 2 preauthentication mechanism bootstraps the layer 2 authentication process in the neighboring networks and completes most of the handoff-related operations before the mobile moves to the new network. Thus the layer 2 authentication delay during handoff is reduced to the time needed to complete the four-way handshake operation which generates the pairwise transient key (PTK).

There are the following two key advantages of the proposed techniques:

- The proposed mechanism takes care of the shortcomings of the existing layer 2 preauthentication mechanisms (e.g., 802.11i) by supporting preauthentication across subnets and administrative domains while providing equivalent performance to IEEE 802.11i. The proposed mechanism

reduces the authentication delay to only the four-way handshake delay, resulting in an average of 16 ms authentication delay. This delay is comparable to the performance offered by IEEE 802.11i-based preauthentication.
- By supporting preauthentication across heterogeneous access networks, the proposed mechanism eliminates the dependence on layer 2 access for authentication. By preauthenticating with the target network using the current interface, the mobile does not need to turn on its secondary interface for authentication, and thus saves battery power.

We describe related work, details of the proposed mechanisms, and experimental results in Section 6.4.

6.1.3 Layer 3 Configuration

The IP address acquisition and duplicate address detection processes are two main components that contribute to a mobile's layer 3 configuration delays in a new network. The IP address acquisition process involves signaling exchanges between the mobile and a DHCP server, and the mobile waits for a random period of time to complete the duplicate address detection before it can assign the new address to its interface.

In the following, we describe two of the proposed mechanisms that reduce the IP address acquisition delay and duplicate address detection delay using *proactive* and *reactive* techniques, respectively:

1. By using the proposed *proactive* discovery mechanisms, the mobile discovers the DHCP server or router in the target network and communicates with it to obtain the IP address from the target network over a secured handover tunnel that is set up between the mobile and the target router. The mobile then checks the uniqueness of the IP address and caches the IP address of the target network locally until it moves to the new network, where the mobile assigns the address to its interface.
2. As part of the optimized *reactive* address configuration process, we have proposed a router-assisted duplicate address detection mechanism, where the router multicasts the ARP (address resolution protocol) cache at periodic intervals so that the mobile does not need to initiate an address resolution process to determine the uniqueness of the address.

The proposed proactive layer 3 configuration mechanism eliminates the signaling exchange between the client and the server completely by obtaining the IP address over a secured proactive tunnel before the mobile hands over to the target network. This proactive IP address configuration process can work in conjunction with the preauthentication mechanism to securely obtain the IP address. The proposed proactive mechanism reduces the layer 3 configuration delay to about 100 ms, which is equivalent to the time taken to assign an IP address statically to a mobile's interface. Compared with existing mechanisms (e.g., FMIPv6), the proposed mechanism is client-assisted, works for interdomain mobility, and preconfigures the IP address securely.

In situations where proactive layer 3 configuration is not possible, the proposed reactive technique can be used to reduce the delay for duplicate address detection. Compared with other available mechanisms that reduce the duplicate address detection delay (Forte et al., 2006a), the proposed reactive mechanism does not need any additional elements in the network to detect the uniqueness of an IP address. The proposed mechanism reduces the duplicate address detection delay from 4 s to a few hundred milliseconds, a time that is dependent upon the router advertisement interval.

We describe related work, details of the proposed mechanisms, and experimental results in Section 6.5.

6.1.4 Layer 3 Security Association

When the IP address of either of the communicating end points changes, a new security context needs to be established, which requires generation of new keys. This process results in additional signaling exchanges, giving rise to additional handoff delays and media interruption during layer 3 security association.

We have proposed two different mechanisms that can reduce the delays due to layer 3 security association during handover. These mechanisms can be categorized as *reactive* and *proactive*:

1. The proposed *reactive* mechanism maintains the network layer identifier address of the mobile by introducing an additional anchor agent (e.g., a home agent) into the network. This allows the security context to be maintained by avoiding a rekeying process when the mobile's IP address changes. Compared with the traditional nonoptimized mechanism, the proposed mechanism reduces the handoff delay and packet loss during security association.
2. The proposed *proactive* mechanism uses a preregistration technique to establish the security association in the target network. Using preregistration with the outbound SIP proxy server in the target network and home subscriber server, the mobile establishes the security context ahead of time by generating the cipher key (CK) and the integrity keys (IK) proactively during the AKA (authentication and key agreement) phase.

The layer 3 security binding between two communicating hosts can be maintained either by way of establishing a security context proactively or by hiding the IP address change in a reactive manner with the help of an additional network element in an environment such as a mobile VPN (virtual private network), when proactive optimization is not feasible. While the mechanism proposed by Miu and Bahl (2001) cannot operate in a heterogeneous access network, the proposed reactive mechanism can reduce the packet loss to zero and still work for both homogeneous and heterogeneous handover. Unlike another proactive proposal by Bargh et al. (2004), our proposed proactive technique reduces the security risk by avoiding the domino effect (Housely and Aboba, 2007) that results when the security context is transferred between the end points. The proposed proactive mechanism reduces the layer 3 security association delay to zero, but it depends upon the mobile's ability to discover the outbound proxy server in the neighboring network.

We describe related work, details of the proposed mechanisms, and experimental results in Section 6.6.

6.1.5 Binding Update

A longer distance between the mobile node and the correspondent node or home agent delays the binding update, resulting in an overall handoff delay and packet loss, since the media are repeatedly forwarded to the previous network until the binding update is complete.

We have proposed two mechanisms, namely *hierarchical binding update* and *proactive binding update*, that reduce the binding update delay and as a result minimize the packet loss:

1. The proposed hierarchical binding update mechanism is a reactive mechanism that uses a two-level hierarchy of addresses (e.g., a local care-of address and a global care-of address) and introduces an anchor agent called a mobility agent (MA) into the network to limit the global signaling update during the mobile's mobility within a domain. This mechanism reduces the global signaling update by 70% for a network with 10 subnets per domain.
2. The proposed proactive binding update mechanism sends the binding update to the home agent and correspondent node before the layer 2 handover over a secured tunnel, and uses the IP address that is obtained as part of the proactive configuration process from the target network as the new care-of address. Using this mechanism, the mobile eliminates the binding update delay completely after its move to the network.

Unlike MIP-RR (Perkins, 2002c), the proposed reactive mechanism works for both network layer (e.g., MIP) and application layer mobility (e.g., SIP) protocols and supports dynamic load balancing and fast handoff. The proposed proactive mechanism avoids the binding update completely, as the mobile does not need to send a new binding update after the handover.

We describe related work, details of the proposed mechanisms, and results in Section 6.7.

6.1.6 Media Rerouting

During the media rerouting process, transient data may be lost because of handoff operations in several layers or be delayed because of operations such as encapsulation, decapsulation, tunneling, and buffering in the network.

In the following, we summarize the proposed mechanisms that reduce the data traversal delay by optimizing the media rerouting process:

1. *Reactive forwarding* of data from the previous network. This mechanism uses the reactive forwarding mechanism to redirect the in-flight data from the previous network using an application layer mobility proxy in the case of a longer binding update delay.
2. *Proactive multicasting*. This mechanism proactively multicasts the in-flight data to the neighboring candidate networks and reduces in-flight packet loss.
3. *Mobile-controlled proactive buffering*. The proposed mobile-controlled buffering mechanism provides a per-mobile packet buffer at the edge router that controls the buffering period dynamically, based on the handoff duration during a proactive handoff. The proposed techniques can be categorized into two categories, namely *time-limited buffering* and *explicit buffering*.

The proposed *reactive* mechanism is the first forwarding technique of its kind that uses an application layer mobility proxy to forward in-flight data from the previous network. The proposed *proactive* multicasting mechanism avoids the need for any additional network elements in the neighboring networks and proactively multicasts data to the neighboring networks where the mobile is about to move to. Unlike the existing scheme by Tan et al. (1999), this mechanism uses a single multicast address for all the mobiles that are in one MA's domain, where the MA encapsulates the unicast data destined for any particular mobile with the appropriate multicast address.

Compared with the existing buffering techniques by Khalil et al. (1999) and Moore (2004) that depend upon extensions of Mobile IPv4 and Mobile IPv6, respectively, the proposed mechanism is an independent protocol and can be used with any type of mobility protocol (e.g., SIP or Mobile IP). For example, we have experimented with the proposed buffering control protocol for both MIPv6 and SIP-based mobility protocols without making any changes to these protocols. Both of our approaches, when applied to proactive handover mechanisms, reduce the packet loss to zero.

We describe the details of reactive forwarding techniques, proactive multicasting techniques, proactive buffering techniques, related work, and the experimental results in Sections 6.8 and 6.9.

6.1.7 Route Optimization

The end-to-end transport delay of media traffic affects the performance of real-time communication. The media transport delay is also affected when signaling is delayed because there is a long route between the mobile node, home agent, and correspondent node. Several mobility-related operations, such as triangular routing and encapsulation and decapsulation processes, add further delays to the media transport. Thus, it is essential to optimize the route between the correspondent node and the mobile node so that both media transport delay and signaling transport delay are minimized after handover. Although MIPv6 (Carpenter, 2000) supports route optimization techniques, Mobile IPv4 and its variants such as MIP-LR (Jain et al., 1999) and Proxy MIPv6 (Gundavelli et al., 2008) suffer from a route optimization problem.

We have designed the following route optimization techniques that minimize the length of the signaling and media routes between the mobile node and correspondent node:

1. We have designed an interceptor-assisted packet modifier that is used at the end hosts and helps maintain a direct media path between the mobile node and correspondent node by modifying the source and destination addresses as needed. This technique can be applied to both the MIP-LR (Jain et al., 1999) and the MIPv4 (Perkins, 2002b) mobility protocols. SIP-based mobility management (Schulzrinne and Wedlund, 2000a) is an existing approach that provides route optimization using a direct media path, but this technique can be used to support real-time traffic only.
2. We have designed a route optimization technique that uses a packet interceptor and a forwarding module at the mobile's outbound SIP proxy. This technique takes care of routing indirection of the SIP signaling caused by the underlying network layer mobility protocol Mobile IP in an IMS (IP Multimedia Subsystems) environment.
3. We have designed a binding-cache-based technique that uses proxy binding update to establish a mapping at the local anchor agent (media access gateway) and routes packets locally instead of routing them via a local mobility agent (LMA) in the Proxy MIPv6 environment.

The interceptor-assisted packet modifier is an application layer technique and reduces the end-to-end delay of the media traffic by 60% for large packets (e.g., 1024 bytes). Its benefit becomes more effective when the mobile is further away from the home network. In an IMS environment, the proposed intercepting proxy-assisted mechanism reduces SIP registration delays by 20% and SIP INVITE delays by 30%. By using the binding-cache-based route optimization technique, the end-to-end media delay is minimized and does not change even if the distance between the mobile and the LMA is increased.

We describe the details of these route optimization mechanisms, related work, and experimental results in Section 6.10.

6.1.8 Media-Independent Cross-Layer Triggers

Posthandoff detection mechanisms and detection mechanisms in layer 2 (e.g., in the access point) and layer 3 (e.g., in the router) points of attachment work independently of each other, causing additional delays during handover. Handoff-related functions are spread across different layers of the protocol stack and are executed independently. There is also no existing mechanism that allows control information to be exchanged across layers. However, for efficient network communication, it is essential for a protocol layer to utilize cross-layer information. Thus, it is useful to have a set of abstract primitives that can be used to pass on information across layers in order to expedite the handoff operations.

We have proposed a set of abstract primitives that can pass information across layers and work independently of the access mechanism (e.g., CDMA or 802.11). Some of these abstract primitives were used to develop the media-independent handover functions that have recently been standardized in the IEEE 802.21 standards. Unlike other proposals, these primitives can be applied to support handover among heterogeneous access networks such as 802.11 and CDMA. These triggers can be categorized as *information service*, *command service*, and *event service* triggers. Using these primitives, the mobile can quickly detect new networks and the loss of old networks. Section 6.11 provides a detailed description of these triggers and the various implementation steps. As part of the proposed mechanisms, we have designed the following cross-layer triggers that expedite the network detection process and trigger upper-layer handoff operations such as IP address acquisition and binding update:

1. *Proactive triggers.* In order to prepare for an impending handoff and perform some of the handoff operations ahead of time, we have developed cross-layer triggers such as `MIH_Link_Going_Down`, `MIH_Link_Handover_Imminent`, and `MIH_Link_Parameters_Report` that will trigger many of the handover-related upper-layer operations, such as application layer discovery and network-layer-assisted layer 2 preauthentication.
2. *Reactive triggers.* We have developed link layer cross-layer triggers such as `MIH_Link_Up`, `MIH_Link_Down`, `Link_Detected`, and `MIH_Link_Get_Parameters` events to expedite upper-layer handoff operations in an access-independent manner such as handoff between 802.11 and CDMA networks. Unlike other event triggers, the proposed cross-layer triggers work across different access mechanisms.
3. *Cross-layer triggers.* Using these mechanisms, layer-3-related information (e.g., subnet prefix and default router address) is passed during access point discovery. This is accomplished by modifying the layer 2 access point beacon and stuffing it with layer 3 information.

We have used these triggers to build a media-independent proactive handoff system as described in Chapter 11. We describe related work, details of the proposed mechanisms, and experimental results in Section 6.11.

In the rest of the chapter, we describe the details of the proposed mechanisms for each of the handover components that we have summarized, related work, and results from an experimental prototype that we built using these optimization techniques.

6.2   Introduction

In order to experiment with the key optimization techniques, we implemented an Internet mobile multimedia test bed and the associated functional components, with which we demonstrated several of the mobility functions. In particular, we implemented a configuration agent, signaling agent, mobility agent, home agent, authentication agent, and authorization agent using IETF-based protocols, namely DHCP (Droms, 1997), SIP (Rosenberg et al. 2002), MIP, PANA (Jayaraman et al. 2008), and Diameter (Calhoun et al. 2003a), over heterogeneous access networks including IEEE 802.11 and CDMA. We have described the details of the implementation of the multimedia test bed in Dutta et al. (2004a). In the rest of the chapter, we describe the optimization techniques associated with some of the primitive operations of the handoff event as described in Chapter 3, namely discovery, authentication, security association, configuration, media delivery, and buffering. In addition, we explain how cross-layer triggers can help expedite the handoff-related operations and reduce the delay.

In the following sections, for each of the handoff components, we follow a systematic approach in which we describe the performance parameter (e.g., handoff delay or packet loss) that is optimized, highlight the fundamental principles and techniques that were used to optimize these parameters, demonstrate an experimental system that validated these techniques, and compare the results obtained by applying these core techniques with results obtained with the nonoptimized version.

6.3   Discovery

As discussed in Chapter 3, experimental results show that network discovery and resource discovery processes in IEEE 802.11 networks contribute to a large amount of delay during handoff. During a handoff between heterogeneous access networks involving Wi-Fi and cellular networks, discovering a cellular network such as a GSM network also takes time (Rahnema, 1993; Steele et al. 2001), depending upon the channel assignment strategies and the type of handover scenario, as described in Chapter 2. In this section, we propose an application layer network discovery mechanism that can discover the network elements and resources in the neighboring networks independently of the underlying access technology. Using this discovery technique, the mobile can proactively discover many of the layer 2 and layer 3 network resources, namely the channel number, default router's address, and authenticator in the target network. This proactive operation will help to reduce the handoff delay as many of the discovery-related operations, namely layer 2 scanning, router solicitation, and server discovery, do not need to be performed after the handoff.

In this section, we first describe the general principles that are needed to optimize the delays contributed by discovery operations in several layers. Then, we cite related papers that describe attempts to optimize the discovery-related delay at the expense of other systems resources such as network bandwidth and CPU cycles. We then introduce the proposed application layer discovery technique and describe its advantages over the existing discovery mechanisms. Finally, we illustrate the experimental results obtained in a test bed environment.

6.3.1   Key Principles

The following are the key principles that govern the optimization of the discovery process. This optimization process is aimed at optimizing the delay during discovery with respect to other network resources such as processing power at the end hosts and network bandwidth.

1. Limiting the number of signaling exchanges between the mobile and the centralized server needed to discover the network resources.
2. In the case of passive scanning, an increase in the rate of beacon advertisement reduces the time to discover the new point of attachment at the cost of additional network bandwidth and processing at the end hosts.
3. Caching of neighboring network resource parameters before the mobile moves to the new network.
4. Use of a media-independent application layer discovery protocol to discover network resources to support handover in heterogeneous access networks without depending upon any access-specific technology.

6.3.2   Related Work

In cellular networks such as GSM and CDMA, the pilot signals of the mobile, namely the BCCH (broadcasting channel) and the sync channel, report the details of the neighboring networks to the serving MSC (mobile switching center). Serving MSCs use this information to decide the target networks for the mobile. Recently, for IP-based networks, some efforts have been under way to design discovery protocols that provide service discovery and network discovery in different layers. Here, we highlight some of the related work in the area of network discovery and optimization techniques for it.

Several task groups in the IEEE 802.11 standards groups have proposed network discovery mechanisms in layer 2 and the application layer. The IEEE 802.11u (Gast, 2005) working group has proposed methods of network selection, along with methods for other external networks such as cellular networks. The IEEE 802.11k (Stallings, 2004) working group has proposed methods that enable the access points (APs) to query mobile devices for location and neighbor information. This group has proposed several new request/response measurement mechanisms, namely measurement pilots, neighbor reports, link measurement, station statistics, and location configuration information, so that the mobile can obtain information about its neighbors and make appropriate decisions to achieve a fast transition. However, the IEEE 802.11k-based discovery mechanism is limited to 802.11 access networks and works in layer 2 within the same ESS (extended service set).

Several service discovery protocols and architectures exist today, including SLP (Service Location Protocol) (Guttman et al. 1999), JINI (Waldo, 1999), UPnP (http://www.upunp.org), Salutation (Miller and Pascoe, 2000), and LDAP (Lightweight Directory Access Protocol) (Johner and Corporation, 1998). However, these focus mostly on how a user retrieves service-related information assuming that the information is already available in databases. The service-related information and hence the servers that host the information can be organized into a hierarchy, for example in a way similar to the Internet Domain Name System (DNS). The service-related information can either be preconfigured or be provisioned dynamically on the servers. The information can then be updated either by human administrators or automatically by the servers themselves exchanging updates with each other. However, none of these protocols provide support for discovering information about the neighboring networks in higher layers, for dynamic construction of discovery databases, or for determining what information to collect and provide to mobiles. Instead, the existing service discovery mechanisms focus on how to retrieve information already existing in databases. These mechanisms rely on all of the local network providers implementing service information servers, which are usually not deployed in public networks. Recently, the IEEE 802.21 working group has finalized an information service mechanism that provides information discovery in the application layer. Some of the techniques developed as part of this work, such as application layer discovery mechanisms using RDF (Resource Discovery Framework) (Lassila et al. 1999), have contributed to the development of the information server (IS) components of IEEE 802.21. We describe the details of these mechanisms in Section 6.3.3.

A representative example of a discovery protocol in layer 3 is the Candidate Access Router Discovery (CARD) protocol (Liebsch et al. 2005), which provides a network discovery mechanism in layer 3. A candidate access router is an access router in a neighboring network that a mobile device may move into. CARD is designed to be used by a mobile device to discover a candidate access router before the mobile performs IP layer handoff to a neighboring network. With CARD, the mobile listens to layer 2 identifiers such as IEEE 802.11 BSSIDs broadcast from the radio access points in neighboring networks prior to making a decision about IP layer handoff. The mobile then sends these layer 2 identifiers to the access router in its current network, which in turn maps the layer 2 identifiers to the IP addresses of the candidate access routers in the neighboring network and then sends the candidate router addresses back to the mobile. In order to use CARD to support network neighborhood discovery, the routers in the network need to be upgraded. This also needs security and trust between the neighboring routers and thus may not work if it involves handoff between two administrative domains.

There are a few related papers that describe attempts to reduce the network discovery time in an IEEE 802.11 environment. Shin et al. (2004b) adopted a selective scanning and caching strategy to reduce the handover latency in an IEEE 802.11 environment. However, this method is more applicable to an environment where the mobile has associated with the neighboring APs in the past, and cannot be applied if the target access point is a new AP. Montavont et al. (2005) proposed a periodic scanning method, where the mobile scans different channels periodically and builds up a list of neighborhood APs. However, this mechanism generates more traffic, and as a result consumes more energy. Velayos and Karlsson (2004) provided techniques to reduce the layer 2 discovery process by reducing the beacon interval time and performing the search phase in parallel with data transmission. Brik et al. (2005) proposed the use of a second interface to scan while communicating with the first interface, thereby avoiding a scanning delay during communication. Most recently, Forte and Schulzrinne (2007) have developed discovery mechanisms using cooperative roaming techniques that are suitable for working in an infrastructureless environment.

6.3.3 Application Layer Discovery

As part of the work on minimizing the handoff delay due to the discovery component of the handoff process, we have developed an access-independent information-server-based application layer discovery mechanism that helps to discover the network parameters and resources of the target network (Dutta et al. 2006c). Unlike the existing network discovery mechanisms, this application layer discovery mechanism does not depend on any access-specific discovery technique such as IEEE 802.11u.

This application layer discovery technique can be applied in both infrastructure-assisted and end-system-assisted scenarios. We have analyzed how this discovery mechanism can be effective in a collaborative environment using an *end-system-assisted* approach (Zhang et al. 2005), where each end system can act as a source of information about the neighborhood. As part of the *infrastructure-assisted* scheme, the information server stores the details of the networks and the associated resources in a generic format, in an access-independent manner, that can be queried by a mobile client at any time. The client communicates with the information server and discovers the neighboring network elements, such as the access router, authentication agent (IEEE 802.11i authenticator), configuration agent (e.g., DHCP server), and authorization agent (e.g., AAA server), and communicates with these entities prior to its handover to these networks. By discovering the details of the target access points prior to handoff, the mobile keeps the MAC address and channel number of the access point in its cache and avoids some parts of the scanning procedure, such as channel probing during 802.11-based handover. We have described how proactive discovery of routers and authentication servers helps to complete other handoff-related functions, such as authentication and configuration prior to handoff (Dutta et al. 2006a). The proposed information-server-assisted discovery technique has been adopted as one of the discovery mechanisms for the Media Independent Information Service (MIIS) function of IEEE 802.21. An evaluation of a complete system using a network-assisted discovery scheme is described in Chapter 11. We describe the details of the architecture and proposed schema below.

Currently, no database-querying mechanism allows one to obtain detailed information about a neighboring network given a property such as the network type or the GPS coordinates of the mobile. Such detailed information might include the MAC addresses of the neighboring APs, the channel numbers associated with those APs, and the IP addresses of the DHCP server, the router, and the AAA server. Currently, DHCP provides a DHCP option mechanism (Droms, 1999) whereby a client can discover a specific server and the geo-coordinates of the nearby access points (Polk et al. 2004). However, a DHCP server usually stores information specific to a subnet and cannot provide services to a mobile that is not located in the same subnet without the help of a relay agent, namely a DHCP relay agent. Thus, the DHCP-based discovery mechanism is limited to a specific subnet and cannot span over multiple networks. The query mechanism should also be extensible and should accommodate proprietary vendor definitions. Thus, it is desirable to design a query mechanism that can support schema-based (or subschema-based) access and can cover networks beyond a specific subnet.

The proposed approach is based on a new architecture called Application-layer Information Service (AIS) that supports network discovery, including methods to solve the discovery database construction problem and methods for mobiles to discover information regarding neighboring networks. AIS is designed to be extensible enough to support current and future types of network information that may be needed by mobile nodes. AIS leverages existing protocols as much as possible. Although information about the network elements can have multiple uses, we focus on how the mobile can use this discovery information to support secured and proactive handoff. Some of the key design factors that need to be looked into when one is designing a discovery architecture include constructing the information, retrieving the information, and the format of the information stored in the information server.

In the following, we describe a sample implementation of the query and response processes that are part of the network discovery mechanism. To query information related to a specific network interface (e.g., 802.11 or CDMA), a mobile needs first to know which information attributes are supported by a network interface. Thus, a query–response mechanism may use two steps: the first query provides the metadata information (i.e., the attribute names) and the second query provides the values of the attributes the mobile is interested in or requires.

The information on the information server should be stored in a standard, easy-to-access manner. We used an RDF-based schema (Lassila et al. 1999) to describe and store the information regarding networking elements and their characteristics on the information server. RDF is a framework that describes a language for representing information about World Wide Web resources. It is intended for representing metadata, such as the title, author, and modification date of a Web page, and copyright of Web resources.

RDF provides a common framework for expressing information so that it can be exchanged between applications without loss of meaning. It is intended for describing information that needs to be processed by applications, rather than being only displayed to people. Therefore, RDF-based query and response mechanisms provide a suitable way for mobiles to report to and retrieve information from the application information server. They allow a mobile to query specific elements of information about a network by providing the characteristics of the information elements in a granular manner.

The characteristics of these network information elements may be the SSID (service set identifier), location information (geo-coordinates), or the layer 2 (L2) security information. The RDF schema defines the structure of the information elements, as well as the relationship between these elements. An RDF schema is usually partitioned into two schema types: a *basic schema* and an *extended schema*. The basic schema is static and includes media-independent classes and properties. The extended schema includes properties that are dynamic in nature, such as bandwidth.

Figure 6.1 shows a very simple view of an RDF-based tree illustrating how these network entities are constructed in a hierarchical manner. It shows the network elements in the neighborhood networks and their interdependency and shows how the location information, L3 information, L2 information, and network types are constructed in a hierarchical manner.

We present the schema for the information service in Appendix A. Here, we briefly present the architecture and describe the functional components used in the information query and update

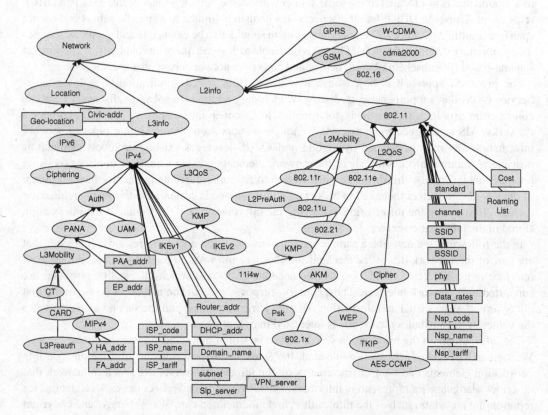

Figure 6.1   Interdependency chart for network elements

process. At the information server end, we used Joseki (Kosugi and Davies, 1973) to interpret the RDQL (Seaborne, 2004) and send appropriate responses to the client. We used Jena (McBride, 2002) to form the RDQL. Jena is a Java framework for building Semantic Web applications. It provides a programmatic environment for RDF, RDFS (McBride, 2004), and OWL (Web Ontology Language) (McGuinness et al. 2004), including a rule-based inference engine. The implementation in Jena was coupled to relational database storage so that an optimized query was performed over the data held in a Jena relational persistence store. We used a Joseki server to publish RDF models on the Web. These models were represented by URLs and could be accessed by queries using HTTP GET.

6.3.4   Experimental Results and Analysis

Figure 6.2 shows a possible deployment architecture where this information discovery scheme could be useful. Initially, the mobile is in network 1, and is connected to access point AP1. Network 2, Network 3, and Network 4 are the neighboring networks. The information server stores information about these networks and the associated network elements, namely the authentication server, configuration server, authorization server, and access point identifier.

We have implemented both a mechanism for population of the information database by the end clients and a process for network discovery during handover. Although there are several ways an information server can be populated (Dutta et al. 2006c), we implemented an end-system-assisted population scheme in the current experiment. As the mobile moves from one network to another

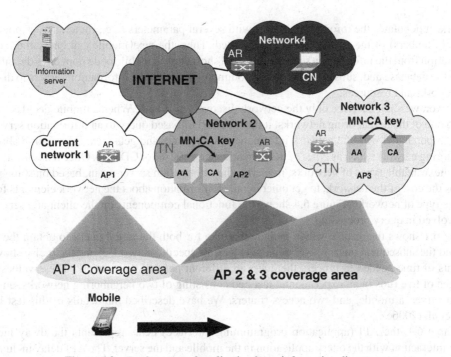

Figure 6.2   Deployment of application layer information discovery

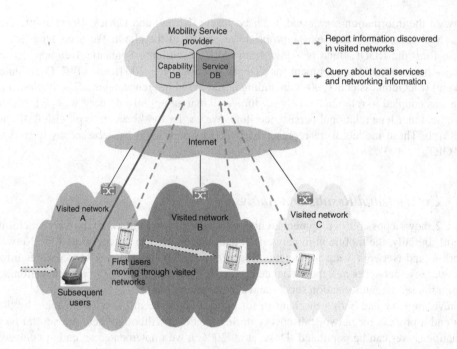

Figure 6.3 Information population and query process

network, it populates the information server with several parameters (e.g., router, access point, and channel numbers) of the network it has just visited. Thus, the next mobile can query the required information from the information server. Figure 6.3 shows how a mobile node populates the information in the database and, subsequently, how it communicates with the information servers to discover the networks and resources.

However, we shall describe only the network discovery part here. When a mobile decides to hand over to one of the neighboring networks, it makes an RDF-based query to an information server that has been populated with the network information beforehand, and gets meta-information about the neighboring networks such as types of the networks (e.g., 802.11 or CDMA). Once it has information about the available types of networks, it queries the information server again, based on some policy such as the cost of the network, to get other detailed information about the network elements for one specific type of networks. Figure 6.4 shows the functional components on the client and server that are involved in query processing.

Table 6.1 shows the results, which include the time for both the initial query to obtain the meta-data and the subsequent queries to obtain the values for specific network elements. It also shows the amounts of time needed to process the query in different parts of the network. These values show averages of five runs in an experimental test bed consisting of two neighboring networks, an information server, a mobile, and two access routers. We have described the details of this test bed in Dutta et al. (2006c).

In Table 6.1, the API (application programming interface) delay represents the delay incurred during interaction with the query application in the mobile and the server. The API delay includes the interaction with the database and is dependent upon the implementation language, such as Java or C.

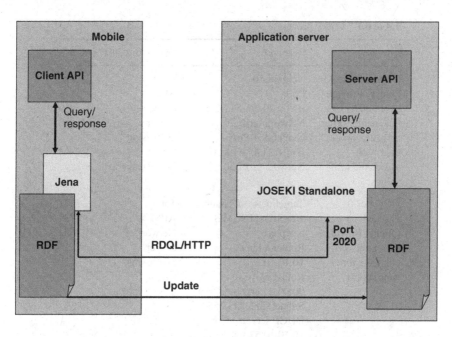

Figure 6.4 Interaction among the functional components

The network layer delay includes the delay due to the TCP layer transaction. The processing delay at the client and the server includes the time spent on HTTP processing at each of these nodes. In this experiment, during the transaction for query 1, the mobile sent 1288 bytes of data and obtained 1684 bytes in response. Query 2 involved 1713 bytes of data sent by the client and 1335 bytes of data sent by the server. The queries and responses were carried back and forth in chunks to accommodate the maximum segment size (MSS), thus adding to the delay. Since these queries and responses were carried as part of HTTP messages, TCP was the chosen transport method. The transport delays could be significantly reduced if UDP were used as a transport protocol instead. During the database update procedure by the client, we observed that the average update delay by the client was 353 ms with a standard deviation of 153 ms. This information update time will of course depend upon the amount of data being updated and the network bandwidth. However, query update time is not critical for handover decisions.

Optimizing the query delay is important for mobiles with a high mobility rate as the mobile needs to make a decision ahead of time based on the query delay. Increased processing power at the end clients and improved transport methods will help optimize the delay associated with the queries and responses. We observed that the delay due to the second query was less than that for the initial query because of the additional ARP performed during query 1. The response to the first query is the metadata, where the mobile finds the relevant networks, which were of type 802.11 here, and using a tariff as the policy it chooses a specific network type and decides to get more information about other network elements such as the access router, PANA server, and DHCP server. A more detailed breakdown of the stack level delays for each of the IS primitives is presented in Chapter 11.

The IEEE 802.21 working group has included both XML (Extensible Markup Language) and the TLV (time–length–value) format as part of the Media Independent Information Service. The proposed mechanism contributed to the XML format that was included. RDQL uses an XML

Table 6.1 Query process

Query type	Response	Processing delay (ms)	
Current PoA: AP **Query:** Provide list of 802.11 type neighboring networks and their with associated tariff values	Neighbor 0 PoA: ID:00:20:A6:53:B2:5E, Network tariff:20 Neighbor 1 PoA: ID:01:23:45:67:8:AB, Network tariff:50	Total API Network Server processing Client processing	2292 1291 919 18 64
Neighbor 0 selected **Query:** Provide list of network elements for Neighbor 0	Target network channel: 10 SSID: ITSUMO newpoa1 Router address: 10.10.10.52 Router MAC: 00:00:39:e6:8b:ee Subnet: 255.255.255.0 DHCP server: 10.10.10.52	Total API Network Server processing Client processing	1473 991 451 13 18

format for query and response. We did a preliminary performance comparison between the XML and TLV formats. The sizes of the queries and responses obtained via RDQL were much larger than in the case of TLV. However, if the basic schema were changed to a flatter structure, then the size of the queries and responses would be reduced. On the other hand, XML-based queries provide more extensibility and flexibility in terms of their ability to query a specific network element. Since the number of bytes going over the air is a concern, we also used a compressed version of XML for the query response. By using this compressed version of XML during information querying, we reduced the overall discovery time. Another approach to reducing the query time is to use a combination of XML and TLV where the mobile makes an XML query but obtains the information in TLV format.

6.4 Authentication

In Chapter 3, we defined the authentication and authorization processes that are needed during mobility. We also illustrated how, during a mobility event, authentication and authorization processes add to the disruption in communication and packet loss. Figure 6.5 shows a basic Internet roaming scenario where two different administrative domains that are managed by different wireless service providers establish business agreements between them in order to provide roaming services for their customers. In particular, these business relationships allow users who belong to one domain (the home domain) to access the network and services in other domains (e.g., Domain A or Domain B in Figure 6.5). A domain here is defined as an administrative domain. There may be several subnets within an administrative domain. These agreements are enforced by means of the AAA infrastructure deployed (e.g., the AAAv (visited AAA) or the AAAh (home AAA)) in each domain. In a nutshell, the home AAA domain (the domain that the user belongs to) is equipped with a home AAA and

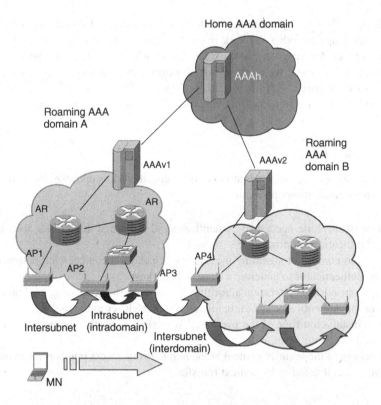

Figure 6.5 Illustration of roaming environment

each roaming AAA domain is equipped with an AAA proxy (AAAv) that contacts the home AAA infrastructure in order to verify the roaming user's credentials. Figure 6.5 also highlights three different types of movement: intrasubnet, intersubnet, and inter-AAA-domain ("interdomain" hereafter). Link-layer handoff is the common scenario in this roaming architecture. Thus, optimization of the establishment of security during link-layer handoff deserves some attention.

In general, authentication and authorization take place in the target network after the mobile moves to the new network. For example, in IEEE 802.11-based networks, the authentication mechanism requires an IEEE 802.1X message exchange with the authenticator in the target network, such as an access point that can initiate an EAP (Extensible Authentication Protocol) (Aboba et al. 2004) exchange with the authentication server. Following successful authentication, a four-way handshake with the wireless access point derives a new set of session keys to encrypt the data. The handover latency introduced by this authentication mechanism has proved to be larger than what is acceptable for some handover scenarios involving interdomain handover. Hence, improving the handover latency due to authentication procedures is a necessary objective in such scenarios. Various standards organizations, such as the IEEE 802.11i and 802.11r working groups, 3GPP, and the WiMAX forums, are developing access-specific techniques to reduce the authentication delay. However, these mechanisms are designed to work at link-layer level, which has some implications and limitations for intertechnology and intersubnet handoff, such as the inability to preauthenticate.

In this section, we first describe the key principles that should be considered when optimizing authentication delay against other network resources, namely bandwidth, processing power, and battery power. We then describe some of the related work where attempts have been made to reduce the authentication delay. Then, we describe our proposed authentication mechanism and highlight the key differences compared with the existing techniques. Finally, we describe our experimental test bed and analyze the measurements that validated the proposed optimization mechanism.

6.4.1 Key Principles

The following are the key principles that need to be considered to optimize delay and processing power during an authentication operation:

1. Minimization of the time needed to authenticate and authorize the mobile after each handoff during the reauthentication procedure.
2. Reduction of the number of signaling messages that need to be exchanged between the mobile node and the authenticator to generate a shared secret key.
3. Use of an appropriate key generation algorithm that reduces the processing load on the end hosts.
4. Placement of the authenticator and authentication server closer to the mobile.
5. Reduction in installation time of the preshared keys (PSKs) on the authenticator in the case of proactive authentication.
6. Proactive caching of the security context at the neighboring access points prior to handoff, either by proactive authentication or by context transfer.

6.4.2 Related Work

IEEE 802.11i and IEEE 802.11r (O'Hara, 2004) provide link-layer handoff optimization mechanisms that attempt to reduce the delay due to link-layer authentication during a node's mobility. IEEE 802.11i was conceived to provide stronger security for IEEE 802.11 WLANs. It relies on IEEE 802.1X for the authentication and access control of IEEE 802.11 stations (STAs).[1] As part of 802.1X, a successful authentication allows both the STA (mobile node) and the AP to generate a pairwise master key (PMK). Typically, the AP relies on a backend authentication server (AS) such as an AAA server acting as a termination point of an EAP (Extensible Authentication Protocol) authentication method, in order to verify the authentication credentials of a peer and deliver the PMK to the AP after the verification is successful. In the preshared-key mode, the STA and AP preshare a 256-bit key that is used as the PMK. Therefore, no EAP authentication is needed. Moreover, the four-way handshake protocol uses the PMK to mutually authenticate the STA and AP and establish fresh pairwise transient keys (PTKs) to protect the link-layer frames. However, IEEE 802.1X authentication can last from several hundred milliseconds to several seconds (Bargh et al. 2004). Hence, each time an STA moves from one AP to another, this delay and the associated packet loss during the handoff affect real-time applications such as VoIP. In order to overcome this problem, IEEE 802.11i introduces a mechanism of preauthentication, where the STA starts a new EAP authentication with the target AP that it is likely to hand off to, through its currently associated AP. After the EAP authentication has completed successfully, the generated PMK is stored properly at the target AP. When

---

[1] "STA" and "mobile node" are used interchangeably here.

the STA finally roams to the target AP, both parties engage in a four-way handshake using a specific PMK. Therefore, EAP authentication is not performed after the handoff. By decoupling the authentication and network access control operations from the handoff, IEEE 802.11i preauthentication reduces the handoff delay. However, 802.11i also has some drawbacks and limitations that are worth mentioning:

1. Each IEEE 802.11i preauthentication involves a full EAP authentication. Consequently, it implies a lot of signaling with the authentication server during each movement.
2. The mechanism does not work when the APs involved belong to different distribution systems, where a distribution system is used to interconnect a set of basic service sets and integrated local area networks (LANs) to create an extended service set (ESS). For example, intersubnet and interdomain preauthentication is not possible.
3. The full association and four-way handshake are still required to be finished after the movement.

IEEE 802.11r overcomes most of these problems by introducing a three-level key hierarchy (started either from a master session key (MSK) generated during an EAP authentication or from a PSK) and a supporting architecture that allows the STA to perform a fast transition between APs within the same mobility domain without the need to run EAP authentication during each movement. Additionally, IEEE 802.11r allows one to perform part of the four-way handshake and some resource reservation at the target AP before the STA moves. When the STA finally hands off, it only needs to reassociate with the target AP to complete the handoff. Thus, IEEE 802.11r reduces the handoff delay compared with IEEE 802.11i. However, the IEEE 802.11i and IEEE 802.11r mechanisms do not work when the APs involved belong to different distribution systems, which is the case for intersubnet and interdomain handoffs. Basically, the reason is that the 802.11i and 802.11r handover optimization mechanisms are based on link-layer frames, which cannot operate across different subnets.

IEEE 802.11f, a trial-use recommended practice, defined a context transfer and caching mechanism to transfer some of the 802.11i keying-related information between neighboring APs. It used the Inter Access Point Protocol (IAPP) to transfer keys between the access points. However, IEEE 802.11f has been administratively withdrawn since 2006 because of security concerns due to communication between the access points.

The problem of applying link-layer handoff optimization mechanisms between different subnets has also been addressed by the research community. However, most of the solutions are based on context transfer mechanisms (Bargh et al. 2004; Duong et al. 2004; Georgides, 2004). The optimization is achieved by transferring the security context (keys and related parameters) created by the STA and the previous AP to the new AP between subnets. Consequently, the STA does not need to run a full EAP authentication to create a new PMK, and only the four-way handshake is required after the handoff. For example, Bargh et al. (2004) explained how to transfer an IEEE 802.11i context between two APs with different networks by using a combination of Context Transfer Protocol (CxTP) (Loughney et al. 2005) and CARD (Liebsch et al. 2005). Georgides (2004) extended Cellular IP to signal a context transfer between two base stations (BSs) with two different gateways (GWs). Here, the new GW contacts the previous GW to recover the security context from the previous BS. Duong et al. (2004) also proposed an optimized solution based on CxTP and CARD in which a context is proactively transferred when a move of the mobile node (MN) move is imminent. From a security perspective, it is not always a good idea to transfer cryptographic keys between different network entities. For example, Housely and Aboba (2007) have raised a warning about security context transfer. Additionally, to achieve a secure context transfer, one needs to have certain security associations and strong trust relationships between the policy enforcement points such as APs, which are not always possible. Finally, such a transfer only allows handoff between technologies of the same type, such as 802.11 (homogeneous handoff).

Mishra et al. (2004) and Pack and Choi (2002) completely avoided the use of context transfer by preinstalling keys into APs before the STA moved to the target network. In general, the methods described by those authors are based on algorithms that steer the key installation process based on the movement of the mobile node. These solutions assume that an AAA server or trusted third party is in charge of predistributing keys to different APs that the MN could potentially associate with. This implies that the AAA server has knowledge about the location of the APs. This may work when a single wireless service provider is considered. However, in roaming scenarios, the home AAA server needs to know the location of the APs in the visited domain. Unfortunately, this is not always possible, since usually the visited domain does not want to reveal details about its internal network deployment for reasons of privacy, even when a roaming agreement has been defined. Additionally, the assumption that an AAA server is able to store the key after EAP authentication is not always true (e.g., in the case of RADIUS). Ruckforth and Linder (2004) proposed a different approach, where a combination of Fast Mobile IPv6 (Koodli, 2005) and IEEE 802.11i frames is used to inform the user's home domain AAA server about the next IPv6 router and the next AP to which the STA may move. With this precise information, the AAA server creates a new PMK and sends it to the AP and AR. However, this solution is restricted to IPv6 networks because of the MIPv6-related messages between the access routers. Forte and Schulzrinne (2007) proposed a cooperative roaming approach to authenticating the mobile, but its use is limited to a domain only.

The handover process often requires authentication and authorization for acquisition or modification of resources assigned to the mobile device. In most cases, these authentications and authorizations require interaction with a central authority in some realm. In some cases, the central authority may be distant from the mobile device. The delay introduced by such an authentication and authorization procedure adds to the handover latency, and consequently affects ongoing application sessions. The HOKEY (Handover Keying) working group in the IETF has defined two types of authentication models, namely EAP-based early authentication (Ohba et al. 2010) and efficient reauthentication (Clancy et al. 2008), to reduce the delay due to authentication. In EAP early authentication, AAA-based authentication and authorization for a candidate access point (CAP) is performed while ongoing data communication is in progress via the serving access network, so that the mobile can complete AAA signaling for EAP before the mobile device moves. The applicability of EAP early authentication is limited to scenarios where candidate authenticators can be discovered and an accurate prediction of movement can be easily made. In addition, the effectiveness of EAP early authentication may be less significant for some particular intertechnology handover scenarios where simultaneous use of multiple technologies is not a major concern. There are two types of early authentication model, namely *direct* and *indirect*. In the *direct* model, the service access point (SAP) is not involved in the EAP exchange and only forwards the EAP preauthentication traffic as it would any other data traffic. The direct preauthentication model is based on the assumption that the mobile device can discover candidate authenticators and establish direct IP communication with them.

In the *indirect* preauthentication model, it is assumed that a trust relationship exists between the serving network (or serving AAA realm) and the candidate network (or candidate AAA realm). The SAP is involved in EAP preauthentication signaling. This preauthentication model is needed if a peer cannot discover the candidate authenticator's identity or if direct IP communication between the mobile device and the CAP is not possible owing to security or network topology issues. Figure 6.6 illustrates the two types of authentication modes, direct and indirect.

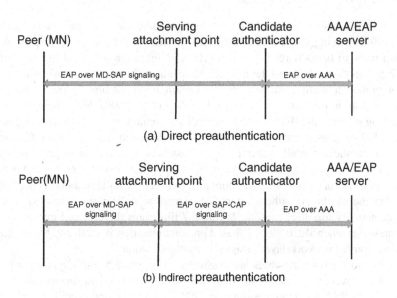

Figure 6.6   Direct and indirect preauthentication

Ohba and Yegin (2010) described how the Procotol for carrying Authentication for Network Access (PANA) can be used to support preauthentication. That RFC defines an extension to the PANA protocol (Forsberg et al. 2008) used for proactively executing EAP authentication and establishing a PANA security association between a PANA client (PaC) resident on the mobile node in an access network and a PANA authentication agent (PAA) in another access network to which the PaC may move. The extension to the PANA protocol is designed to realize the direct preauthentication defined by Ohba et al. (2010).

6.4.3   Network-Layer-Assisted Preauthentication

In order to deal with the limitations of the existing mechanisms, we have proposed a network-layer-assisted link-layer preauthentication mechanism (Dutta et al. 2010; Lopez et al. 2007) that can deal with many of the drawbacks of the existing approaches. Our proposed mechanism is aimed at reducing the link-layer handoff latency when the existing link-layer handoff optimization mechanisms cannot be applied in cases involving interdomain and interaccess technologies. It uses preauthentication in the network layer to assist link-layer handoff optimization techniques by allowing a fast transition even when the APs involved in the handoff do not share same link layer. Although this mechanism can work independently of the link-layer access technology, our studies and experiments were focused on 802.11-based access networks. The proposed mechanism also preserves the security criterion required by the IETF by not allowing context transfer between APs. In this section, we describe the architecture of this mechanism, provide experimental results from a test bed implementation, and compare these with IEEE 802.11i preauthentication.

In an interdomain mobility scenario, the authentication process is followed by an authorization process. In addition to reducing the delay due to layer-3-related authentication and authorization, our proposed mechanisms can reduce the authentication delay in the link layer when existing preauthentication mechanisms (e.g., 802.11i-based preauthentication) cannot be applied to take care of handoffs involving interdomain, intersubnet, and interaccess technologies. A successful authentication prior to handoff results in proactive configuration and establishment of a security association between the mobile and the network elements in the target network. We have discussed two types of preauthentication, namely *direct preauthentication* and *indirect preauthentication*, in the preauthentication problem statement draft (Ohba et al. 2009) that is being discussed in the HOKEY working group in the IETF. In the case of direct preauthentication, the serving authenticator forwards the EAP preauthentication traffic as it would do for any other data traffic, or there may be no serving authenticator at all in the serving access network. In indirect preauthentication, it is assumed that a trust relationship exists between the serving network (or serving AAA domain) and the candidate network (or candidate AAA domain). Indirect preauthentication is needed if the peer cannot discover the candidate authenticator's IP address or if IP communication is not available for security or network topology reasons. Figure 6.7 illustrates the protocol interaction among the network components when IEEE 802.11i-based preauthentication is used, and Figure 6.8 shows the protocol interaction for network-layer-assisted preauthentication.

Both roaming and nonroaming cases are illustrated in Figure 6.5. Initially, during the discovery phase, the MN discovers by some means (e.g., the 802.21 information service) the target AP and the IP address of the PAA that manages the target AP. Then the MN preestablishes a PANA security association (SA) (preauthentication phase) with the candidate target network (CTN) via its serving network, by performing an EAP exchange between the MN and the PAA. In the example shown, EAP-TLS (Aboba and Simon, 1999) is used as the EAP method for the authentication. The PAA

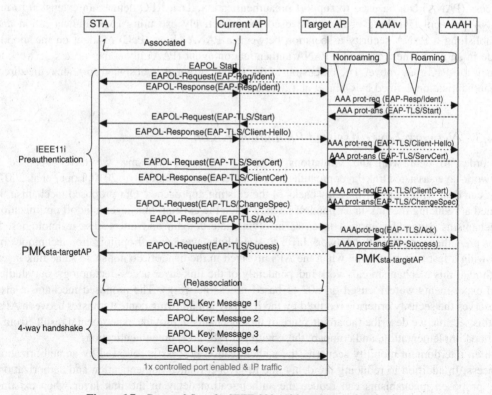

Figure 6.7   Protocol flow for IEEE 802.11i-based preauthentication

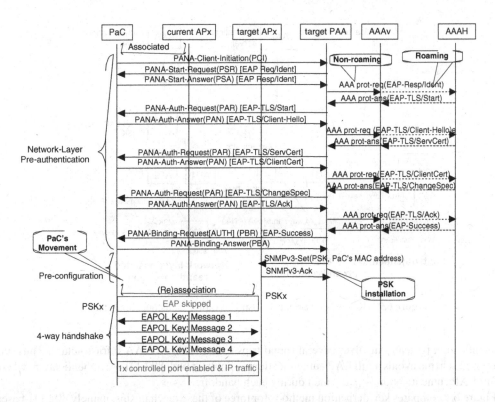

Figure 6.8 Protocol flow for network-layer-assisted layer 2 preauthentication

can rely on a backend AAA server to carry out an EAP authentication. From the MSK generated during the EAP authentication, the PAA can derive a distinct PSK for each AP. The PAA installs these keys in these APs (preconfiguration phase), and provides the MN with the required information (e.g., the APs' MAC addresses) to generate the same PSKs. Then the MN moves to the new AP and, after association, runs a four-way handshake by using the specific PSK, PSK_{ap}, generated during PANA preauthentication. At this point, the handoff is complete. Thus, by preauthenticating and preconfiguring the link, the establishment of a security association during handoff is reduced only to a four-way handshake.

In comparing the IEEE 802.11i preauthentication presented in Figure 6.7 with the PANA-based network layer preauthentication shown in Figure 6.8, one may notice that both of these schemes reduce the delay induced by the authentication process during the handover between access points. In particular, the delay is reduced to the time for the four-way handshake required to establish a security association between the PaC and the target AP in both cases. Therefore, in terms of handoff delay, both schemes result in comparable values. However, the proposed mechanism obtains the same reduction even when the APs belong to different subnets, which may be part of different administrative domains. Thus, it deals with the limitations imposed by the regular IEEE 802.11i preauthentication mechanism. Another interesting advantage of the current proposal is that a PAA can control and distribute PSKs to several APs through a single EAP authentication, the one performed during the preauthentication shown in Figure 6.8. This means that although two messages are required for key installation, when the mobile running the PaC moves between APs covered by the same PAA's area, it avoids additional EAP authentication. As depicted in Figure 6.7, EAP

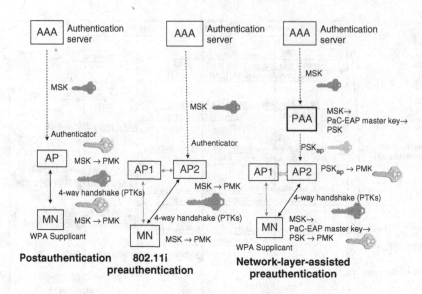

Figure 6.9   Key generation mechanisms in various authentication schemes

authentication typically involves several round trips to the backend AAA infrastructure. Thus, the proposed scheme avoids a full EAP authentication, in contrast with 802.11i preauthentication, where a full EAP authentication is performed during each handoff.

Figure 6.9 compares key derivation methods for three of these mechanisms, namely 802.11i-based reauthentication, 802.11i-based preauthentication, and network-layer-assisted layer 2 preauthentication. However, we describe in the following only the key derivation and key installation procedures that are part of our proposed network-layer-assisted preauthentication mechanism.

6.4.3.1   Derivation of Preshared Key

During PANA-based preauthentication, a master session key is generated after EAP authentication. The MSK is used to derive a master key for the PAc and the enforcement point (PaC–EP master key), specific to both the AP and the mobile node. In turn, the PaC–EP master key is used to derive the PSK. Since the PSK is dynamically derived from the PaC–EP master key, it has an associated lifetime. In PANA, the lifetime of the PaC–EP master key (and thus the PSK lifetime) is bounded by the PANA security association lifetime, which, in turn, is bounded by the MSK lifetime. Since each EAP reauthentication generates a new MSK, a new PaC–EP master key and PSK are derived. For security reasons, when a new PSK is installed in the AP, a four-way handshake must subsequently be run. This allows new, fresh PTKs to be generated from the new PSK. It is worth mentioning that, in general, the PaC–EP master key can be used to bootstrap link-layer security at policy enforcement points (PEPs) for any type of link layer (e.g., 802.11 or CDMA), which allows the MN to roam among multiple PEPs with different link-layer types without additional EAP execution if the PEPs are controlled by the same PAA.

6.4.3.2 Key Installation Process

The PAA installs the PSK in the target access points. We consider two key installation methods, namely *preemptive* and *on-demand*. As part of the preemptive installation process, the PAA installs PSKs in a preemptive way in all target APs. However, this introduces scalability and resource consumption problems when many APs are under the control of one PAA or many MNs are connected to APs served by one PAA. Since it provides the needed PSK for a particular MN and AP before the MN is attached, it reduces the time required to start the four-way handshake.

Alternatively, an AP may inform the PAA when an MN has associated with it. This mechanism is referred to as on-demand key installation for the AP. Although this mechanism can save systems resources, it introduces a delay before network access is gained because both the MN and the AP need to wait for the PSK provisioning.

In order to take advantage of these two methods and minimize some of their disadvantages, algorithms such as those proposed by Mishra et al. (2004) and Pack and Choi (2002) could be used. These algorithms determine the most probable APs that the MN may move to, so that the PAA can install PSKs only at those APs selected by the algorithm, as part of the preemptive key installation. However, if the prediction fails and the MN finally moves to another AP where a PSK has not been installed, on-demand key installation may be used instead. Depending on the number of APs and the number of users, a wireless service provider may decide to use one or other of these techniques or even a combination of both.

6.4.4 Experimental Results and Analysis

We implemented the proposed network-layer-assisted preauthentication mechanism in a test bed as shown in Figure 6.10. Here, we illustrate several different scenarios and demonstrate how network-layer-assisted preauthentication can provide link-layer handoff optimization. In particular, we describe the application of the preauthentication mechanism over IEEE 802.11 networks and compare the results with the existing preauthentication mechanism for IEEE 802.11i. Figure 6.11 shows the interaction among several functional components and the protocols used between each pair of these components.

In this experimental test bed, we used the HostAP daemon (Malinen, 2005a) and the MADWiFi driver (MADWiFi Driver 2013) and configured three Linux systems to act as access points. Two of these access points (AP1 and AP2) worked as IEEE 802.11i APs. Both of these APs could work in either PSK (when network layer preauthentication was used) or 1X EAP mode. There was also inbuilt RADIUS client functionality within the AP (for cases where network layer preauthentication was not enabled). Each AP implemented an SNMPv3 (Simple Network Management Protocol) agent that allowed it to set PSKs and associated parameters such as key lifetimes. Finally, the last access point (AP0) was configured with open authentication. The MN was a laptop equipped with WPA supplicant software (MADWiFi Driver 2013) that provided 802.11i functionality, a MADWiFi driver, and Open Diameter's PANA client implementation (http://sourceforge.net/projects/diameter/). The PANA agent was based on an Open Diameter implementation that also provided an inbuilt Diameter client. We used Open Diameter and FreeRADIUS (http://www.freeradius.org/) as the AAA protocol implementations.

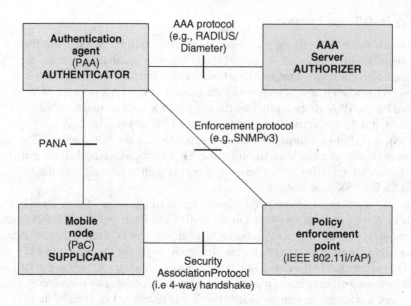

Figure 6.10   Experimental test bed for preauthentication

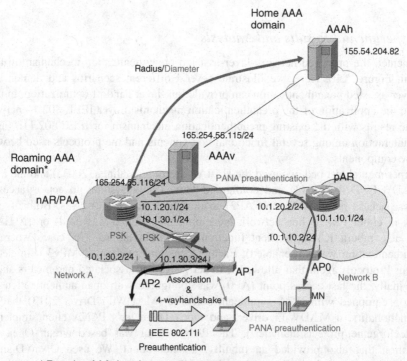

* Roaming AAA domain in roaming case.
For nonroaming case, it acts as MN's home AAA domain.

Figure 6.11   Interaction among the functional components

We experimented with three types of movement scenarios, involving both roaming and nonroaming cases. In the roaming case, the mobile node was visiting an administrative domain different from its home domain. Consequently, the AAAh, which was placed in a different continent in our experiment (e.g., at the University of Murcia, Spain), needed to be contacted. In the nonroaming case, we assumed that the MN was moving within its home domain and only the local AAA server (AAAv) was contacted.

The first scenario did not involve any preauthentication. The MN was initially connected to AP0 and moved to AP1. Because neither network-layer authentication was enabled nor IEEE 802.11i preauthentication was used, the MN needed to engage in a full EAP authentication with AP1 to gain access to the network after the move (postauthentication). This experiment shows the effect of delay when there is no preauthentication.

The second scenario involved 802.11i preauthentication and involved movement between AP1 and AP2. The MN was initially connected to AP2, and started IEEE 802.11i preauthentication with the target access point AP1. This is an ideal scenario for comparing the values obtained from 802.11i preauthentication with those obtained from the proposed network-layer-assisted preauthentication. Both the first and the second scenarios used RADIUS as the AAA protocol, with the APs implementing a RADIUS client.

The third scenario took advantage of the proposed network-layer-assisted link-layer preauthentication. It involved movement between two APs (e.g., between AP0 and AP1) that belonged to two different subnets where 802.11i preauthentication was not possible. Here, Diameter was used as the AAA protocol, where the PAA implemented a Diameter client.

In this third movement scenario, the MN was initially connected to AP0 in Figure 6.10. The mobile node started PANA preauthentication with the PAA, which was co-located with the AR in the new candidate target network (nAR in network A), from the current associated network (network B). After authentication, the PAA installed two preshared keys, PSK_{AP1} and PSK_{AP2}, in AP1 and AP2, respectively, by using a preemptive key installation method. Finally, because PSK_{AP1} was already installed, AP1 immediately started a four-way handshake upon the mobile's arrival in network A.

As illustrated, we used the same target access point AP1 to perform the handover in all three scenarios. Therefore the measurement of the four-way handshake time was always taken at this access point (i.e., AP1). In the first scenario, the mobile node was initially attached to AP0 because we were trying to demonstrate the case where the 802.11i preauthentication cannot be executed since the two access points are connected to two different subnets. This happened when the target AP (AP1) was not placed in the same distribution system as the current AP (AP0). In the second scenario, both AP1 and AP2 were configured with 802.11i support, so that we could simulate 802.11i-based network protection. Therefore, in order to initiate a handoff to AP1, the MN started the test attached to AP2 after running an initial EAP authentication. Finally, in the third scenario, the MN was initially attached to AP0 and the handoff was performed to AP1. In this case, we simulated the scenario in which layer-2-based 802.11i preauthentication cannot be performed but network layer preauthentication can be used instead.

The MN used the application layer discovery mechanism discussed in Section 6.2 to discover the PAA's IP address and all required information about the target APs, namely AP1 and AP2 (e.g., channel, security-related parameters, and MAC address), at some point before the handoff. This avoided scanning during link-layer handoff. Because the focus here is on reducing the time spent on

Table 6.2 Experimental results (ms) for preauthentication

	Type of authentication					
	Postauthentication		802.11i preauthentication		Network-layer-assisted preauthentication	
Type of movement	Nonroaming	Roaming	Nonroaming	Roaming	Nonroaming	Roaming
T_{auth}	61	599	98	638	177	831
T_{conf} 2 AP	–	–	–	–	16	17
$T_{assoc+4-way}$	18	17	16	17	15	17
Total	79	616	114	655	208	865
Time affecting handover	79	616	16	17	15	17

the authentication part during handoff, we shall not discuss the details of how we reduced the layer 2 scanning time. We have described the details of how scanning was optimized in Dutta et al. (2005e). The MN could also use any of the existing techniques that reduce layer 2 scanning, as described in Section 6.3.

Table 6.2 shows the average timings (rounded off to the most significant digit) associated with some of the handoff operations that we measured in the test bed. We briefly explain each of these timings in the following:

- T_{auth} refers to the execution of the EAP-TLS authentication procedures. This time does not distinguish between whether this authentication was performed during preauthentication or during a typical postauthentication.
- T_{conf} refers to the time spent during PSK generation and installation after EAP authentication is complete. When network layer preauthentication is not used, this time is not considered.
- $T_{assoc+4-way}$ refers to the time dedicated to the completion of association and the four-way handshake with the target AP after the handoff.

We have shown the total time taken by the process by adding these components. Finally, we have also highlighted the time that affects the handoff in each case.

Each of these timings may safely be considered as independent for each experiment. Thus, the authentication phase, the configuration phase, and the association or four-way handshake can be considered as independent events. In fact, the time $T_{assoc+4-way}$ seems to be similar in value regardless of the movement scenario. Also, independent of whether PANA was run in the roaming or nonroaming case, the value of T_{conf} remains the same.

The second and third columns in Table 6.2 show the results for the nonroaming and roaming cases, respectively, when no preauthentication was used. The fourth and fifth columns depict similar cases when IEEE 802.11i preauthentication was used. Finally, the last two columns show the results when network layer preauthentication was used. When preauthentication is used, only $T_{assoc+4-way}$ affects the handoff time. When no preauthentication is used, the time affecting the handoff includes T_{auth} (the complete EAP-TLS authentication) plus $T_{assoc+4-way}$. These results illustrate how network-layer-assisted layer 2 preauthentication can provide results comparable to 802.11i-based preauthentication and at the same time can support intersubnet and interdomain mobility, which cannot be supported by IEEE 802.11i.

In Chapter 11, we illustrate how the proposed preauthentication mechanism can interwork with other handoff-related operations and how application layer and network layer mobility protocols can be used to build a complete handoff system.

6.5   Layer 3 Configuration

In Chapter 3, we defined a mobile's configuration processes during a mobility event. We also illustrated how the mobile's layer 3 configuration processes affect the handoff delay and contribute to packet loss. During layer 3 configuration, the mobile acquires an IP address and assigns to its interface so that the mobile can communicate using the newly obtained IP address. Before assigning the IP address, the client usually performs duplicate address detection (DAD) by way of ARP or neighbor discovery in an IPv4 or an IPv6 network, respectively. For an IPv4-based network, this detection procedure may take 4–15 s (Vatn and Maguire, 1998). The DAD-related delay for stateless address configuration of an IPv6 address identifier may be up to 1500 ms and depends on a random value that determines the neighbor solicitation interval (Narten et al. 1998).

In this section, we first analyze the effect of layer 3 configuration on handoff delay for both IPv4 and IPv6 networks. Then, we describe the key principles that need to be considered in order to optimize the delay due to configuration. We introduce a few related studies in which the configuration-related delay was optimized. Then, we describe the proposed techniques that expedite part of the layer 3 configuration process at the cost of additional signaling messages. Finally, we highlight the experimental results from our test bed.

As part of the investigation into layer 3 configuration optimization techniques using DHCPv4 and MIP, we verified that with the ARP enabled, the IP address acquisition took an average of 15 s, but when the ARP was suppressed, the average time for IP address acquisition was 436 ms. We conducted several experiments to analyze the effect of two factors on IP address acquisition, namely duplicate address detection (Thomson and Narten, 1998) and router selection, on the disruption of real-time voice traffic over an IPv6 network. We experimented with both Mobile IPv6 from USAGI (Tuominen and Petander, 2001) and SIP-based terminal mobility (Wedlund and Schulzrinne, 1999).

DAD confirms the uniqueness of the IPv6 address on the link. During the DAD process, the new address is called a tentative address. According to RFC 2462 (Thomson and Narten, 1998), a tentative address is not allowed to be used by a node. This means that the MN cannot send packets with a tentative address as a source IPv6 address and has to discard all inbound packets to a tentative address during the DAD phase. This imposes an additional delay on any mobility binding update, such as a re-INVITE in the case of SIP. With default values as described in Thomson and Narten (1998), the average delay caused by DAD was 1500 ms.

Router selection also plays an important role during handoff. According to RFC 2461 (Narten et al. 1998), a host needs to perform certain steps before switching to another access router. These additional steps, such as routing table update and neighbor unreachability detection (NUD), contribute to the delay due to the router selection process. We describe these two processes in the following.

Routing Table Update

To perform rapid handoff, hosts in an IPv6 environment should attach to the new access router whose RA (router advertisement) is the most recent. However, commonly used Linux hosts do not always select the new access router quickly. If the routing table has other routes, a host may select a different router. In this case, an IPv6 host performs NUD against old router to confirm unreachability, and after confirmation of unreachability, the host is allowed to switch to another router to connect.

Neighbor Unreachability Detection

Neighbor unreachability detection verifies that two-way communication with a neighboring node exists. The host sends a neighbor solicitation to a node and waits for a solicited neighbor advertisement. If a solicited neighbor advertisement is received, the node is considered reachable. During a handoff operation, an IPv6 host must confirm unreachability to an old access router before switching to a new access router by using a NUD mechanism in the absence of an aggressive router selection mechanism. Our study showed that NUD with default values could impose more than 8 s delay on the configuration process without a mechanism such as aggressive router selection.

We now briefly explain how NUD contributes to the configuration delay. In NUD, each neighbor has a reachability state. When a host confirms that a neighbor is reachable, the reachability state of that neighbor is called REACHABLE. Then the host waits for REACHABLE TIME (measured in milliseconds) before the state goes to STALE. On receiving an RA, it can also go to the STALE state. During the STALE state, nothing happens until the host sends new packets. After the host sends a packet, active reachability confirmation starts in the state of DELAY. During this state, the host waits for another DELAY FIRST PROBE TIME (in seconds) and goes to the PROBE state. In this state, the host uses neighbor solicitation to confirm reachability with a predefined number of retransmissions (MAX UNICAST SOLICIT). The host does not receive any neighbor advertisement from the target neighbor, and the reachability state of the neighbor goes to NULL. The amount of delay introduced by the NUD process depends upon the time when the host gets a new RA and the NUD state of the old access router at that moment. If a host detects unreachability of an old access router before getting a new RA, the NUD operation may not introduce any additional delay into the configuration process.

In order to study the effect of DAD and NUD on the configuration delay, we modified the Linux kernel to avoid the DAD process and enabled an aggressive router selection procedure in the kernel module, which helped the mobile to communicate with the new router quickly without doing NUD (Narten et al. 1998).

Figure 6.12 shows the IPv6 test bed where we experimented with SIP-based mobility to study the effects on handoff delay due to DAD and NUD. This IPv6 test bed had one home network (N1) and two visited networks (N2 and N3). The experiments involved three movement scenarios: (i) movement between the home network and a visited network (from N1 to N2), (ii) movement between visited networks N2 and N3, and (iii) movement between visited network N3 and the home network N1. The experimental results shown in Table 6.3 demonstrate how DAD- and NUD-related delays affect both the signaling and the media redirection delays in the case of SIP-based terminal mobility.

The delays shown in Table 6.3 are not inclusive of layer 2 access delays such as 802.11 scanning delays. Two different scenarios were considered: (a) SIP mobility without aggressive router selection, and (b) SIP mobility with aggressive router selection. Both the handoff-related signaling delay and the media delay are shown when the mobile moves between the home network and two visited networks, namely visited network 1 and visited network 2. H12 denotes when the mobile moves from the home network to visited network 1, H23 denotes when the mobile moves from visited network 1 to visited network 2, and H31 denotes when the mobile moves from visited network 2 back to the home network. The values demonstrate how by avoiding DAD and adopting an aggressive router selection technique to reduce the effect of NUD, we could reduce the signaling delays to 200 ms and

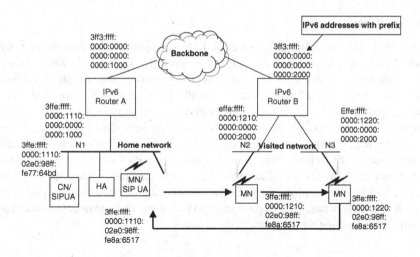

Figure 6.12   Experimental IPv6 test bed for handoff

Table 6.3   Effect of duplicate address detection (IPv6) on handoff

Handoff case	Signaling delay (ms)		Media delay (ms)	
	SIP with DAD and NUD	SIP w/o DAD and NUD	SIP with DAD and NUD	SIP w/o DAD and NUD
H12 (home–visited 1)	3829	171	3854	421
H23 (visited 1–visited 2)	3932	161	4188	419
H31 (visited 2–home)	1935	161	1949	408

media interruption to less than 500 ms for SIP-based mobility (Wedlund and Schulzrinne, 1999). We have published the details of this experiment (Nakajima et al. 2003).

6.5.1   Key Principles

The following are some of the key principles that can help optimize the time taken for IP address acquisition during the layer 3 configuration process:

1. Reduction of the number of signaling messages exchanged between the mobile node and the DHCP server during stateful IP address acquisition.
2. Minimizing the time taken to verify the uniqueness of the IP address of the mobile.
3. Performing the address uniqueness checking ahead of layer 3 handoff.
4. Prefetching and caching of the new IP address reduces the time taken for IP address acquisition after the handoff.
5. Performing the address resolution by mapping between the IP address of the target router and the MAC address before the mobile has moved to the new network.

6.5.2   Related Work

For IPv6 networks, Moore (2006) and Han et al. (2003) proposed some optimization techniques to carry out DAD optimization for IPv6 clients. Optimistic DAD (Moore, 2006) ensures that the probability of address collision is not increased and thus improves the resolution mechanisms for address collisions. There are a few proposals from the IETF, such as Passive DAD (Forte et al. 2006a) and the DHCP rapid commit option (Park et al. 2005), that try to expedite the IP address acquisition for IPv4 networks. We have proposed and implemented two optimization techniques that help expedite the IP address configuration process, namely router-assisted duplicate address detection and proactive IP address configuration. Compared with the existing techniques, the first approach does not need any additional agent in the network and the router assists in reducing the time taken for IP address acquisition. The second approach reduces the delay at the cost of additional resources such as tunnels between the target router and the mobile and additional network bandwidth. In the following, we describe these two methods in detail.

6.5.3   Router-Assisted Duplicate Address Detection

We have designed and implemented a router-assisted duplicate IP address detection mechanism that reduces the layer 3 configuration time by expediting duplicate IP address detection. It adopts the general principle of the network doing the duplicate address detection instead of the mobile itself. In this mechanism, an upstream router keeps the list of IP addresses configured in a specific subnet in its neighbor cache. A router in each subnet acts like a reporting agent and sends a list of IP addresses that are currently in use via a scope-based multicast address. A scope-based multicast address could be a multicast address with some TTL (time-to-live) value that can work over a range of subnets. An upstream router can send the list of IP addresses in use in the neighboring subnets periodically using a scoped multicast address. A TTL-scoped multicast address can be used to limit the number of subnetworks the router can cover. For example, a router in a subnet can use a TTL of one, whereas an upstream router can use a TTL that is higher than one and can cover multiple subnets. Figure 6.13 shows how the mobile obtains the list of used IP addresses that could be used in its own subnet or in neighboring subnets from the router, and thus does not need to perform an ARP before it assigns the address.

Thus, a mobile can obtain the list of addresses that are currently in use in its own subnet or in the neighboring subnets from the router without performing an ARP and having to wait for the ARP reply. Unlike other approaches, the proposed approach does not need any new elements in the network and does not need changes in the DHCP server. However, there is a trade-off between the frequency of router advertisements and the load on the network. This technique also needs some modifications to the router, and the neighbor cache entry in the router needs to be rebuilt in the case of a power failure. We have presented details of the proposed duplicate address detection mechanism in Dutta et al. (2006b).

6.5.4   Proactive IP Address Configuration

We have designed a proactive configuration technique that can work independently or in conjunction with the preauthentication mechanism described earlier. It adopts the general principle of proactive caching. The proactive configuration mechanism consists of several steps, namely proactive address

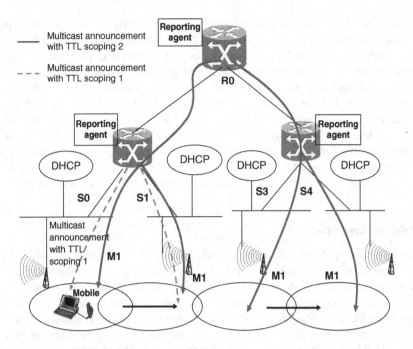

Figure 6.13   Router-assisted duplicate address detection

acquisition, proactive duplicate address detection, and proactive address resolution. In the following, we describe these steps in detail.

6.5.4.1   Proactive IP Address Acquisition

Although FMIPv6 (Johnson et al. 2004) can proactively acquire an IP address by obtaining the router prefix from the next access router, it expects the adjacent routers to cooperate and discover each other. Thus, the FMIPv6-based fast-handoff mechanism does not work for interdomain mobility. The proposed technique is client-assisted, and can be applied in both intradomain and interdomain mobility scenarios. In the proposed technique, the client obtains the IP address of the target network while the mobile is still in the current serving network. The client assigns this proactively obtained address to a virtual interface and performs a subsequent proactive binding update to the home agent or correspondent node. Alternatively, the mobile can store the address in a local cache and assign the address later on. This avoids the delay due to the signaling messages needed during the address acquisition process after the handover.

6.5.4.2   Proactive Duplicate Address Detection

When the DHCP server dispenses an IP address, it updates its lease table so that the same address is not assigned to another client for a specific period of time. At the same time, the client also keeps a lease table locally so that it can renew when needed. In some cases, where a network consists of both DHCP- and non-DHCP-enabled clients, there is the possibility that another client in the LAN may have been configured with an IP address from the DHCP address pool. In such a scenario, the server detects a duplicate address based on an ARP or on IPv6 neighbor discovery for an IPv4 or an IPv6 network, respectively, before assigning the IP address. This detection procedure may take from 4 to 15 s (Vatn and Maguire, 1998) and will thus contribute to a larger handover delay.

In the proposed method, the mobile node performs duplicate address detection ahead of time, while it is still in the previous network, thus reducing the IP address acquisition time. This is performed by a DHCP relay that is co-located with the next target router. In the case of stateless address configuration, the proactive duplicate address detection over the candidate target network is performed by the previous access router (PAR) on behalf of the mobile at the time of proactive establishment of the handover tunnel, since duplicate address detection over a tunnel is not always performed.

6.5.4.3   Address Resolution

The address resolution process has been defined in Chapter 3. Through the address resolution process, one can obtain a mapping between the MAC address and IP address. Having prior knowledge of the IP-address-to-MAC-address mapping, the neighboring first-hop router and the mobile do not need to discover each other in layer 2 after the mobile has moved to the new network. For example, if the MAC-to-IP address mappings are known to the mobile ahead of time, the mobile can communicate with nodes in the target network after attaching to the target network without waiting for an ARP broadcast or neighbor solicitation process. The mobile communicates with the access router, authentication agent, configuration agent, and correspondent node after the handover.

Here, we describe several possible ways of proactively performing address resolution to obtain a MAC–IP address mapping:

1. An information service mechanism (e.g., IEEE 802.21) can be used to resolve the MAC addresses of the nodes. This requires that each node's network information (e.g., IP address, channel address, and authentication scheme) is populated in the information server database. This information can be entered using approaches discussed by Dutta et al. (2006c)
2. The authentication protocol that helps to preauthenticate the mobile or the configuration protocol that is used for preconfiguration can piggyback the MAC address of the network entities during the preauthentication or preconfiguration process. The mobile can thus keep this MAC address in its cache and avoid the address resolution process after it is handed over to the target network. For example, if PANA is used as the authentication protocol for preauthentication, PANA messages may carry AVPs (attribute–value pairs) that can be used to carry the MAC address. In this case, the PANA authentication agent in the target network may perform address resolution on behalf of the mobile node and carry the related network parameters to the mobile node before the handover.

When the mobile node attaches to the target network, it installs the proactively obtained address resolution mappings without necessarily performing address resolution queries for the nodes in the target network. On the other hand, the nodes that reside in the target network and are communicating with the mobile node need also to update their address resolution mappings for the mobile node as soon as the mobile node attaches to the target network. These proactive address resolution methods could also be used for those nodes to proactively resolve the MAC address of the mobile node before the mobile node attaches to the target network.

In order to expedite the address resolution process, a mobile could trigger the address resolution process as soon as it detects a new network. This is based on the mobile gratuitously performing address resolution (Johnson et al. 2004; Perkins, 2002b), in which the mobile node sends an ARP request or an ARP reply in the case of IPv4 or a neighbor advertisement in the case of IPv6 immediately after the mobile node attaches to the new network, so that the nodes in the target network can quickly update the address resolution mapping for the mobile node.

6.5.5   Experimental Results and Analysis

We have demonstrated proactive address acquisition for both IPv4 and IPv6 networks using PANA. Independently of the preauthentication mechanism, we have also used stand-alone protocols, such as GIST (General Internet Signaling Transport) (Schulzrinne and Hancock, 2008) and IKEv2 (S. Eronen, 2006), to configure the mobile proactively. We now briefly describe these experiments. We have described the details of these techniques in Dutta et al. (2005e).

These experiments verified that the number of messages exchanged between the client and the network nodes (e.g., router or server) during IP address acquisition, the processing time at the end systems, and the network load are some of the key factors that contribute to the layer 3 configuration delay. Proactive caching of the IP address at the client and router and the server-assisted proactive duplicate address detection technique reduce the layer 3 address acquisition delay at the cost of additional resource usage at the mobile.

6.6   Layer 3 Security Association

In Chapter 3, we defined a security association and illustrated how reestablishment of a security association affects the handoff delay and packet loss during a mobility event. The security association between two communicating nodes can exist in multiple layers. IPSec (Kent and Atkinson, 1998a) provides security association in layer 3. A layer 3 security association is uniquely identified by an SPI (security parameter index), destination IP address, and ESP (Encapsulating Security Payload). Thus, when the IP address of any one of the communicating hosts changes, a new security association needs to be reestablished between the pair of nodes. During repeated handoff of a mobile, the security association between the mobile and the communicating host over the secured channel needs to be reestablished when the end-point identifier (e.g., IP address) changes. The process of reestablishing the security association requires an exchange of messages to derive a new key and processing at the end hosts, and thus contributes an added delay during handoff.

In this section, we describe the key principles that need to be considered when optimizing the delay due to reestablishment of a security association. We then describe related work where attempts have been made to optimize the delay due to security association during handoff. We describe our proposed techniques that optimize the delay due to security association at the cost of additional resources such as an additional home agent in the network and additional tunneling operations. Finally, we illustrate the experimental results obtained in our test bed.

6.6.1   Key Principles

The following are the key principles that should be considered when one is aiming to minimize the delay due to security association:

1. Maintain the security binding between the two communicating end points.
2. Avoid signaling exchanges between the peers in order to generate the encryption keys.
3. Maintain the security context by way of reactive or proactive context transfer.
4. Maintain constant end-point connection identifiers.
5. Hide the change of IP address of the end points by using an additional home agent.

We have designed an optimization technique that is based on some of these principles and experimented with it in a test bed. We have published the details of this optimization technique and the results in Dutta et al. (2005d, 2007d).

6.6.2   Related Work

Miu and Bahl (2001) described an architecture that helps to maintain the security association when a mobile moves between the public Internet and a private enterprise network. However, this solution is limited to movement between homogeneous networks (e.g., 802.11b). Rodriguez et al. (2004) introduced the concept of a mobile router, where end clients with multiple access technologies connect to the mobile router's downlink interface. In this case, the end clients do not change their IP addresses; rather, the mobile router continually changes the external IP addresses as it moves around and connects to different access networks, such as GPRS, CDMA, and 802.11b networks. This router uses NAT (network address translator) functionality to shield the clients from reinitiating the sessions.

6.6.3   Anchor-Assisted Security Association

In this section, we describe a proposed mechanism that optimizes the handoff delay due to security association at the cost of an additional home agent in the network. While the handoff delay is reduced, this mechanism introduces a tunneling overhead because of the additional home agent in the network. This optimization technique is based on the key principles 1, 2, and 4 listed in Section 6.6.1. In Chapter 11, as part of our discussion of systems evaluation, we demonstrate a handoff optimization technique in IMS (IP Multimedia Subsystem) that uses principle 3 in conjunction with other optimization techniques.

The proposed mechanism uses an anchor agent that acts as a home agent and maintains the security association during the handoff, thereby reducing the handoff delay and packet loss (Dutta et al. 2005d). The key principle introduced in this technique is maintaining the security association with the end client even when the end-point identifier changes. This avoids the delay due to reestablishment of the security association during the mobile's handoff. We have experimented with this technique using both network layer mobility and application layer mobility protocols. In the experiment, a mobile with two interfaces moved back and forth between an enterprise network equipped with 802.11, a cellular network with CDMA1XRTT access, and a hotspot equipped with 802.11. By introducing an anchor such as a secondary home agent into the network, one can achieve secured seamless communication without the need to reestablish the IPSec (Kent and Atkinson, 1998a) tunnels during each subnet move.

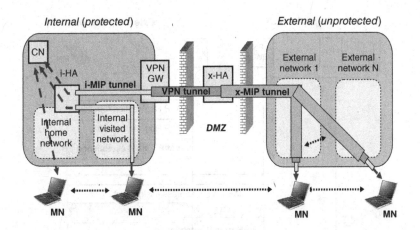

Figure 6.14   Anchor-agent-assisted security association

Figure 6.14 illustrates a scenario in which security reassociation is avoided by introducing an additional home agent, x-HA. By using the external home agent x-HA, the mobile does not need to set up a new IPSec association when it moves between subnets or domains. An internal home agent (denoted by i-HA) inside the intranet supports mobility inside the intranet. The external home agent, placed in the DMZ (demilitarized zone),[2] handles the mobile's mobility outside the enterprise and ensures that the security association with the mobile does not break when the mobile changes its IP address.

Figure 6.15 illustrates how Mobile IP tunnels and VPN (IPSec) tunnels are set up during a mobile's movement from an enterprise network to an external network. If the mobile uses Mobile IP's reverse tunneling, the data from the mobile will flow to the correspondent host in the reverse direction to the path shown in Figure 6.15. These tunnels are the additional systems resources expended when the handoff delay is reduced by avoiding reestablishment of the security association.

We have described the details of the architecture, implementation, and experimental verification in Dutta et al. (2004d). We briefly describe the techniques and associated results here.

The i-HA and x-HA collectively ensure that the packets received by the i-HA can be forwarded to the mobile when it is on an external network. A mobile has two MIP home addresses: an internal home address, i-HoA, in the mobile's internal home agent and an external home address, x-HoA, in the external home agent. The mobile's care-of address registered with its i-HA is referred to as its internal care-of address and is denoted here by i-CoA. The mobile's care-of address registered with the x-HA is referred to as its external care-of address, denoted by x-CoA. The instance of MIP running between the mobile and its i-HA is referred to as the internal MIP, or i-MIP. The instance of MIP running between the mobile and the x-HA is referred to as the external MIP, or x-MIP. After successful establishment of a VPN (e.g., after a successful IPSec security association), the mobile obtains an address, denoted by TIA (tunnel inner address), from the VPN gateway (VPN-GW).

When a mobile moves into a cellular network, setting up a connection with the cellular network can take a long time. For example, in our experiment, we routinely experienced 10–15 s delays in setting up PPP (Point-to-Point Protocol) connections to a commercial CDMA2000 1xRTT network.

---

[2] "DMZ" is defined in the List of Abbreviations.

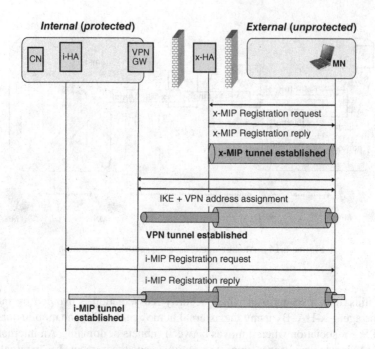

Figure 6.15   Mobile IP and VPN tunnels

In addition, establishing an IPSec connection to the mobile's enterprise network could also lead to excessive delay. To enable seamless handoff, handoff delays need to be significantly reduced.

Therefore, we applied handoff preprocessing and make-before-break techniques to reduce the handoff delay. In particular, in such techniques, the mobile anticipates the need to move out of a currently used network, based on, for example, the signal-to-noise ratio in the network. When the mobile believes that it will soon need or want to switch to a new network, it starts to prepare the connectivity to the target network while it still has good radio connectivity to the current network and the user traffic is still going over the current network. Such preparation may include the following steps:

1. Activate the target interface if the interface is not already on (e.g., a mobile may not keep its cellular interface always on if it is charged by connection time).
2. Obtain an IP address and other IP-layer configuration information (e.g., the default router address) from the target network.
3. Perform required authentication with the target network.
4. Establish the network connections needed to communicate over the target network (e.g., a PPP connection over a CDMA2000 network).

Although both interfaces are turned on at the same time, the decision to switch over from one interface to another will depend upon a local policy, which can be client-controlled or server-controlled. In this case, handover anticipation was based purely on the signal-to-noise ratio (SNR) of the 802.11 interface. But this handoff decision could be based on any other specific cost factor. When the mobile decides that it is time to switch its application traffic to the target interface, it takes the following steps:

1. It registers its new care-of address acquired from the target network with the x-HA.
2. It establishes a VPN tunnel (IPSec association) between its x-HoA and the VPN gateway inside the DMZ of its enterprise network.
3. It registers the gateway end of the VPN tunnel address as its care-of address with the i-HA. This causes the i-HA to tunnel packets sent to the mobile's home address to the VPN gateway, which then tunnels the packets through the VPN tunnel and the x-MIP tunnel (a Mobile IP tunnel formed by the external home agent) to the mobile.
4. When the mobile moves back to the enterprise network, the VPN and MIP tunnels are torn down. Tearing down the VPN tunnel takes up to a few seconds, owing to negotiation between the end points. Thus, some in-flight packets may be lost or arrive at a later time, leading to out-of-order packet delivery. Most of today's applications are capable of reordering out-of-sequence packets (e.g., out-of-sequence RTP packets).
5. When the mobile moves from one external network to another external network and acquires a new local care-of address (x-CoA), the mobile's x-HoA remains the same. Therefore, the mobile's existing security association does not break. The mobile needs only to register its new local care-of address with the x-HA so that the x-HA will tunnel the VPN packets to the mobile's new location.

6.6.4   Experimental Results and Analysis

We experimented with the proposed technique for both CBR (constant bit rate) traffic (audio) and VBR (variable bit rate) traffic (video) and analyzed the packet loss, delay, and jitter during the hand-off. Figure 6.16 shows the experimental test bed where we conducted this experiment. It shows an enterprise network, two home agents (external home agent and internal home agent), a VPN gateway, a cellular network, and another external Wi-Fi network. The mobile moved back and forth between the enterprise network, the cellular network, and the Wi-Fi hotspot. The mobile set up an IPSec connection with the VPN gateway.

In the absence of the proposed optimized technique, the mobile experienced packet loss due to the delay associated with IPSec tunnel setup and tear-down every time it changed its point of attachment. Without any optimization, layer 2 configuration took about 10 s in a CDMA network, and layer 3 address acquisition took about 3 s in an 802.11 network. The two binding updates, namely external and internal MIP registration, took about 300 and 400 ms, respectively, to complete. The IPSec-based security association took about 6 s. As the mobile moved back to the home network, it took around 200 ms for Mobile IP deregistration. These signaling exchanges resulted in degradation of real-time services due to the associated delay and packet loss. However, using a combination of an anchor-based security association technique that helped to maintain the security binding and a make-before-break technique, we could obtain zero packet loss during handover from the 802.11 network to the cellular network and vice versa. Although there was no packet loss in the optimized case, the mobile received a few out-of-order packets during its movement back from the cellular network to the 802.11 network as the transit packets on the slow cellular link arrived later than the initial packets that arrived via the 802.11 interface.

Figure 6.17 shows the interaction between different network components (e.g., the correspondent node (CN), MN, i-HA, x-HA, and VPN-GW) during the mobile's movement from the 802.11 to the CDMA network and vice versa when the optimization technique was deployed. Figure 6.17(a) shows how the mobile receives the data traffic while in the 802.11 network and how it prepares to hand over to the CDMA network at a certain threshold signal value, S1, by setting up PPP connections and

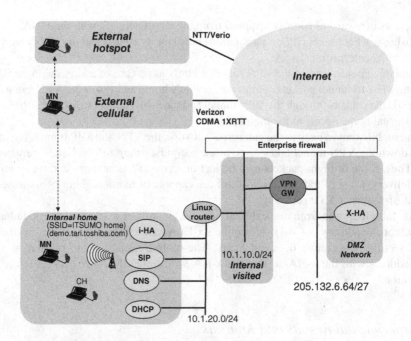

Figure 6.16   Experimental test bed for security association

establishing xMIP and IPSec tunnels. Figure 6.17(b) shows how, at an SNR value of S2, the mobile updates the internal home agent with the tunnel inner address. At this point the data flows directly to the CDMA interface using triply encapsulated tunnels. Figure 6.17(c) shows the signaling sequence when the mobile goes back from the CDMA network to the 802.11 network.

Figure 6.18 shows the results for packet loss with and without optimization of the security association. Figure 6.18(a) shows the packet loss due to reestablishment of the security association. Figure 5.15(b) shows how packet loss is avoided by introducing an additional home agent as the anchor point. Although no packet loss was observed, the mobile received out-of-order packets when it moved from the cellular to the Wi-Fi network. When the mobile is in the cellular network, the slope of the RTP traffic is less steep, meaning that the packets are subject to buffering delay in the CDMA base station and the output rate is less than the input rate. If after the handoff a packet is delayed beyond a certain threshold (e.g., the interpacket delay between the last packet before handoff and the first packet after the handoff is larger than 300 ms), then the packet may be considered lost for certain applications, such as VoIP.

Figure 6.19 illustrates the packet transport delay in the 802.11 and CDMA networks and the jitter introduced during handoff between the 802.11 network and the CDMA network for CBR traffic such as VoIP. We used the audio application RAT (Robust Audio Tool) (http://www-mice.cs.ucl.ac.uk/multimedia/software/rat/) to generate the audio traffic. Figure 6.20 illustrates the packet transport delay in the 802.11 and CDMA networks and the jitter introduced during handoff between the 802.11 network and the CDMA network for VBR traffic such as video over IP. We used the video conferencing application VIC (http://www-nrg.ee.lbl.gov/vic/) to generate the video traffic. Both RAT and VIC are open source software.

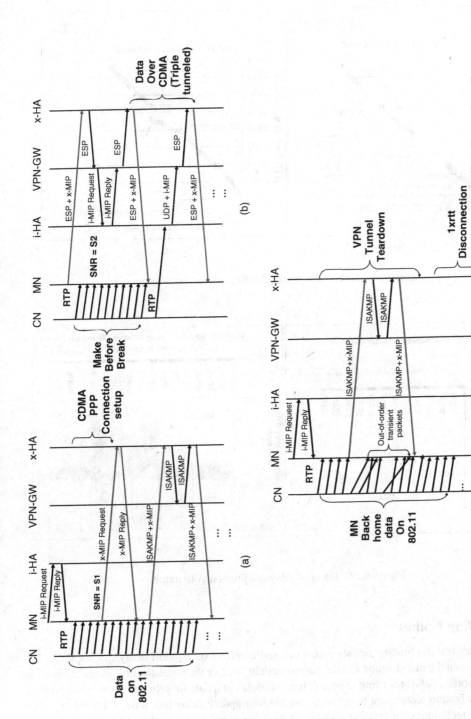

Figure 6.17 Interaction between network components during handoff

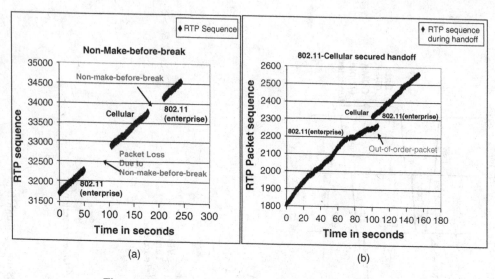

Figure 6.18   Effect of security rebinding and its optimization

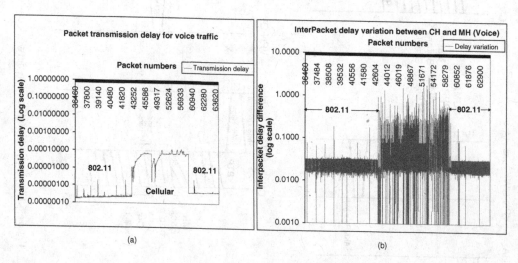

Figure 6.19   Effect of delay and jitter on VoIP traffic

6.7   Binding Update

We have described the binding update procedure and its effect on handoff delay in Chapter 3. The distance between the mobile node and the correspondent node or the home agent (HA) contributes to the binding update delay, resulting in overall handoff delay and data loss. In this section, we propose several optimization techniques to optimize the binding update delay and reduce the effect of this delay. These techniques for binding update can be categorized as *hierarchical binding update* and *proactive binding update*.

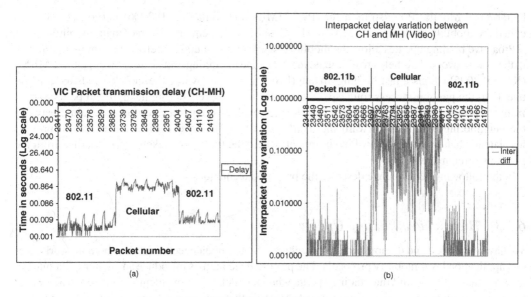

Figure 6.20   Effect of delay and jitter on video traffic

We first describe the key principles that should be considered when one wishes to optimize the binding update delay or reduce the effect of binding update. We describe some of the related work where attempts have been made to reduce the binding update delay. Then we introduce our proposed techniques that use some of these principles to optimize binding update delay at the cost of additional resources in the network. We have validated these binding update optimization techniques using experiments.

6.7.1   Key Principles

The following are some of the key principles that should be taken into account when optimizing binding update delay:

1. Limiting the traversal of the binding update closer to the mobile after every handoff.
2. Use of two levels of binding update by using an anchor agent between the home agent and the mobile node.
3. Applying the binding update proactively in the previous network before the mobile has moved to the new network.
4. Simulcasting the data to help reduce the data loss due to a longer binding update delay. This can probably be achieved by using a localized multicast approach.

6.7.2   Related Work

There have been enhancements to the layer-3-based mobility protocols to reduce the binding update delay when the CN and MN are far apart. MIP regional registration (Fogelstroem et al. 2007) provides hierarchical Mobile IP registration for IPv4. HMIPv6 (Soliman et al. 2006) introduced an agent called the mobility anchor point (MAP) to localize the management of intradomain mobility.

Proactive binding updates allow the mobile to send a binding update before the mobile has moved to the new network. This helps to eliminate the delay due to binding update after the handoff. FMIPv6 (Koodli, 2005) adopts a fast binding update (FBU) technique, where the mobile sends the binding update to the previous access routers so that the in-flight packets during handoff can be forwarded from the previous access router to the mobile. However, this requires additional signaling between the neighboring routers to forward the data. Malki (2007) also described techniques to provide low-latency handoff in an MIPv4 environment, where the transient packets are forwarded from the previous foreign agent.

In the following sections, we describe the proposed techniques.

6.7.3   Hierarchical Binding Update

We have developed and demonstrated hierarchical binding update techniques for both network layer and application layer mobility protocols. The proposed techniques introduce an anchor point into the network that helps to limit the binding update when the mobile's movement is confined to a domain, where a domain is defined to be a set of subnetworks that are controlled by a mobility agent. This technique helps to optimize binding update delay and reduces the network load, at the cost of an additional network element such as a mobility agent. We have applied these techniques to both the application layer mobility protocol and the network layer mobility protocol.

We have presented details of the implementation and experimental analysis for the above two cases in Das et al. (2002) and Dutta et al. (2004c), respectively. These techniques were developed around the same time as the other related hierarchical mobility management techniques. We describe the two techniques and experimental results in the following.

6.7.3.1   Network-Layer Mobility-Agent-Assisted Technique

We have designed a network-layer-based intradomain mobility management protocol (Das et al. 2002) by adopting a similar approach, where a mobility agent acts like an anchor point. Figure 6.21 shows how an anchor agent, called a mobility agent, can be used to provide hierarchical binding update when the mobile moves within a domain.

The mobile assigns two addresses: a *local care-of address* and a *global care-of address*. The first time the mobile moves to a domain, it sends two binding updates, one to the mobility agent with its local care-of address and one to its home agent with the address of the mobility agent, which is same as the global care-of address. Thus, any packet from the home agent is intercepted by the local mobility agent first. The local mobility agent first decapsulates the original packet, then encapsulates it again with the local care-of address, and then sends it to the mobile. For every subsequent move within the domain, the local binding update is sent to the anchor agent only and is not propagated to the home agent. Although this technique reduces the delay due to binding update, the traffic is subject to additional processing delay due to encapsulation and decapsulation at the mobility agent. Figure 6.22 shows the call flow when the MN first moves into a new domain managed by a mobility agent. Figure 6.23 shows the call flow during subsequent intradomain movement.

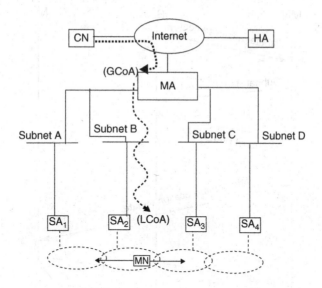

Figure 6.21   Functional architecture for hierarchical mobility agent

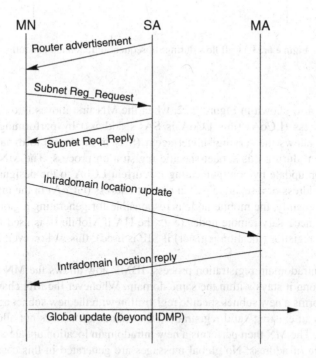

Figure 6.22   Initial intradomain location update

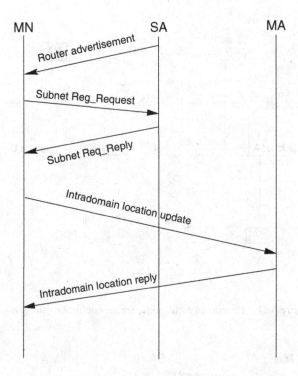

Figure 6.23 Call flow during subsequent intradomain handoff

According to the flow shown in Figure 6.22, when the MN first moves into a domain, it obtains a local care-of address (LCoA) (this LCoA is SA_2's address) by performing a subnet-specific registration. IDMP allows the serving subnet agent (SA_2 in this case) to dynamically assign the MN a mobility agent during this subnet-specific registration process. The MN then performs an intradomain location update by communicating its current LCoA to the designated MA. The MA includes either its address or a separate global care-of address (GCoA) in the intradomain location update reply. Subsequently, the mobile node is responsible for generating a global location update (registration) to the necessary remote nodes (e.g., the HA if Mobile IP is used for global mobility management or the registrar (location register) if SIP is used); this is, however, independent of the IDMP specifications.

After the initial intradomain registration process, IDMP now allows the MN to retain its global care-of address as long it stays within the same domain. Whenever the MN changes subnet within this domain, it performs a new subnet-specific registration with the new subnet agent. Since the MN indicates that it has an existing valid registration, the subnet agent does not allocate it a new MA address in this case. The MN then performs a new intradomain location update and informs its MA of its new local care-of address. No global messages are generated in this case, since the global care-of address remains unchanged. As with other hierarchical mobility management schemes, the localization of intradomain mobility significantly reduces the latency of handoffs across subnets within the same domain and also decreases the frequency of global signaling traffic.

Figure 6.23 shows the call flow during the subsequent intradomain movement.

6.7.3.2   Application Layer Anchoring-Agent-Assisted Technique

In the case of an application layer mobility protocol, we used a back-to-back SIP user agent (B2BUA) as the anchor point, possibly closer to the mobile node. A B2BUA consists of two SIP user agents where one user agent receives a SIP request, possibly transforms it, and then has the other part of the B2BUA reissue the request. A B2BUA in each domain needs to be addressed by the mobile host (MH) in the visited domain. The B2BUA issues a new request to the correspondent host (CH) containing its own address as the media destination and then forwards the packets, via RTP translation or NAT, to the MH. We have described the details of how a B2BUA can be used as an anchor point to reduce the binding update delay in Dutta et al. (2004c). Figure 6.24 illustrates the functional architecture of the use of a B2BUA to reduce the binding update delay. Figure 6.25 shows the detailed protocol flow for a B2BUA-based binding update.

6.7.4   Experimental Results and Analysis

This section describes the basic prototype implementation of IDMP. We first describe the functional components of the implementation. The mobility agent handles local registration requests from MNs that are currently in its domain, and provides temporary bindings to the MNs as long as they remain in that domain. As far as the handling of such registration (or location update) requests is concerned, there is little functional difference between the HA and the MA. Unlike the HA, which has a permanent list of mobility bindings for each MN associated with its home network, the MA maintains a dynamic list of mobility bindings for currently registered MNs. The major functional difference between the HA and the MA is in terms of packet forwarding to the MN. When the MN is away from the home network, the HA is responsible for collecting all of the packets directed to the MN's permanent IP address and tunneling those packets to the global care-of address (which is also the IP

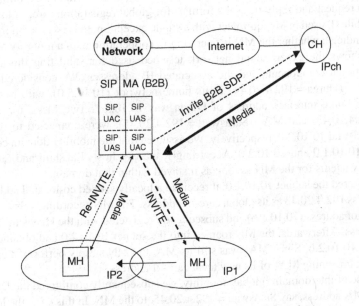

Figure 6.24   B2BUA-based hierarchical binding update

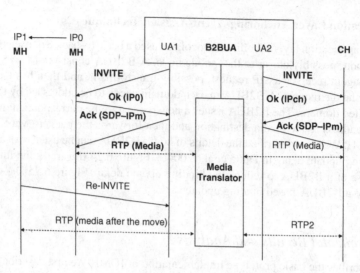

Figure 6.25　B2BUA-based flow for hierarchical binding update

address of the MA interface). The task of the MA is simpler; it receives the packets automatically, and after decapsulating the packets, redirects the inner IP packet to the MN's local care-of address.

In fact, the HA can potentially be unaware of the use of IDMP and the presence of the MA. As in conventional Mobile IP, it simply has to intercept all packets intended for the MN from the home network, encapsulate them, and forward them to the care-of address specified in the MN–HA registration message. We used MosquitoNet Mobile IP (Baker et al. 1996) in our experimental test bed. The registration request and reply message formats for global registrations are, in fact, identical to those in the Mobile IP used in MosquitoNet with a single exception: the reserved bit in the flags field is now used to indicate whether the MN is operating in cooperation with a mobility agent.

Figure 6.26 shows the experimental network test bed used for validating this mechanism. It shows the functional components and the associated IP addresses. We considered a single MN served by its HA (Durga = 192.4.20.44) in its home network (10.10.5.0), with home IP address 10.10.5.10. The home interface address of Durga was 10.10.5.1. Two MAs, for example MA_1 (Lakshmi = 192.4.20.43) and MA_2 (Saraswati = 192.4.20.45), were connected to routers serving subnets 10.10.1.0 and 10.10.2.0, respectively. We assumed that the mobility domain comprised both of the subnets 10.10.1.0 and 10.10.2.0. Accordingly, the two hosts Lakshmi and Saraswati could serve as mobility agents for the MN as long as it stayed within that domain.

As the MN entered the subnet 10.10.1.0, it received a locally scoped co-located address 10.10.1.6 and the IP address 192.4.20.43 as its global care-of address. The MN accordingly first informed MA_1 of its local care-of address (10.10.1.6) and subsequently registered with the HA using 192.4.20.43 as its care-of address. Afterwards, the MN roamed into the subnet 10.10.2.0 and obtained a new local care-of address 10.10.2.6. Since MA_1 was still its MA, the MN simply performed an intradomain location update, informing MA_1 of its new local care-of address.

To test the case of interdomain (global) mobility, we subsequently configured the DHCP server to provide a new MA address, say Saraswati = 192.4.20.45, to the MN. In this case, the MN performed both the intradomain and the interdomain registrations.

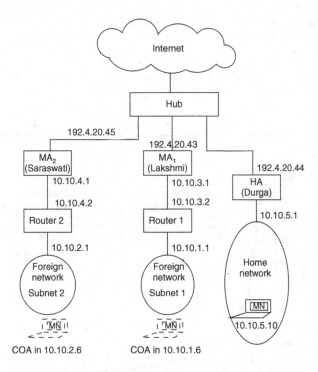

Figure 6.26 Experimental test bed for hierarchical mobility

6.7.4.1 Analysis of Results

In this section, we compare the signaling overhead associated with MA-assisted mobility management with that of base Mobile IP. We use the following parameters to express the signaling overhead:

- $L_g = 46$: size of global registration packet (in bytes).
- $L_1 = 50$: size of local registration packet (in bytes). (Note that $L_g \leq L_1$, since the global registration request does not contain the local care-of address field.)
- T_s: average duration for which the MN remains in a subnet (seconds/subnet).
- T_d: average duration for which the MN remains in a domain (seconds/domain).
- N: average number of subnets in a domain.
- $N_{MA} = 2$: average number of hops from the MN to the MA when the MN is in a foreign network.
- $N_{HA} = 5$: average number of hops from the MN to the HA when the MN is in a foreign network. (The numbers 2 and 5 are arbitrary.)

Clearly, T_s and T_d depend on the network and topology and on the mobility pattern of the MN. For the sake of simplicity, we assume $T_d = N \times T_s$ in our analysis. Table 6.4 displays the expressions for the signaling overhead in basic Mobile IP and under hierarchical mobility management involving an MA. In each expression, the factor of 2 is due to the fact that each registration attempt involves the exchange of a registration request and a corresponding reply message.

The global and local signaling overheads per hop in the MA-assisted hierarchical mobility architecture (referred to as DMA (dynamic mobility agent) in the figure)) are plotted against T_s for different

Table 6.4 Expressions for signaling overhead

Architecture	Signaling overhead (bytes/s)		
	Local per hop	Global per hop	Total in network
Mobile IP	0	$2L_g/T_s$	$2N_{HA}L_g/T_s$
Mobility agent	$2L_l/T_s$	$2L_g/T_d$	$2N_{HA}L_g/T_d + 2N_{MA}L_l/T_s$

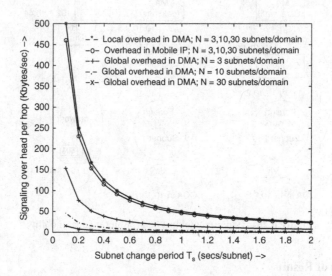

Figure 6.27   Global and local signaling overhead for IDMP

values of N (3, 10, and 30) in Figure 6.27. These numbers were chosen arbitrarily in order to reflect an increase in the number of subnets per domain. As expected, the global signaling overhead in the proposed architecture is significantly less than the corresponding local overhead. Also, the signaling overhead decreases as the MN stays longer in a subnet (and domain). As the number of subnets in a domain increases, the global signaling overhead reduces, whereas the local signaling overhead remains unchanged. In other words, the global signaling overhead in basic Mobile IP and the local overhead with a hierarchical MA do not depend on N. Since global signaling messages travel over a larger number of hops (and hence consume a larger portion of the network resources), hierarchical mobility management has an advantage over Mobile IP in terms of total network capacity (aggregated over all hops). From the plots in Figure 6.27, it is clear that hierarchical mobility management results in a significant reduction in the network signaling overhead, especially when mobiles change subnets more frequently and when a larger number of subnets form a single domain. As N_{HA} increases, the reduction in signaling overhead in the proposed scheme becomes more significant. For example, if we use hierarchical mobility management (with a DMA) in a network with 30 subnets per domain instead of 3 subnets per domain, the percentage gain in terms of signaling overhead is approximately 14% when the subnet mobility rate is kept constant.

6.7.5   Proactive Binding Update

We have designed a proactive binding update technique that works in conjunction with the preauthentication and proactive configuration techniques described earlier in Section 6.3 and Section 6.5, respectively. As part of the proposed technique, the mobile uses a proactive handover tunnel and sends the binding update proactively with the cached IP address that the mobile has obtained from the target network. Thus, any packet destined for the new address is picked up by the next access router and is tunneled to the mobile before it moves to the new point of attachment. After the mobile has moved to the new network, the IP–IP tunnel is deleted and the new IP address is assigned to the physical interface, but no new binding update is necessary (Dutta et al. 2010). This effectively removes the need to send another binding update after the mobile has moved to the new network. Thus, this technique completely eliminates the delay associated with the binding update. This technique uses the general optimization principle of reducing the number of signaling messages exchanged after the handoff to establish the new identifier. However, it introduces the additional complexity of managing a transient tunnel between the mobile and the router in the target network.

We have experimented with the proposed proactive binding update technique for both an application layer mobility protocol based on SIP and a network layer mobility protocol, namely MIPv6. In Chapter 11, we will describe how the proactive binding update mechanism works in conjunction with other handoff optimization techniques, namely preauthentication and proactive IP address acquisition, for both the network layer and the application layer mobility protocols.

6.8   Media Rerouting

In Chapter 3, we defined media rerouting as one of the handoff components. Media rerouting is the final step in the handover process before the data path is reestablished between the communicating nodes. The media rerouting process may include several elementary operations, such as encapsulation, decapsulation, tunneling, buffering, and store-and-forward. There is an overhead associated with the encapsulation, decapsulation, and tunneling operations. During the media rerouting process, transient data may be lost or delayed because of these operations. However, the in-flight data can be captured and redirected to the new point of attachment. Redirecting the transient media during handoff reduces the in-handoff packet loss. There are several ways the transient data can be forwarded from the CN to the mobile. Here, we describe a few candidate protocols or mechanisms that can be used to forward the traffic so that the loss of in-flight data is minimized.

In this section, we describe the key principles that should be considered when designing optimization techniques to reduce the media delivery delay and packet loss. We then highlight some related work where the media delivery was optimized to reduce the in-flight media delay and packet loss. We describe a few redirection techniques that we have developed based on these key principles. These are (1) data redirection using a forwarding agent, (2) mobility proxy-assisted time-bound data redirection, and (3) time-bound multicasting. These techniques help mitigate the effect of binding update delay by forwarding the in-flight data to the new point of attachment. Compared with the previous related work, the proposed techniques do not need any changes in the existing networking infrastructure. Finally, we demonstrate these techniques with some experimental results from our test bed.

6.8.1 Key Principles

The following are some of the key principles that help reduce the packet loss due to the redirection of in-flight data during handoff. We have applied some of these key principles in designing the optimization techniques presented here.

1. Simultaneous binding of both of the care-of addresses (the care-of addresses at the current network and the target network) to the home agent or CN reduces the data loss due to media redirection.
2. Forwarding of in-flight data from the previous network during handoff reduces the packet loss. Forwarding from the previous network can be done using either reactive tunnels[3] or an application layer forwarding technique. The forwarding technique helps to forward those packets from the previous network that would have been lost owing to the delay in the binding update. However, it cannot avoid packet loss completely (i.e., that due to L2 and L3 handoff delay). Nevertheless, it could be useful for low-latency applications, as there is no network buffer delay.
3. A combination of buffering and forwarding can be applied without doing a simultaneous binding update. Buffering techniques can be applied to any part of the network, such as the edge of the network, the core of the network, or the source.
4. Bicasting or localized multicasting of data at the edge of the network helps to reduce data loss.

6.8.2 Related Work

There is a relatively small amount of related work, namely RFC 4881 (Malki, 2007) and Perkins and Wang (1999) and Calhoun et al. (2000), on helping to reduce the transient data loss during a handoff by redirecting the data from the previous network. Koodli (2005) proposed reactive and proactive handover mechanisms that allow the in-flight data to be forwarded from the previous network and buffered in the target router, respectively. We have experimented with these techniques and have presented the results in Chapter 3. Vakil et al. (2001) designed a virtual soft-handoff method for CDMA-based wireless IP networks using a localized multicasting technique. However, this scheme works for CDMA networks only and does not provide a generalized solution suitable for other types of access network, such as 802.11. Tan et al. (1999) proposed a fast-handoff scheme for wireless networks that used a hierarchical mobility approach and used a multicasting technique to reduce packet loss during intradomain handoff. However, this approach requires that each mobile node is assigned a dedicated multicast address.

We have developed a few mechanisms that help to reduce the packet loss due to binding update delay during handoff. These techniques are protocol-independent and could be applicable to both network layer and application layer mobility protocols. These mechanisms are dependent upon packet-forwarding and localized multicasting techniques. Unlike the mechanisms proposed by Koodli (2005), these mechanisms do not need any cooperation between the previous access router and the next access router and work across administrative domains. Unlike the mechanisms proposed by Vakil et al. (2001) and Tan et al. (1999), the proposed localized multicasting technique is access-independent and does not need to assign a multicast address to each mobile node. We present these techniques in detail in the following sections.

---

[3] "Reactive handover tunnel" (RHT) is defined in the List of Abbreviations.

6.8.3 Data Redirection Using Forwarding Agent

The forwarding agent takes care of capturing the in-flight data during handoff and sends it to the new destination. The placement of the forwarding agent in the network determines the amount of in-flight data that can be forwarded during the handoff. In the most ideal scenario, it is best to place the forwarding agent as close as possible to the access networks.

Figure 6.28 shows a scenario where a mobile moves from network 1 to network 2. Forwarding of data from the current network (e.g., network 1) to the target network (e.g., network 2) can be established reactively by setting up a transient tunnel between the router in the current network and the mobile in the new network or by applying any application layer forwarding technique. Similarly, proactive forwarding of data from networks 2 and 1 is established by setting up a transient tunnel between router 2 and the mobile in the previous network. We categorize these types of transient tunnels into two basic categories, namely *proactive* and *reactive*. Depending upon the nature of the tunnel and the type of data forwarding, the tunnels can be defined as a proactive handover tunnel (PHT) or a reactive handover tunnel (RHT). We now define the functionalities of these tunnels in more detail.

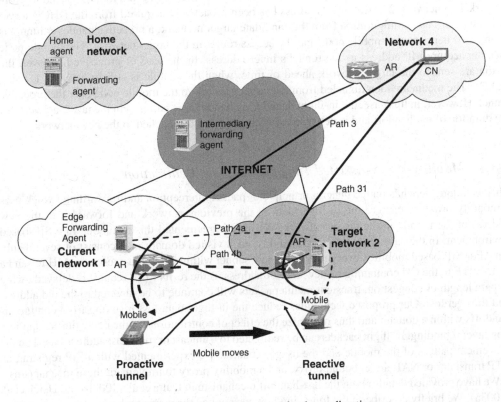

Figure 6.28 Forwarding agent for data redirection

6.8.3.1   Reactive Handover Tunnel

According to Figure 6.28, a reactive handover tunnel is established between the mobile and router 1 in the previous network after the mobile has moved to the new access network. The reactive handover tunnel helps to forward the in-flight data traffic from the previous network until a new path has been established between the mobile and the correspondent node. Path 4a in Figure 6.28 shows how the in-flight data is redirected from network 1 to network 2 until path 5 is established between the mobile and the correspondent node. In-flight data is sent over this tunnel and is received by the mobile in network 2.

6.8.3.2   Proactive Handover Tunnel

A proactive handover tunnel is established between the mobile in network 1 and the router in network 2 before the handover. Path 4b in Figure 6.28 shows the data that is redirected from network 2 to network 1 after path 3a has been established with the new network before the handover. This mechanism is useful during proactive handover. In such a case, data path 3a is established before the mobile has moved to network 2. In this case, data is forwarded from network 2 to network 1 while the mobile is still in network 1 and data is buffered at router 2 during the mobile's handoff from network 1 to network 2. After an IP address has been proactively acquired from the DHCP server or via stateless autoconfiguration from the candidate target network, a proactive handover tunnel is established between the mobile node and the access router in the target network. The mobile node uses the acquired IP address as the tunnel's inner address. In the case of proactive handover, the media are sent to the target network ahead of time, when the mobile is still in network 1, using path 3a. The media are then tunneled from the target network to the mobile node over the proactive tunnel. However, in this case, the in-flight data during handover is buffered in the target network for the duration of the handoff and is delivered after the mobile has attached to the new network.

6.8.4   Mobility-Proxy-Assisted Time-Bound Data Redirection

This technique depends on the general principle of packet interception and forwarding, which uses a mobility proxy to capture the in-flight data in the previous network and forward it to the new address of the mobile in the target network. We have implemented this technique in a SIP-based environment. In the case of intradomain mobility, each visited domain may consist of several subnets. For SIP-based mobility, every move to a new subnet within a domain causes the MH to send a re-INVITE to the CH containing its new care-of address. If the re-INVITE request is delayed owing to path length or congestion, transient media packets will continue to be directed to the old address and thus get lost. Our proposed techniques reduce the in-flight data loss resulting from continuous handoffs within a domain and thus minimize the effect of contributions to the delay during application layer rebinding. In-flight packets can be redirected to a unicast or multicast address based on the movement pattern of the mobile and the usage scenario. We experimented with a SIP registrar, an RTP translator or NAT, an outbound proxy, and a mobility proxy to implement these mechanisms.

We have provided details about the fast-handoff mechanism in Dutta et al. (2003b) and Hsieh et al. (2003a). We briefly describe in the following how we applied these two techniques.

Figure 6.29 shows the basic framework for mobility-proxy-assisted media redirection. In this specific framework, the visited network has an outbound proxy. We enhanced this proxy with the ability to temporarily register visitors (Schulzrinne, 2001). Here, the mobile node in the visited network

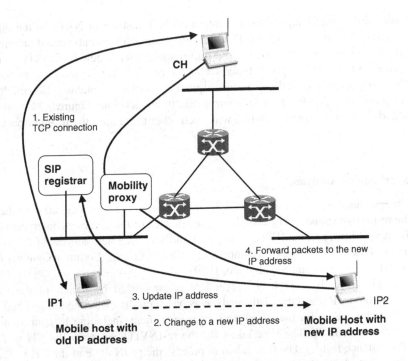

Figure 6.29 Media redirection using SIP-based mobility proxy

obtains a temporary, random identity from the visited network and uses it as its new address-of-record to register with the registrar in the visited network. The hierarchical registration speeds up the registration process, but does not address the "delayed binding update" issue that occurs when SIP's re-INVITE feature is used if the CH is very far away. We have taken care of the effect of delayed binding update using a mobility-proxy-assisted technique.

In the experiment, each subnet within a domain was equipped with a mobility proxy that had the ability to intercept a packet destined for the mobile's old address and forward it to the new destination. The RTP translator (http://www.cs.columbia.edu/IRT/software/rtptools/) (Schulzrinne et al. 2003) provided an application-layer forwarding technique that could forward the RTP packets for a given address and UDP port to another network destination. SIP requests typically traverse a SIP proxy in the visited network, the *outbound proxy*. As the mobile moves to a new network, it sends both re-REGISTER and re-INVITE messages via the outbound proxy. This outbound proxy can be configured as a visited registrar. Thus, the visited-network registrar receives the registration updates from the MH that has just moved, and immediately sends a request to the mobility proxy in the network that the MH just left. The request causes the mobility proxy to intercept the packets, and the RTP translator forwards any incoming packets to the new address of the MH. After a set time interval or after no media packets have been received by the RTP translator, the mobility proxy relinquishes this old address and removes the forwarding-table entry, assuming that the re-INVITE has reached the CH.

Alternatively, the outbound proxy can use the data in the MH-to-CH re-INVITE to configure the mobility proxy in the previous network. The advantage of this approach is that the outbound proxy usually has access to the Session Description Protocol (SDP) information containing the MH media address and port, thus simplifying the configuration of the translator or NAT. On the other hand, this outbound proxy has to remember the INVITE information for an unbounded amount of time and become call stateful, since it needs the old information when a new re-INVITE is issued by the MH. We verified the mobility-proxy-based technique by using two different tools, namely rtp-trans (http://www.cs.columbia.edu/IRT/software/rtptools/) and Linux iptables (Herrin, 2000), that help direct the transient traffic from the previous subnet to the new one. Figure 6.29 shows how the SIP registrar and the mobility proxy interact with each other to forward the in-flight packets to the new network.

6.8.4.1 Experimental Analysis

In the experimental test bed shown in Figure 6.30, RTP translators are associated with the mobility proxies in the respective subnets. The mobility proxy in each of these subnets intercepts the traffic meant for the mobile host, and the RTP translator forwards it to the new address of the mobile host after capturing it. This was achieved by a combination of SIP-CGI (SIP Common Gateway Interface) and SIP REGISTER (Lennox and Schulzrinne, 1999).

We now provide an analysis of how packet loss is minimized by means of the forwarding mechanism using the RTP translator. This analysis is based on the experimental test bed shown in Figure 6.30. The time taken for the complete subnet movement, including IP address acquisition and layer 2 movement, is T_s. The time taken for the re-INVITE to reach the CH is T_i (mostly decided by the distance factor). The time taken to process the re-INVITE at the CH is T_p, the time taken to register at the SIP proxy is T_g, and the time taken for the SIP registrar to forward a packet after capturing it is T_f. The packet generation rate at the CH is P_r packets per second. Thus, the

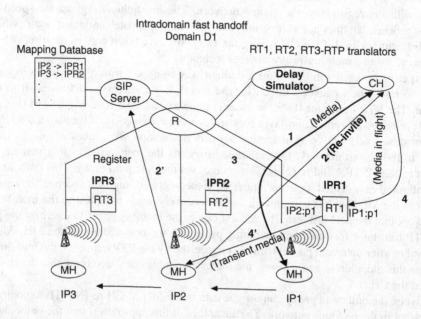

Figure 6.30 Experimental test bed for mobility proxy

total number of packets lost during a handoff using SIP registration and the RTP translator rtptrans is $P_{rt} = (T_s + T_g + T_f) \times P_r$.

As part of the experiment, we delayed the re-INVITE signal to simulate the distance between the CH and MH after the mobile had moved to the new network. The tools VIC and RAT (Sasse et al. 1995) were used to measure the delay performance of audio and video streaming traffic, respectively. We delayed the traversal of re-INVITE signals by 100, 200, and 500 ms and 1, 2, and 3 s to emulate the distance between the visited network and home network. This technique also shows how the RTP translator helps the media redirection and mitigates the packet loss during the mobile's movement.

In an earlier experiment (Dutta et al. 2001), we measured the processing time for the re-INVITE at the CH to be about 100 ms. Complete SIP registration took about 150 ms. It took about 200 ms to complete the subnet movement and IP address acquisition, including the layer 2 detection. In the current experiment, we measured the packet forwarding delay due to redirection at the registrar to be less than 1 ms when an iptables-based NAT approach was used, whereas an RTP translator approach added 4 ms of delay. The additional delay was due to the application layer redirection used by rtptrans. In the current 802.11-based experimental environment, the mobile lost about 15 packets owing to layer 2 delay, IP address acquisition delay, re-INVITE processing delay, registration, and packet forwarding delay.

Figure 6.31 compares the efficiency of the SIP-based optimized handoff approach using a combination of a mobility proxy and an RTP translator with a SIP-based mobility protocol without a fast-handoff technique. As the figure shows, the relative packet gain at the mobile node increases as the distance between the CH and MH, measured by number of hops, increases for a given packet generation rate.

6.8.5 Time-Bound Localized Multicasting

A locally scoped multicast technique allows in-flight data to be multicast to neighboring networks during handoff. This helps to avoid packet loss when the MH can predict that it is about to move to one of the new subnets within the neighboring network. We have applied this mechanism to reduce the packet loss during media delivery for both network layer and application layer mobility protocols. These mechanisms are described below.

6.8.5.1 Network-Layer-Based

We have applied the proposed technique to reduce packet loss for IDMP and have described the details in Das et al. (2002). Figure 6.32 shows the architecture in which we applied this time-bound multicasting mechanism to support fast handoff for network layer mobility.

In this case, the mobility agent encapsulates the unicast packets in a multicast address and sends them to the neighboring base stations proactively. The base stations buffer these packets and, upon the mobile's arrival, the unicast packets are delivered to the client. We now briefly describe below how fast handoff is achieved by applying these optimization techniques in the architecture shown in Figure 6.32.

It is assumed that a layer 2 trigger will be available (either to the MN or to the old BS) indicating an imminent change in connectivity. A layer 2 trigger is an indication from the lower layer to trigger an action in layer 3. As shown in Figure 6.33, the MN moves from SA_2 to SA_3. To minimize the service interruption during the handoff process, the mobile node or the old subnet agent (SA_2) generates a *MovementImminent* message to the MA serving the MN. Upon reception of this message, the MA

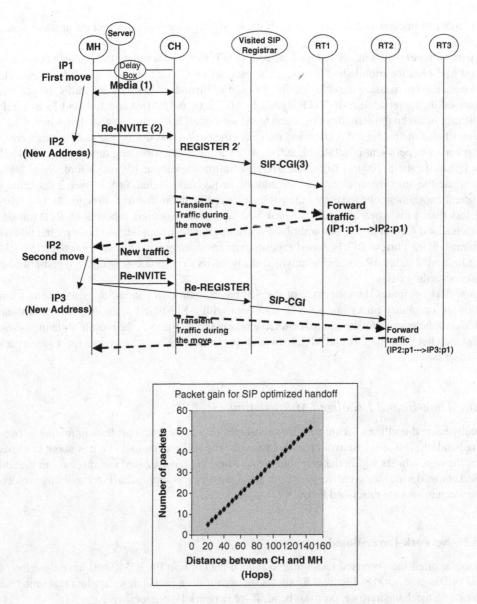

Figure 6.31   Reducing packet loss with localized media redirection

multicasts all inbound packets to the entire set of neighboring subnet agents (SA$_3$ and SA$_1$ in this case). Each of these candidate subnet agents buffers such arriving packets in per-MN buffers, thus minimizing the loss of in-flight packets during the handoff transient. When the MN subsequently performs a subnet-level registration with SA$_3$, this subnet agent (SA$_3$) can immediately buffer all such buffered packets over the wireless interface without waiting for the MA to receive the corresponding intradomain location update.

We now highlight some key benefits of the proposed techniques compared with other existing localized multicasting techniques.

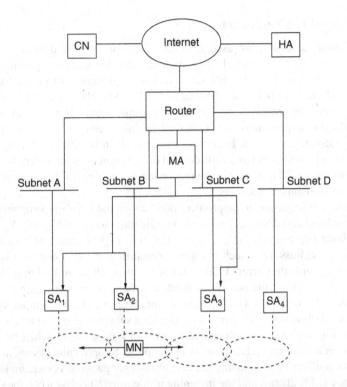

Figure 6.32   Time-bound multicasting for IDMP

The proposed technique uses a network-controlled, network-initiated, or mobile-initiated handoff technique. It is the MA that decides the set of target BSs to which in-flight packets are multicast. This is especially useful in scenarios where the MN may be in contact with multiple BSs and is unable to specify the future point of attachment exactly. While current cellular networks use a network-controlled handoff technique (where the base station controller (BSC) determines the candidate BS based on link-layer measurements supplied by the MN or BS), the IP mobility model is typically MN-driven, with the MN selecting a foreign agent (FA) from a list announced via agent advertisements. The proposed technique preserves the network-controlled handoff model for future IP-based cellular networks, without compromising the MN's ability to select such fast-handoff support.

Unlike other multicasting-based fast-handoff approaches, the proposed multicasting scheme prevents unnecessary wastage of wireless bandwidth, since base stations do not unilaterally transmit all arriving multicast packets over the wireless interface. Such proactively multicast packets are temporarily buffered by the BS in per-user buffers and forwarded to the MN over the wireless interface only if the mobile happens to register at that BS. If MN does not register at a particular subnet agent, the buffered packets are discarded after a specified maximum time interval.

We shall now briefly describe the approaches to implementation. We have described the details of the mechanism and its pros and cons compared with other approaches in Misra et al. (2002). The use of a locally scoped multicast is only effective if the MH can quickly acquire a multicast address and there is a multicast infrastructure available. Additional encapsulation overhead is an associated trade-off.

6.8.5.2   Fast-Handoff Implementation

For a prototype implementation, we used IP multicast to proactively distribute such packets to possible points of attachment. This mechanism requires only one multicast group per neighbor set; all BSs that are neighbors of a specific BS are members of this multicast group. Since a single BS can be a neighbor of multiple BSs, each BS can in fact be a member of multiple multicast groups. This approach does not require the establishment of dynamic multicast groups for individual MNs. The membership of the neighborhood set is also not dynamic: given a fixed network topology, the set of neighboring BSs stays constant. Each BS is thus permanently subscribed to one or more multicast groups, each of which always has a well-defined distribution tree. Accordingly, the fast-handoff scheme does not require a BS to dynamically join or leave a group, and hence does not suffer from any transient tree-establishment latencies.

Figure 6.33 shows the sequence of the protocol flow among the network components and how the packets are encapsulated and decapsulated as the mobile moves from SA_{old} to SA_{new}. On receiving a *MovementImminent* message, the MA encapsulates an in-flight packet and then tunnels it to the appropriate multicast address. (For such multicast forwarding, the MA does not perform the conventional tunneling towards the current LCoA.) On receiving such a tunneled multicast packet, each subnet agent first decapsulates the outermost header. It then buffers the decapsulated packet in a per-user buffer, using the destination address in the inner header (which is unique to a specific MN) as an index. When a mobile node subsequently performs a subnet-specific registration with a subnet agent (say SA_3 in Figure 6.32), the subnet agent can then forward any cached packets to the MN before the intradomain location update process is complete. Simple calculations indicate that even a small user buffer is effective in reducing the loss of in-flight packets. For example, if the intradomain update latency (L) is 200 ms and the incoming traffic rate (R) is 144 Kb/s, then a buffer size of

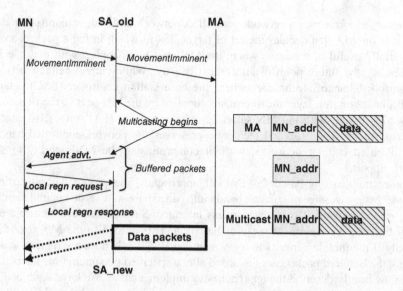

Figure 6.33   Flow of scope-based multicast data redirection

$(L * R) = 3.6$ Kbytes is able to protect against buffer overflow due to multicast packets transmitted during the handoff.

6.8.5.3   Application-Layer-Based Approach

In the case of application layer mobility, the mobile informs the visited registrar or B2BUA of the temporary multicast address of its contact address in the SDP. Once the MH has arrived in its new subnet, it updates the registrar or B2BUA with its new unicast address, while it continues to receive the in-flight data over the multicast address. We have described the details of how a locally scoped multicast address can be used to reduce packet loss during handoff in Dutta et al. (2004c). Multicast agents may be co-located with the first-hop router or coexist with the B2BUA or SIP proxy. Using scoped multicast is only effective if the MH can quickly acquire a multicast address that can be used as part of the SDP to update the B2BUA. Figure 6.34 shows how this forwarding technique can be applied to support fast handoff using application layer mobility. In this case, the mobile sends a re-INVITE to the B2BUA with the locally scoped multicast address in the SDP when the handoff is imminent. Thus, media traffic is multicast to both of the neighboring subnets. After the handover to the new network, the mobile sends a new re-INVITE with the unicast address in the SDP. This unicast address is the care-of address of the mobile after it has moved to the new network. Thus, the effect of binding update delay is minimized by redirection of the in-flight packets during handoff by way of reducing the packet loss.

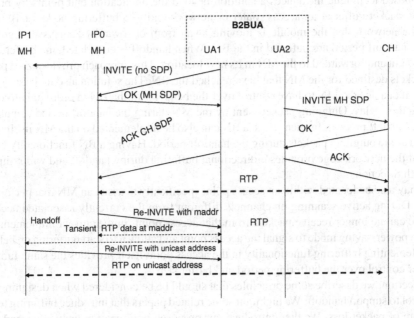

Figure 6.34   Data redirection using a multicast agent

There is a likelihood that duplicate packets will be received during the mobile's movement between subnets. RTP packets have their own sequence numbers associated with them, and thus these packets can be reordered. However, mechanisms similar to those described by Perkins and Wang (1999) can be adopted to take care of duplicate non-RTP-based traffic.

6.9   Media Buffering

In Chapter 3, we introduced buffering as one of the subprocesses of the media-routing process during handoff that help to reduce the packet loss at the cost of added packet delivery delay. Although media redirection techniques help redirect the in-flight packets caused by delayed binding update, some packets may be lost during link-layer handover. Thus, it is essential to mitigate the effect of the link-layer handover delay by reducing the packet loss. Bicasting or buffering the transient packets in different parts of the network can be applied to minimize or eliminate packet loss. However, bicasting alone cannot eliminate packet loss if the link-layer handover is not seamless. Although buffering can mitigate the effect of layer 2 handoff by reducing or eliminating packet loss, it introduces an additional one-way delay for the in-flight packets. While this additional end-to-end delay may not affect streaming traffic, interactive traffic such as VoIP applications cannot tolerate the jitter resulting from variable one-way delay. An ability to control the buffer dynamically provides a reasonable trade-off between delay and packet loss that is within the threshold limit for supporting real-time communication. We have developed a dynamic buffer control protocol (BCP) (Dutta et al. 2006e) that can provide a dynamic buffering mechanism based on the duration of the handoff and on the placement of buffers at the edges of the networks.

The proposed technique introduces a solution beyond the application end points by providing a per-mobile packet buffer at an access router or network entity (a buffering node, or BN) near the edge of the network that the mobile is moving away from or towards. Packets that are in flight during the handoff period are buffered in the BN. When handoff is completed, the buffered packets are flushed out and forwarded to the MN in its new location. This approach provides zero packet loss for all packets destined for the MN that have reached the BN. The solution also includes a buffering scheme that enables the MN to have control over the behavior of the BN in order to help reduce the overall handoff delay. Outgoing packets sent by the MN during the handoff period can also be lost during the handoff process. In such cases, a BN can also be implemented on the MN itself to provide buffering for the outgoing packets during the handoff period. Having a BN functionality both in the MN and at the network edge provides bidirectional buffering during handoff and will reduce packet loss in both directions.

A BN may also be located within an access point specifically to assist an MN that performs active scanning. During active scanning on channels different from the currently associated access point, the mobile can no longer receive packets from that access point. In the current implementation, the MN uses a power-saving mode to signal the access point and allow it to start buffering on behalf of the MN. Implementing buffering functionality in the access point itself provides the same functionality with better control over the buffering period and buffer size.

In this section, we describe some principles that should be considered when designing a buffering protocol to support handoff. We highlight some related papers that introduce buffering techniques to take care of packet loss. We then introduce our proposed optimization techniques and elaborate on a dynamic buffer control protocol that reduces the packet loss at the cost of added delay. Finally, we present experimental results obtained using two different buffer control approaches that we have proposed.

6.9.1 Key Principles

The following are the key principles that need to be considered in designing a buffer control protocol for handoff of a mobile:

1. Buffering in-flight packets during handoff can eliminate packet loss, inclusive of the effects of layer 2 handoff delay.
2. The added delay due to network buffering may not be suitable for low-latency applications, as "delayed packets" beyond a certain threshold are considered to be lost. However, buffering can help TCP-type traffic without compromising the data throughput and streaming traffic (e.g., IPTV).
3. The buffering period can be adjusted based on the handoff interval.
4. Media can be buffered at any part of the network, such as at the source, at the edge routers, in the core network, or at the mobile.
5. In most cases, buffering is useful for proactive handoff, where packets are buffered before the handoff begins and are flushed after the handoff is over.
6. The packet generation rate at the source, the handoff period, the time taken to signal the buffer to be flushed, and the packet transmission time are some of the parameters that affect the optimal buffer length at the edge router.
7. While the overall buffering period is influenced by the handoff delay, buffering affects the end-to-end delay, the number of packets delayed, and the jitter.
8. The jitter due to buffering of packets at the router node can be compensated by use of a playout buffer at the mobile.

6.9.2 Related Work

Moore (2004) and Krishnamurthi et al. (2001) developed buffering techniques for Mobile IPv6. Khalil et al. (1999) described a Mobile IPv4 buffering protocol that resembles the proposed method described here. These proposals define extensions to the Mobile IPv4 and Mobile IPv6 protocols to support buffering in the network during a handover period. Moore (2004) described the use of a P-bit in the mobility header of the binding update and LBU (local binding update) messages. The proposal by Krishnamurthi et al. (2001) is very similar to the proposal by Khalil et al. (1999). However, that technique adds a discovery feature to provide buffering capability and takes advantage of IPv6 router advertisements to check the buffering capability of a network.

There are alternative mechanisms that reduce packet loss without the use of any buffer management protocol but depend heavily on the cooperation of the end clients. Most multimedia applications resort to playout buffers, FEC (forward error correction) (Rosenberg and Schulzrinne, 1999), RTCP-based feedback (Ott et al. 2006), and other stream repair techniques (Perkins and Hodson, 1998) to minimize the effects of packet loss or to reduce jitter. However, the existing end-system-assisted solutions may not be appropriate to a wireless medium where the bandwidth is limited and the end hosts are separated by a long distance. These mechanisms have not been applied to take care of packet loss during handoff.

In layer 2, there is an existing method that uses the power management functionality of IEEE 802.11 to avoid packet loss while the MN is actively scanning (Roshan and Leary, 2003) channels. In this method, the MN signals to the current access point that it is entering sleep mode, and the access point attempts to buffer packets for the MN until the MN wakes up. However, this method cannot be used to buffer packets during a handover, because the method assumes that the MN will continue to be associated with the access point after it has woken up in order to stop buffering, and its applicability is limited because the method does not carry additional information such as traffic flow identification information, the buffer size, and the buffering period, which might be required to meet particular quality-of-service requirements of the mobile.

The existing proposals are tightly coupled to specific mobility management protocols such as Mobile IPv4 and Mobile IPv6. In contrast, the proposed buffering method can work with any mobility management protocol by allowing the buffering control mechanism to be defined as a separate protocol. In the existing proposals, the location of buffering nodes is limited to mobility agents such as home agents and mobility anchor points. In contrast, the proposed method provides more flexibility in the location of buffering nodes. In the existing proposals, the forwarding of buffered packets to the mobile node after completion of the handover period depends on the forwarding behavior of the mobility agent that is part of the mobility protocol. In contrast, the proposed technique defines its own mechanism for establishment of the tunnel used for forwarding buffered packets to the mobile node, to provide complete independence from the mobility management protocol. The proposed method also defines detailed queuing and forwarding mechanisms for the buffered packets, and the detailed behavior in error situations is defined, whereas such details are missing in the existing proposals.

6.9.3   Protocol for Edge Buffering

We have designed and implemented a dynamic buffering scheme (Dutta et al. 2006e) that ensures zero packet loss at the cost of an additional end-to-end delay that can be controlled dynamically. Figure 6.35 shows four different scenarios that illustrate how buffering techniques can be applied in different parts of the network. The figure shows how buffering can be applied in the previous access router, in the next access router, at the source, or at the destination. The buffering scheme can be used in conjunction with existing mobility protocols, or as an independent network or link-layer access mechanism.

The ability to control the buffer dynamically provides a reasonable trade-off between delay and packet loss within the threshold limits for real-time communication. We have experimented with two kinds of buffering schemes, namely *time-limited buffering* and *explicit signaling buffering*. In the time-limited buffering technique, the mobile node and buffering node can negotiate a buffering period that is conveyed to the buffering node during the initial setup signal. The buffering node buffers the packets for the duration of the buffering period as defined in the initial control message. In the case of explicit signal buffering, the buffering period is equivalent to the total handoff period and the additional time taken to flush the buffer after the handoff has taken place. We have experimented with both types of buffering schemes, using different traffic rates and buffering periods in conjunction with our Media Independent Preauthentication mechanism. The average interpacket delay during handoff, the packet loss, and the average number of packets buffered were calculated for each case. In the case of explicit signaling, the number of packets buffered was dependent upon the handoff delay and packet generation rate. Time-limited buffering, on the other hand, introduced a higher probability of packet loss resulting from buffer overflow.

6.9.3.1   Protocol Details

In this section, we briefly describe the details of the protocol for the dynamic buffering mechanism. The buffer control protocol is used by the MN to request buffering services at the BN. This is a

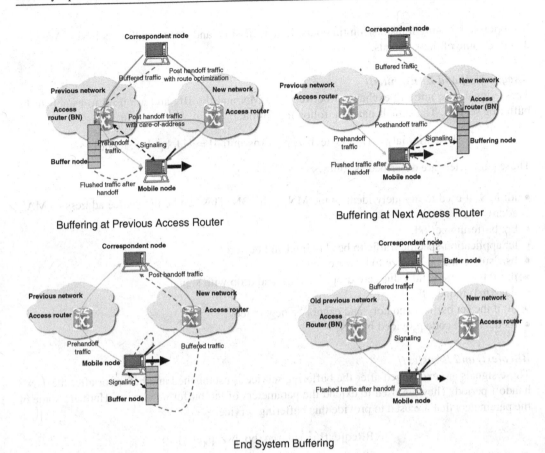

Figure 6.35 Alternative ways of using buffering

simple and reliable messaging system composed of pairs of request and answer signals. The BCP may be defined as a new protocol or as an extension to an existing protocol such as PANA, SIP, Mobile IPv4, Mobile IPv6, or a link layer protocol. For example, in Mobile IPv6, it may be possible to define a new mobility option in the binding update or acknowledgement message exchange that carries the BCP in TLV format. In PANA (Forsberg et al. 2008), it is possible to define BCP AVPs that can be appended to the PUR (PANA update request) and PUA (PANA update response) message exchange. Other methods may be employed as long as the requirements of the BCP signaling can be accommodated. In all these cases, the delivery and encoding of BCP signals may become specific to the protocol that carries the BCP.

Signaling messages for BCP

In the following, we describe some of the signaling messages that are used to take care of buffering. As a rule, request signals are sent from the MN to the BN and answer signals are sent by the BN to the MN in response to a request signal. The BN should never generate a request signal. Request signals carry parameters regarding the request, and answer signals contain result codes. Reliability is supported by using transmission timeouts, retransmission, and error-handling behavior. We now describe some of these signals.

BReq[initial] and BAns[initial]

These signals are initially exchanged between the MN and the BN and are used to establish the buffering service. The signals have the following format:

$$BReq[initial] = \{id, bp, tc, bsz, p\} BAns[initial] = \{id, bp, bsz, rcode\}$$

These parameters are described as follows:

- id: MN id used to uniquely identify the MN to the BN. This can be the source address or MAC address of the MN.
- bp: buffering period.
- tc: application-specific traffic to be classified and buffered.
- bsz: suggested buffer size to be allocated.
- p: FP for EOS; valid values are drop, forward, and drop with signal.
- flag {m}: request flags.
- m: if the set bsz is mandatory and cannot be negotiated.
- rcode: result code provided by BN.

BReq[ext] and BAns[ext]

These signals are exchanged after the buffering service is established and before or after the MN's handoff period. They are used to extend the parameters of the buffering service. Here are some of the parameters that are used to provide this buffering service:

$$BReq[ext] = \{id, seq, bp, bsz, p, coa\}$$

- id: MN id sent in BReq[initial].
- seq: signal sequence number.
- bp: additional buffering period, maybe zero (0).
- bsz: additional buffer size, maybe zero (0).
- p: new FP for EOS; valid values are drop, forward, and drop with signal.
- coa: current CoA of the MN.

$$BAns[ext] = \{id, seq, bp, bsz, rcode\}$$

- id: MN id sent in BReq[initial].
- seq: signal sequence number, must match BReq[ext].
- hp: new buffering period for this service.
- bsz: new buffer size allocated for this service.
- rcode: result code provided by BN.

BReq[stop] and BAns[stop]

These signals are exchanged to stop the buffering service. The following are some of the parameters:

$$BReq[stop] = \{id, p, coa\}$$

- id: MN id sent in BReq[initial].
- p: termination FP for EOS; valid values are drop, forward, and drop with signal.
- coa: current CoA of the MN.

$$BAns[stop] = \{id, rcode\}$$

- id: MN id sent in BReq[initial].
- rcode: result code provided by BN.

Service Attributes

The BCP also creates service attributes (state information) in the BN. These attributes include the following:

1. MN id (id).
2. Buffering period (bp).
3. Negotiated buffer size (bsz).
4. Traffic classification (tc) parameter.
5. FP, current EOS flushing policy.
6. Last extension request sequence number.
7. Current MN CoA.
8. Previous MN CoA.

The attributes should be allocated during the request phase. The values of the attributes are updated by the BN upon receiving valid request signals or other local events. The lifetime of the attributes is limited to the duration of the service. If a positive or negative EOS is met with, the BN should release resources occupied by these attributes.

The protocol flow shown in Figure 6.36 provides a general view of the sequence of the signal exchanges as it relates to the MN's handoff, traffic classification, and buffering period. The variables $o1$ and $o2$ define the buffering overlap periods before and after the handoff, although the MN and BN may still have connectivity. This will happen if the packets are buffered before the handoff starts and continue to be buffered throughout the handoff period and after the mobile's handoff.

6.9.4   Experimental Results and Analysis

In this section, we describe the experimental results for two types of techniques, *time-limited buffering* and *explicit buffering*. We used this mechanism in conjunction with the Media Independent Preauthentication technique, where the target access router was used as the buffering node. Figure 6.37 shows a typical scenario in which we experimented with this technique using Media Independent Preauthentication. Figure 6.38 shows the protocol flow when PANA was used as the buffer control protocol.

Table 6.5 and Table 6.6 show values of the average packet delay, the average packet loss during handoff, and the average number of packets buffered for the time-limited and explicit buffering techniques, respectively. The average packet delay is defined as the delay between the last packet before the handoff and the first packet after the handoff. The value of x in these tables represents the inherent buffer delay and is assumed to be zero for simplicity in this case, for both time-limited buffering and explicit buffering. The value of y is based on the handover period (hp) and was varied

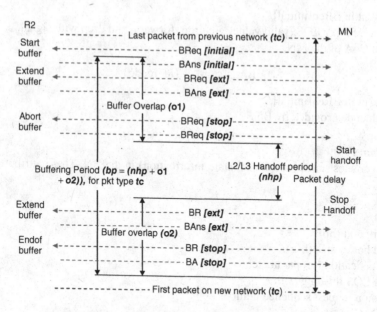

Figure 6.36 Protocol flow for buffer control protocol

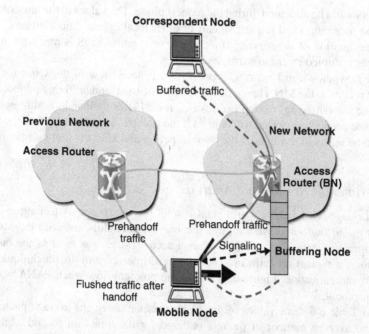

Figure 6.37 Buffering with Media Independent Preauthentication

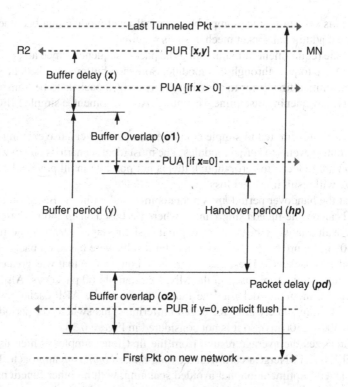

Figure 6.38   Protocol flow using PANA as the BCP

Table 6.5   Results for time-limited buffering

Traffic rate (pkts/s)	x (ms)	y (ms)	Packet delay (ms)	Packet loss	Average packets buffered
70	0	N/A	37	0	2.5
80	0	N/A	42	0	3
90	0	N/A	44	0	3.5
100	0	N/A	45	0	4

in the time-limited-buffering experiment. The value of y is not relevant to the explicit-buffering experiment.

The current solution was implemented using a kernel queue module that hooks into Linux netfilter's QUEUE handler. The new module is called "ipmparb" (IPv4 MPA router buffer). This module has the following advantages:

1. Packet classification is done by iptables, so the module is much simpler. It simply relies on iptable's packet classifier with ipmparb as the target.
2. Implementation is efficient, since packets are routed to the module in *skbuff* objects and so no copying is done. ipmparb simply queues the *skbuff* objects without modification.

3. Implementation is very fast, since ipmparb is a kernel-level module that becomes part of the IP routing stack. No additional socket mechanism is required.

4. It easily meets the requirement of maintaining the packet sequence, since all packets that need to be be buffered have to pass through this module. So, when buffered packets need to be flushed, they can be transmitted first, prior to allowing newly arrived packets to be transmitted.

5. ipmparb can use any queuing discipline we require. At the moment, a simple FIFO queue is used.

User-level interaction is limited to simple control events, so one can use existing user-level commands that pass control events to kernel modules. The most ideal scenario is one in which the overlap periods $o1$ and $o2$ are reduced to zero, though this is not possible in all practical cases. A negative value for $o1$ or $o2$ will result in packet loss.

This means that the handover period hp is not encompassed within y. Based on our experimental results, $o1$ and $o2$ can be fine-tuned using x and y, where y is based on hp and x is based on the average round trip time. An alternative is the use of an explicit flush message instead of fine-tuning y (denoted by "PUR if $y = 0$" in Figure 6.38). The experimental results were based on packet generation rates of 70, 80, 90, and 100 packets/s. These rates were based on a value that was greater than the codec rate of RAT (Robust Audio Tool) used in the MN, for example 60 packets/s. Also, the switchover process to router R2, including deleting the tunnel, updating the ARP cache, and so on, always occurred during $o1$, immediately after R2 began buffering. This switchover period was very small (average about 0.300–0.500 ms), so it is not considered in Figure 6.38.

Table 6.5 summarizes the average results from the first four samples, which used a MADWiFi driver and an IEEE 802.11 Netgear card with the *time-limited buffering* approach. The modification involved only a layer 2 optimization that avoided scanning, with no other functional changes. Four different packet generation rates were used in the experiments. The average handover period was 10–16 ms. This includes delays related to layers 2 and 3 (L2 and L3) that happened in sequence. The average L2 delay was 4–8 ms, followed by an average L3 delay of 6–8 ms. Since the IP address was obtained beforehand, the layer-3-related delay was the delay associated with assigning the previously obtained IP address to its physical interface. The bulk of the handoff delay was avoided because many of the handoff operations were done proactively. The buffering node in the experiment was the next

Table 6.6 Results for explicit buffering

Traffic rate (pkts/s)	x (ms)	y (ms)	Packet delay (ms)	Packet loss	Packets buffered
70	0	30	27	0	19
	0	20	29	0	3
	0	10	12	0	1
80	0	30	28	0	3
	0	20	33	0	3
	0	10	15	0	1
90	0	30	69	0	3
	0	20	46	0	3
	0	10	11	3	1
100	0	30	69	0	10
	0	20	46	0	4
	0	10	11	3	1

access router, which was discovered before the mobile moved to the new network. Because the value of *hp* was very small, *y* could be fine-tuned to its minimum of 10 ms without incurring packet loss. The value of *x* was kept to zero, as it introduces additional delay. When we used RAT as the media agent, it generated an average of 60 pkts/s, and there was no discernible loss in the audio sample and almost always no packet loss.

Table 6.6 shows the experimental results for the same environment (MADWiFi driver and IEEE 802.11 Netgear card) but using the *explicit buffering* approach. The average value of *hp* was 10–16 ms. When explicit signaling between the MN and R2 is used to flush the buffer, it is guaranteed that no packet loss will occur, in contrast to using a value of *y* that gives the possibility of having *o2* < 0, resulting in packet loss. The price is additional delay. As an example, when an ideal *y* value was used at 70 pkts/s, the average delay was only 12 ms, as compared with the case of explicit signaling, for which the average delay was 36 ms. Similarly to the case of time-limited buffering, the experimental results obtained using RAT as a medium did not show any discernible loss in the audio, and it was guaranteed that there would be no packet loss.

The detailed breakdown of the handoff delay in an experiment with 100 packets/s traffic rate, 1024 bytes packet size, and the explicit signaling method is as follows. The total handoff (handover) delay was about 12.5 s. The L2 delay with the MADWiFi driver was about 4.8 ms, and the L3 configuration time was about 0.5 ms. The processing delay (a) for the buffer request at the PAA was 5.699 ms, and the switchover period at the PAA (b) was 0.46 ms, which included tunnel setup and *ioctl* calls. The processing delay at the MN (c) for sending the stop request was 6.788 ms. The processing delay (d) for flush packets from the buffer was 4.626 ms. The flushing period (e) at the PAA was 0.205 ms. Thus the total handover delay *hp* was equal to (L2/L3/(c)) + (d) + (e). Thus, it is apparent that the handover period was a fraction of the total packet delay incurred. The explicit signaling method adds to the total packet delay because of the delay associated with the flushing, whereas time-limited signaling increases the probability of packet loss.

6.9.5 Analysis of the Trade-off between Buffering Delay and Packet Loss

The end-to-end delay of any specific packet and the delay between the last packet at the previous point of attachment and first packet at the new point of attachment are most important. Although the buffering mechanism reduces the packet loss, it also increases both the end-to-end packet delay for the in-flight packets and the handoff delay. A packet that would otherwise have been lost during handoff is buffered in the buffering node for a certain period of time, which is determined by the handoff delay and the time taken to flush packets from the queue at the new point of attachment. Yemini (1983) provided a trade-off analysis between delay and packet loss. That paper also stressed the fact that as soon as the buffer threshold is exceeded, any newly arriving packets cause the first packet in the queue to be lost. Although that paper focused on the queue at the sender side, the analysis in it is applicable in general terms to the theory of the buffering protocol described here. In the case of *explicit signal buffering*, the buffering period is equivalent to the total handoff period and the additional time taken to flush the buffer after the handoff has taken place. The total number of packets stored in the network buffer depends on the buffer length, the transmission time, and the packet generation rate at the source. Packets arrive in the buffer at regular intervals. But when these packets are flushed out of the buffer after the handoff, all of the buffered packets are flushed out at the same time without any interpacket gaps. Although this avoids packet loss, the mobile is subjected to a spike, since these packets arrive at the mobile almost instantaneously. The in-handoff packets that are buffered in the edge router are subject to an increased amount of delay compared with the

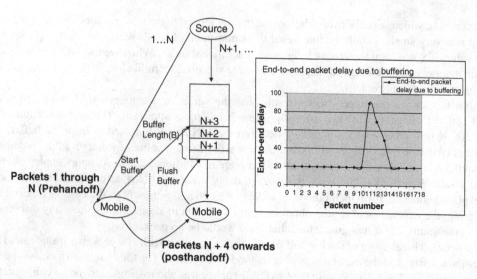

Figure 6.39   Effect of edge buffering on in-handoff packets

prehandoff and posthandoff packets. But each consecutive in-handoff packet is subject to a different amount of delay, since these packets spend different amounts of time in the buffer. Later packets are subject to less delay than the packets that got in first.

The end-to-end delay for in-handoff packets, the delay between the last packet in the old network and the first packet in the new network, and the total number of packets affected by the handoff are the important parameters that need to be considered when one wishes to support real-time communication. The packet generation rate at the source, the handoff period, the time taken to signal the buffer to flush, and the packet transmission time are some of the parameters that determine the optimal buffer length at the router. The overall buffering period is influenced by the handoff delay, and it affects the end-to-end delay, the number of packets delayed, and the jitter. However, the jitter due to buffering at the router node that is observed can be compensated by a playout buffer at the mobile. Figure 6.39 shows how packets are affected during the handoff because of the buffering at the edge router. This shows that during handoff, the packets are subject to jitter because of the increased end-to-end delay due to buffering.

6.10   Route Optimization

In Chapter 3, we defined the different subprocesses that are part of the media rerouting process and illustrated how these affect the handoff delay and packet loss during a mobility event. Triangular routing and the encapsulation and decapsulation processes associated with any mobility protocol affect the performance of real-time traffic, owing to the associated transport delay due to signaling, data, and encapsulation overhead. Route optimization is the process of optimizing the route between the communicating hosts by eliminating encapsulation and decapsulation processes and maintaining a direct route.

In this section, we first describe the key principles that need to be considered for route optimization when one is optimizing packet transport delay and data overhead. Then, we describe related work that describes route optimization techniques for several different mobility protocols. We then describe four route optimization mechanisms that we have developed based on some of these principles, and present some experimental results. These mechanisms optimize the signaling and data traversal by maintaining a direct path between the communicating hosts, and avoid any associated encapsulation overhead. Since Mobile IP inherently suffers from this route optimization problem, we experimented with a few of these optimization techniques; we compare the results with those obtained with MIPv4.

6.10.1 Key Principles

The following are some of the key principles that can be applied to optimize the data path between the end hosts, thereby contributing to reduced end-to-end delay:

1. Maintaining a direct path between the communicating hosts. If a protocol allows the mobile hosts to update each others' identifiers directly without the help of an anchor agent in the middle of the network, it reduces the data traversal path after the handoff.
2. Limiting the media traversal between the communicating hosts within the local domain when both of the communicating hosts are away from home and are visiting in the same domain.
3. Splitting the media and signaling path to avoid data transmission via the home network. Only the mobility-related signaling, such as binding updates, is sent to the home agent, while media traversal is limited to the local domain.
4. Modifying the source and destination addresses at the end host before the data is passed on to the application at the end host. This helps to maintain a direct path.
5. A dual anchoring mechanism that allows the mobile to use different addresses for signaling and media. This method can be applied in IPv6 networks and is applicable when both of the communicating nodes are away from home. Although signaling needs to travel to the home network, it is important to confine the traversal of media to the visited network by avoiding traversal of the longer path to the home network.

6.10.2 Related Work

The IETF has addressed these issues by proposing route optimization support for MIPv6 (Johnson et al. 2004). However, route optimization for MIPv4 was never standardized, although various forms of route optimization have been proposed by others (Wu et al. 2002). Recently, the NETEXT (Network-Based Mobility Extensions) working group in the IETF has included route optimization as one of its working-group charter items and is defining a problem statement and solutions for route optimization for Proxy MIPv6.

We describe below a few route optimization techniques based on the key principles presented earlier, and present the associated experimental results.

6.10.3 Maintaining a Direct Path by Application Layer Mobility

SIP-based terminal mobility (Wedlund and Schulzrinne, 1999) performs binding updates by application layer signaling. It reduces the one-way packet delay by avoiding the triangular routing that is inherently present in Mobile IPv4 (Perkins, 2002b). Both SIP-based mobility and MIPv4 were briefly introduced in Chapter 2 as part of an introduction to mobility protocols. We have augmented the

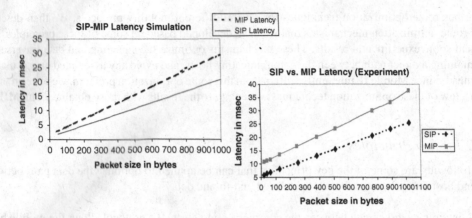

Figure 6.40 One-way data transfer delay for SIP and MIP-based protocols

SIP-based terminal mobility with a complete set of handoff operations to support subnet and domain mobility (Dutta et al. 2001) and compared the effects of triangular routing on packet transmission.

Initially, we compared the latency for SIP-based mobility and Mobile IP using different packet sizes during a subnet handoff. By using a SIP-based mobility protocol, we could obtain a 50% one-way latency improvement for real-time (RTP/UDP) traffic, thus providing a reduction in latency from a baseline of approximately 27 ms to approximately 16 ms for large packets and a 35% increase in utilization due to the avoidance of additional IP-in-IP encapsulation. Figure 6.40 shows the reduction in end-to-end packet delay for SIP compared with MIP for various data packet sizes as obtained from NS2-based simulations (Simulator, 2005) and laboratory experiments. The simulation results demonstrate how a direct signaling path between the CH and MH minimizes end-to-end delay of the data packet and reduces data overhead. These experimental results demonstrate that by maintaining a direct signaling path between the MH and CH, the end-to-end delay of the data packet and the packet loss can be optimized for secured interdomain mobility.

6.10.4   Interceptor-Assisted Packet Modifier at the End Point

Mobile IP with Location Registers (MIP-LR) (Jain et al. 1999) allows the mobile node to register with multiple location registers and avoids triangular routing of the data. We augmented the original MIP-LR with application layer modules, namely a packet interceptor and a packet modifier. The packet interceptor and modifier modules on the sender and receiver sides cooperated with each other to provide route optimization by sending the data directly to the mobile node, and reduced the data overhead by avoiding tunnels. (Interceptor modules intercept outgoing packets, and packet modifier modules change the destination address of the packet on the sender side before it is transmitted.) The packet modifier module on the receiver side changed the destination address back to the permanent address of the mobile before sending it to the application. This way, the application was not aware of the underlying IP address change, and the packets from the correspondent host were sent directly to the mobile's new point of attachment. From the experimental results obtained from our test bed (Dutta et al. 2005a), we verified that one can attain up to 50% reduction in management overhead and up to 40% improvement in end-to-end delay compared with standard Mobile IP in co-located mode.

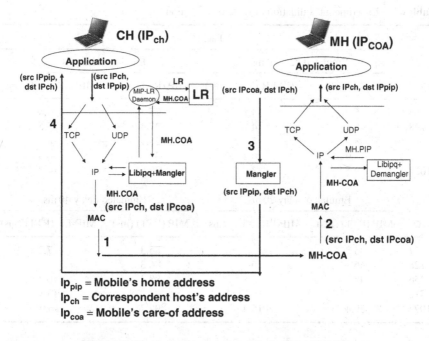

Figure 6.41 Packet interception technique for MIP-LR

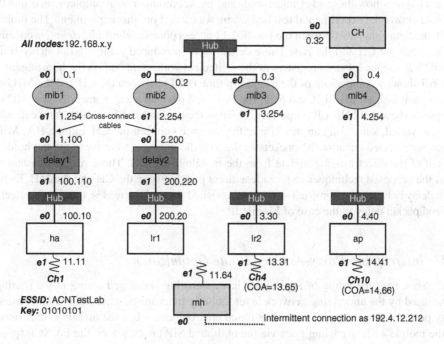

Figure 6.42 Experimental test bed for MIP-LR

Table 6.7 Experimental validation of route optimization

	Packet size			
	Emulated delay 0 ms		Emulated delay 10 ms	
Bytes	MIP (RTT) (ms)	MIP-LR (RTT) (ms)	MIP (RTT) (ms)	MIP-LR (RTT) (ms)
64	5.9	4.1	25.3	5.4
128	6.6	4.7	27.4	5.6
256	8	5.9	28.1	6.2
512	13.9	10.2	32	10.8
1024	19.5	13	39.5	14.2
	Emulated delay 20 ms		Emulated delay 40 ms	
Bytes	MIP (RTT) (ms)	MIP-LR (RTT) (ms)	MIP (RTT) (ms)	MIP-LR (RTT) (ms)
64	45.2	5.9	85.3	7.2
128	46	6.9	86.3	8.6
256	48	7.8	88.2	10
512	51.9	11.4	92	13.8
1024	59.4	15.4	99.7	16.9

Figure 6.41 shows how the packet interceptor and packet modifier were implemented in MIP-LR. Figure 6.42 shows the experimental test bed where we carried out this experiment. The mobile host ("mh" in the figure) moved between the two 802.11 access points. *delay*1 and *delay*2 were emulated delays between the CH and the HA. These delays were introduced using a NIST delay simulator. "lr1" and "lr2" are the location registers in the visited networks, and "ha" is the home agent.

Table 6.7 shows a comparison of the round trip time (RTT) between the CH and the MH between the interceptor-assisted MIP-LR and MIP, for two fixed payload sizes, namely 64 and 1024 bytes. These results show that MIP-LR outperforms MIP as the payload size increases. As the delay factor *delay*1 was varied, simulating an increase in the distance between the CH and the HA, MIP-LR's RTT was not affected because the packets to the MH did not have to traverse via the home agent, as a result of the direct binding update from the mobile to the CH. These results demonstrate the effect of the proposed techniques to provide a direct path between the CH and the MH. Even if the network delay between the home network and the visited network increases, there is no effect on the end-to-end packet delay in the case of Mobile IP.

6.10.5 Intercepting Proxy-Assisted Route Optimization

In some cases, the forwarding of application layer signaling is delayed owing to the routing indirection caused by the underlying network layer mobility mechanism when Mobile IP is used as the mobility protocol. For example, in an IMS-based environment, when the mobile is in a visited network, the mobile's SIP signaling goes via the outbound SIP proxy server, the P-CSCF (proxy call session control function). However, owing to the underlying network layer mobility protocol Mobile IP, the SIP signaling is routed through the home agent even if the SIP proxy server is located very

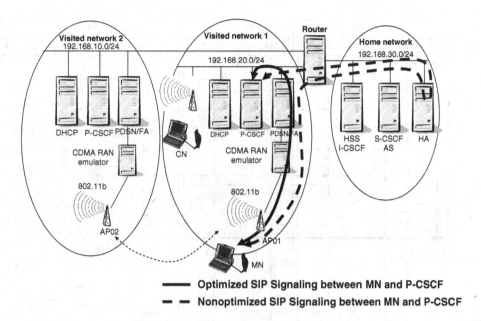

Optimized SIP Signaling between MN and P-CSCF

Nonoptimized SIP Signaling between MN and P-CSCF

Figure 6.43   Route optimization of signaling traffic

close to the mobile. This will delay the SIP reregistration procedure and SIP re-INVITE process after the mobile has handed over to the new network.

For example, in MIPv4, reverse tunneling at the FA forces packets from a mobile node to be routed via the home agent, giving rise to additional route traversal. In the application layer, SIP signaling is usually routed via an outbound proxy (e.g., a SIP server) that is closer to the mobile. However, owing to the indirection imposed by the underlying mobility layer, these packets need to travel to the home network before being directed to the application servers on the edge of the network. This additional routing causes a signaling delay that affects the handoff.

Figure 6.43 illustrates the experimental test bed where we applied the route optimization technique to reduce the traversal delay in the signaling traffic. The solid line shows the optimized path, and the dotted lines show the nonoptimized path routed via the home agent.

RFC 3024 (Montenegro 2001) specifies a means to make use of an encapsulated delivery style to perform selective reverse tunneling. This is intended to support packet delivery to local resources. Packets meant to be reverse-tunneled are sent using the encapsulated delivery style (via the MN–FA tunnel) by the MN. The FA must reverse-tunnel these packets to the HA. Packets not meant to be reverse-tunneled are sent using the direct delivery style (not encapsulated); the FA forwards these and does not use reverse tunneling to send them to the HA. The MN can send all packets meant for the P-CSCF using normal IP routing, and the FA forwards these as regular packets.

This approach solves one part of the trombone routing problem by optimizing the route from the MN to the P-CSCF. However, packets from the P-CSCF to the MN will still be routed via the HA. In addition, this selective reverse tunneling with encapsulated delivery requires changes to be made to the behavior of the MIP protocol. Thus, this feature may not be desirable for already installed MIPv4 infrastructure.

In this section, we present an interceptor-assisted technique to optimize packet delivery from the MN to the P-CSCF and from the P-CSCF to the MN in the visited network. Thus, the cost is

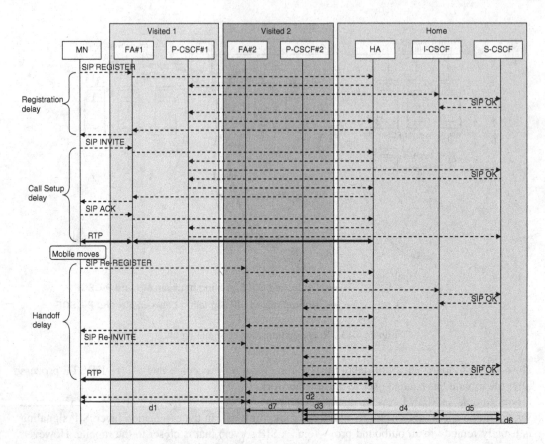

Figure 6.44 SIP signaling flow without route optimization

represented as a delay. The proposed approach provides an encapsulation technique between the FA and the P-CSCF. It installs packet interceptors and forwarding modules in both the FA and the P-CSCF to perform selective tunneling operations in both directions and also establishes an IP–IP tunnel between the P-CSCF and the FA for all of the packets destined for the MN from the P-CSCF. Packets received at the FA via the P-CSCF–FA tunnel are decapsulated at the FA and forwarded to the MN. This is identical to the manner in which encapsulated packets received at the FA via the HA–FA tunnel are processed. In addition, it does not require any changes in the functional behavior of MIP. Figure 6.44 and Figure 6.45 illustrate the SIP signaling flow without and with route optimization, respectively.

6.10.6 Cost Analysis and Experimental Analysis

We have applied this technique to optimize the path for SIP signaling messages, such as re-REGISTER and re-INVITE, in an MIPv4-based mobility environment.

In this section, we provide a simple calculation of the delay based on the cost due to traversal by the signaling messages and the processing cost in the networking nodes. We have not included all of

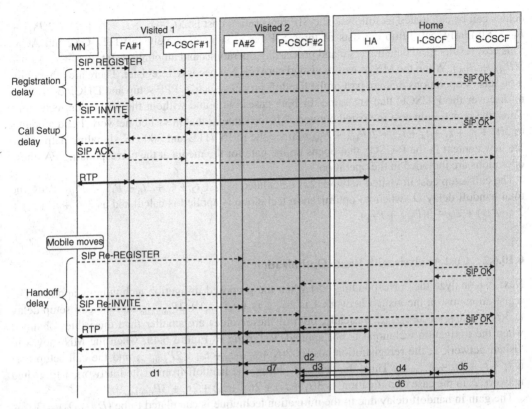

Figure 6.45 SIP signaling flow with route optimization

the processing costs in each networking node in this analysis. We assume that the communication distance between the MN and the FA is d_1, that between the FA and the HA is d_2, that between the HA and the P-CSCF is d_3, that between the P-CSCF and the I-CSCF (interrogating call state control function) is d_4, and that between the I-CSCF and the S-CSCF (serving call session control function) is d_5. The communication distance between the P-CSCF and the S-CSCF is d_6, and that between the FA and the P-CSCF is d_7. Without loss of generality, we assume that d_1, d_5, and d_7 are smaller than the other distances (i.e., the entities are close to each other). The associated costs for traversing these communication distances are t_1, t_2, t_3, t_4, t_5, t_6, and t_7, respectively.

We now analyze the cases with and without route optimization. We assume that the processing costs at the HA and the FA are P_{HA} and P_{FA}, respectively, when the mitigation technique is not applied. On the other hand, when the mitigation technique is applied, there is an additional processing cost due to modifying the packets and the additional lookup at the FA and the P-CSCF. We assume this additional processing cost to be $P_{mitigate}$ for each message. Processing at the HA is completely avoided in the optimized case where mitigation is used, as the signaling does not pass through the HA.

6.10.6.1 Cost Analysis without Route Optimization

First, we discuss the case where no mitigation technique is applied. Before the move, the MN is in visited network 1 and is subject to registration and call setup delays. Referring to Figure 6.44, these delays can be calculated as follows. The SIP registration cost is $2(t_1 + t_2 + t_3 + t_4 + t_5) + 2(P_{HA} + P_{FA})$. Similarly, call setup consists of three SIP-based signals, namely INVITE, OK, and ACK. This cost results from data traversal and processing operations and amounts to $3(t_1 + t_2 + t_3 + t_6) + 3(P_{HA} + P_{FA})$. When the MN moves to visited network 2, it needs to reregister. There are other common sets of operations that are part of the handoff process, such as PPP setup and DHCP operations to discover the P-CSCF, that are same for both cases, with and without the mitigation technique. Thus, the registration cost in visited network 2 is the same as that in visited network 1 and amounts to $2(t_1 + t_2 + t_3 + t_4 + t_5) + 2(P_{HA} + P_{FA})$. Since the re-INVITE and 200 OK signaling help create the new context in the P-CSCF that opens up the gate for the media at the corresponding FA, these operations are included in the operation.

The call setup cost in visited network 2 is calculated as $t_1 + t_2 + t_3 + 2t_6 + P_{HA} + P_{FA}$. Thus, the total handoff delay D_1 when no optimization technique is applied is calculated as $3(t_1 + t_2 + t_3) + 2(t_4 + t_5) + 2t_6 + 3(P_{HA} + P_{FA})$.

6.10.6.2   Cost Analysis with Route Optimization

Next, we analyze the corresponding cost when the proposed mitigation technique is applied. The registration cost in the visited network 1 is $2(t_1 + t_7 + t_4 + t_5) + 2P_{mitigate}$ and the call setup delay is $3(t_1 + t_7 + t_6) + 3P_{mitigate}$. Basically, both of these values are smaller than the values obtained when the mitigation technique is not applied. Referring to Figure 6.45, when the MN moves to visited network 2, the reregistration cost is $2(t_1 + t_7 + t_4 + t_5) + 2P_{mitigate}$ and the call setup cost is $t_1 + t_7 + 2t_6 + P_{mitigate}$. Thus, the delay D_2 during the handoff from visited network 1 to visited network 2, in the case of mitigation, is $3(t_1 + t_7) + 2(t_4 + t_5) + 2t_6 + 3P_{mitigate}$.

The gain in handoff delay due to the mitigation technique is calculated to be $(D_1 - D_2) = 3(t_2 + t_3 - t_7) + 3(P_{HA} + P_{FA} - P_{mitigate})$. The effect of the mitigation technique is felt more when the distance between the visited network and the home network is greater. When the distance is small, the benefit of the trombone routing mitigation is offset by the additional processing time at the FA and the P-CSCF during the packet capture and encapsulation operations.

We increased the communication distances d_2, d_3, d_4, and d_6 by adding an additional 500 ms delay using the NIST delay simulator and measured the performance in the experimental environment. The additional delay between the home network and the visited network emulates a real deployment scenario. Figure 6.46 shows the handoff times both with and without the proposed technique in the IMS test bed environment.

In particular, Figure 6.46 shows the breakdown of the delay into components due to several operations such as layer 2 handoff, PPP link configuration, Mobile IP registration, DHCP INFORM for P-CSCF discovery, and SIP-related signaling. It is important to note that the these route optimization techniques do not affect the delays due to non-SIP-related operations (e.g., PPP, layer 2, and DHCP), and thus remain the same in both cases, with and without mitigation. Thus, we focus on comparison of the reduction in the handoff delay attributed to SIP-related operations only. The amount of delay due to SIP signaling is a large fraction of the total handoff delay. For example, when the proposed mitigation technique was not applied, the delay attributed to SIP signaling was 6.5 s, out of a total handoff delay of 9.4 s, almost 70% of the total delay. When the proposed technique was applied, the delay due to SIP signaling was reduced to 4.9 s. Using this technique, we demonstrated that the SIP signaling packets were not affected by the packet indirection imposed by the underlying mobility protocol. For example, when this technique was applied, SIP reregistration took about 3205 ms,

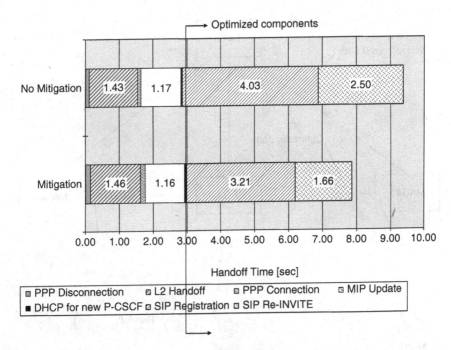

Figure 6.46 Results of route optimization using packet interceptor

compared with 4025 ms in the absence of the optimization technique. Similarly, the SIP re-INVITE took about 1660 ms, compared with 2502 ms in the absence of the mitigation technique.

However, the proposed technique involves additional processing time at the FA and the P-CSCF because of additional packet-capturing and packet-modifying operations. Thus, there is a trade-off between this additional processing time at the FA and the P-CSCF and the reduction in handoff delay due to our proposed technique. The effect of this proposed technique is more pronounced when the SIP server is situated in the visited domain close to the mobile node and the home network is far from the visited network. We have described the implementation details of these techniques in Chiba et al. (2007).

6.10.7 Binding-Cache-Based Route Optimization

In this section, we introduce a binding-cache-based technique that can be applied to Proxy MIPv6 (Gundavelli et al. 2008) to minimize the end-to-end media delay for both intradomain and interdomain mobility. This technique uses principle 1 listed in Section 6.10.1. Many of the terms used in Proxy MIPv6 are defined in Appendix B. Figure 6.47 shows the basic network configuration of a Proxy MIPv6 architecture that involves the intra-LMA movement of an MN. A Proxy MIPv6 domain is equipped with a specific LMA that acts as a home agent. The communicating nodes can belong to the same Proxy MIPv6 domain or different ones. If the LMA is placed too far from the MAG (media access gateway), then the media delivery between the MNs will be considerably delayed. In this specific intra-LMA scenario, MN#1 is anchored at MAG#1 and MN#2 is anchored at MAG#2. MN#2 establishes communication with MN#1 and then performs a handoff to MAG#3. Without any route

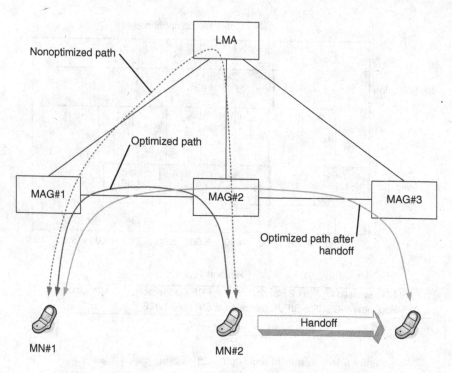

Figure 6.47   Architecture for binding-cache-based route optimization

optimization, before the handoff, data communication between MN#1 and MN#2 goes via MAG#1, the LMA, and MAG#2, and after the handoff it goes via MAG#3. Thus, it is desirable to reduce the media route associated with data traversal before and after handoff. In the intra-LMA scenario, MN#1 and MN#2 operate under the same LMA and the MN's movement is confined to MAGs that are under the same LMA. When the route optimization technique is deployed, the communication path is shortened as the packets bypass the LMA. The dotted line shows the nonoptimized path through the LMA, and the solid lines show the route-optimized paths that bypass the LMA. Similarly, for inter-LMA movement, data traversal via the LMA is avoided and data is forwarded from one MAG to another MAG directly.

Figure 6.48 shows the basic optimization procedure and the flows associated with one of the optimization techniques, which utilizes a binding cache entry (BCE) at the LMA and at the MAG. In this figure, we first show the path optimization from MN#1 to MN#2. Before the handoff, MN#1 attaches to MAG#1 and then MAG#1 sends a PBU (proxy binding update) message to the LMA on behalf of MN#1. Similarly, MN#2 connects to MAG#2, which triggers a PBU message to the LMA on behalf of MN#2. The initial packet from MN#1 to MN#2 is tunneled and is sent via the LMA. As soon as the LMA gets this packet, it figures out how to forward the packet to MAG#2. Then, the LMA sends a new message called a CBU (correspondent binding update) message to MAG#1, notifying it that MAG#2 is the anchoring node for MN#2. After getting this CBU message, MAG#1 keeps a cache that maps MAG#2 to MN#2, and sends a response message called a CBA (correspondent binding acknowledge) message to the LMA. Thus, any subsequent packet from MN#1 destined for MN#2 is intercepted by MAG#1 and forwarded to MAG#2, instead of being forwarded to the LMA. The

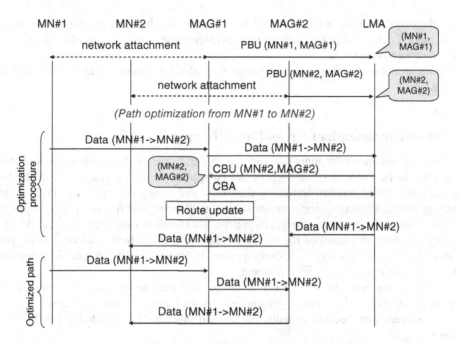

Figure 6.48 Binding-cache-based flow

Table 6.8 Results from route optimization (RO) using binding-cache approach

	Additional round trip delay between MAG and LMA						
	50 ms		100 ms		150 ms		200 ms	
Route optimization	RO	Non-RO	RO	Non-RO	RO	Non-RO	RO	Non-RO
End-to-end media delay (ms)	12	71	12	107	15	167	12	213
SIP REGISTER delay (s)	2.34	2.50	2.84	2.82	3.40	3.41	3.90	4.18

trajectory of the route-optimized packet thus becomes MN#1→MAG#1→MAG#2→MN#2 instead of MN#1→MAG#1→LMA→MAG#2→MN#2, thereby optimizing the route of the data packet from MN#1 to MN#2.

Table 6.8 compares the results obtained with and without route optimization techniques. In particular, these results show the end-to-end media delay between the MNs and the SIP signaling delay for the MNs. In an IMS network, all the SIP-related signals, such as REGISTER and INVITE, traverse all the way to the home network using tunnels between the MAG and the LMA. Thus, the route optimization technique does not reduce the SIP signaling delay, unlike the intercepting proxy-assisted route optimization technique described in Section 6.10.5. In the absence of a route optimization technique, the media traffic flows via the LMA. However, when route optimization is applied, the media traffic bypasses the LMA and flows between the MAGs through the tunnel that is set up between the MAGs. It is evident from Table 6.8 that when the route optimization technique is in place, the end-to-end media delay is not affected even if the delay between the LMA and the MAG is increased. Thus, the handoff delay will be reduced if the route optimization technique is in place after the mobile has moved to the new network.

The details of this route optimization technique have also been presented in an IETF draft (Chiba et al. 2008).

6.11 Media-Independent Cross-Layer Triggers

Cross-layer triggers are useful hints that can expedite the sequential handoff operations that take place in each layer. These handoff triggers can be applied during several stages of the handoff process, namely during handoff initiation, discovery, and configuration. Lower-layer events are generally passed as triggers to upper layers so that mobility-related functions in the upper layers can be expedited. The lower-layer triggers prepare the mobile for an impending handover event by performing different phases of the handover operations proactively. The layer 2 triggers can assist layer 3 operations that rely exclusively on these indicators to perform specific actions such as detection of network attachment or detachment.

In this section, we first describe the key principles that need to be considered when designing handoff mechanisms based on cross-layer triggers. We then describe some related studies in which cross-layer triggers were used to expedite the handoff process. We then define the 802.21-based cross-layer triggers.

6.11.1 Key Principles

The following are some of the principles that should be taken into account when designing cross-layer triggers:

1. Handoff-related functions are spread across different layers of the protocol stack and are executed independently. For efficient network communication, it is essential for a protocol layer to utilize cross-layer information, such as information from layer 2 triggers.
2. Since each protocol layer is also implemented independently in the current operating systems, it is very hard to exchange control information between protocol layers. Thus, it is helpful to have some abstract set of primitives that can pass on information across layers in order to trigger the rest of the handoff operations.
3. Interaction between events across layers expedites the initiation of a specific event in another layer.
4. The handoff initiation process is expedited when the mobile is made aware of an impending handoff. Prior indication regarding an imminent handoff operation helps the mobile to collect information for the upper layers.
5. Triggers from lower layers help initiate many of the upper-layer operations, such as layer 3 discovery, attachment, and configuration processes.

6.11.2 Related Work

There are a few related papers in which the effect of lower-layer triggers during handoff has been demonstrated. Teraoka et al. (2008) proposed unified layer 2 (L2) abstractions for layer 3 (L3)-driven

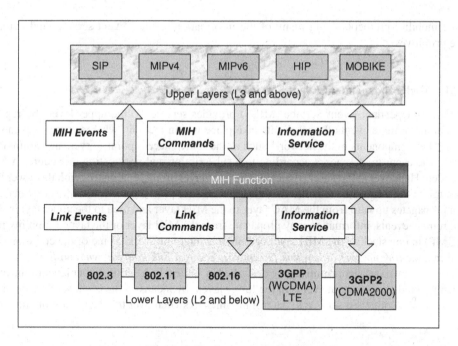

Figure 6.49   Interaction of MIHF

fast handovers. Yokota et al. (2002) described a link-layer-assisted Mobile IP fast-handoff method that uses a combination of a MAC bridge and an 802.11 access point to reduce the handoff period and packet loss during subnet handoff. Tseng et al. (2005) described a topology-aided cross-layer fast-handoff design for IEEE 802.11 and Mobile IP environments. A mobile node can utilize cross-layer topology information, such as the association between 802.11 access points and Mobile IP mobility agents, together with layer 2 triggers, to start layer 3 handoff-related activities, such as agent discovery, address configuration, and registration, in parallel with or prior to layer 2 handoff. However, these triggering techniques are access-specific and do not define any abstract primitives. We have contributed to the development of MIHF (Media Independent Handover Function) (Das et al. 2009), which has recently been standardized by the IEEE 802.21 working group. Unlike other proposals, this work develops abstract primitives that can be applied to support handover between heterogeneous access networks such as 802.11 and CDMA networks. In this section, we describe some of the functional elements of 802.21 services, but we present the experimental results involving mobile-initiated and network-initiated handover in Chapter 11.

6.11.3   Media Independent Handover Function

We have contributed to the design of cross-layer triggers that assist the mobile during its initiation phase or discovery phase by providing useful triggers about an impending movement, attachment to a new network, and disconnection from an old network. Many of these triggers have been standardized as information services, command services, event services as part of MIHF. Some of these primitive services are described in the following. Figure 6.49 shows how MIHF can interwork with the mobility protocols using many of the information service, event service, and command service primitives.

6.11.3.1   Media Independent Event Service

The Media Independent Event Service (MIES) provides services to the upper layers by reporting both local and remote events. Local events take place within a client, whereas remote events take place in other components in the network, such as the router, access point, server, and communicating host. The event model works according to a subscription and notification procedure. A Media Independent Handover (MIH) user (typically an upper-layer protocol) registers with the lower layers for a certain set of events and is notified as those events take place. In the case of local events, information propagates upward from the MAC layer to the MIH layer and then to the upper layers. In the case of remote events, information may propagate from the MIH layer or the layer 3 mobility protocol (L3MP) in one stack to the MIH layer or L3MP in a remote stack. Some of the common events defined include *link up*, *link down*, *link parameters change*, *link going down*, and *handover imminent*. The relevant handover primitives are described in Table 6.9. After the upper layer of the mobile has been notified about certain events in the lower layers by means of the event service, the mobile makes use of the command service to control the links to switch over to a new point of attachment.

6.11.3.2   Media Independent Command Service

The higher layers use the Media Independent Command Service (MICS) primitives to control the functions of the lower layers. MICS commands are used to gather information about the status of the connected links, as well as to pass on the higher-layer mobility and connectivity decisions to the lower layers. MIH commands can be both local and remote. They include commands from the upper layers to the MIH layer and from the MIH layer to the lower layers. Some examples of MICS commands are *MIH Poll*, *MIH Scan*, *MIH Configure*, and *MIH Switch*. These commands instruct an MIH device to poll connected links to learn their most recent status, to scan for newly discovered links, to configure new links, and to switch between available links.

6.11.3.3   Media Independent Information Service

Mobiles need to discover available neighboring networks and communicate with the elements in these networks to optimize the handover process. MIIS defines information elements and corresponding query–response mechanisms that allow an MIHF entity to discover and obtain information about nearby networks. It provides access to both static and dynamic information, including the names and providers of neighboring networks, as well as channel information, MAC addresses, security information, and other information about higher-layer services helpful to handover decisions. This information can be made available via both lower and upper layers. In some cases, some layer 2 information may not be available or the layer 2 information may not sufficient to make intelligent handover decisions. In such scenarios, higher-layer services may be consulted to assist in the decision-making process. MIIS specifies a common way of representing information by using standard formats such as XML (eXtensible Markup Language) and TLV (type–length–value). Having a higher-layer mechanism to obtain information about neighboring networks with different access technologies alleviates the need for a specific access-dependent discovery method. We have

Table 6.9   Sample MIHF primitives

MIH_SAP primitive	Service category	Description
MIH_Capability_Discover	Management	Discover list of events and commands supported by MIHF
MIH_Register	Management	Register with a remote MIHF
MIH_DeRegister	Management	Deregister from a remote MIHF
MIH_Event_Subscribe	Management	Subscribe for one or more MIH events with a local or remote MIHF
MIH_Event_Unsubscribe	Management	Unsubscribe for one or more MIH events from a local or remote MIHF
Link-Detected	Event	Link from a new access network is detected
MIH_Link_Up	Event	L2 connection is established
MIH_Link_Down	Event	L2 connection is lost
MIH_Link_Going-Down	Event	L2 connectivity is predicted to go down
MIH_Link_Handover_Imminent	Event	L2 handover is imminent
MIH_Link_Handover_Complete	Event	L2 link handover to a new access network is complete
MIH_Link_Parameters_Report	Event	Link parameters have crossed specified thresholds
MIH_Link_Get_Parameters	Command	Get the status of the link
MIH_Link_Configure_Thresholds	Command	Configure link parameter thresholds
MIH_Link_Actions	Command	Control the behavior of a set of links
MIH_Net_HO_Candidate_Query	Command	Initiate handover
MIH_MN_HO_Candidate_Query	Command	Initiate MN query request for candidate network
MIH_N2N_HO_Query_Resources	Command	Query available network resources
MIH_MN_HO_Commit	Command	Notify the serving network of the information about the decided target network
MIH_Net_HO_Commit	Command	Network has committed to handover
MIH_N2N_HO_Commit	Command	Notify target network that serving network has committed to handover
MIH_MN_HO_Complete	Command	Initiate MN handover complete notification
MIH_N2N_HO_Complete	Command	Handover has been completed
MIH_Get_Information	Information	Requests to get information from repository
MIH_Push_Information	Information	Notify the mobile node of operator's policies or other information

implemented a version of MIIS based on RDF (Resource Description Framework). Many of the cross-layer triggers that are part of the information services help expedite the initiation of the hand-off process by discovering network components proactively. Two of the primitives of MIIS that are used to discover network services are *Get_Info_Request* and *Get_Info_Response*.

Figure 6.50 shows how the local and remote MIH functions interact with each other. Table 6.9 lists several types of the MIH primitives and their interactions. These are categorized into management, event, command, and information services. There are several scenarios where these triggers could be useful for expediting handoff operations. We have experimented with some of these 802.21-based triggers and demonstrated how these techniques can be used as helpers to many of the existing

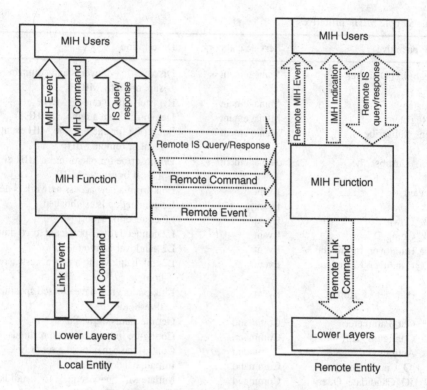

Figure 6.50   Cross-layer triggers with MIHF

mobility protocols such as MIPv6 and SIP-based mobility, and to optimization schemes such as Media Independent Preauthentication. We describe the experimental results in Chapter 11.

6.11.3.4   MIHF Implementation

This section describes the software implementation of MIHF. The software implementation of MIH includes both MIHF and the MIH Information Server. The software is implemented in Java 1.6 and is thus portable across different operating systems. The different components of the MIHF software implementation are shown in Figure 6.51.

This implementation provides an MIH API for the MIH users. The MIH API embodies MIH_SAP and supports both local and remote MIH services. Communications for remote services are realized by the MIH Protocol component, which implements the MIH protocol. The current version of the implementation of MIH Protocol uses UDP as the MIH transport protocol.

The MIH user manager component is responsible for determining the privileges of the MIH users. It enforces coordination between multiple MIH users such that only one particular MIH user is allowed to change the state of network interfaces. This prevents conflicting state changes being made by different MIH users that are employing different handover policies at the same time. An example might be when a network interface is turned on by one MIH user and then turned off by another user.

Network interfaces are managed by the link manager using the MIH LINK API. The MIH LINK API is implemented by the Link Providers components and embodies MIH_LINK_SAP. A distinct

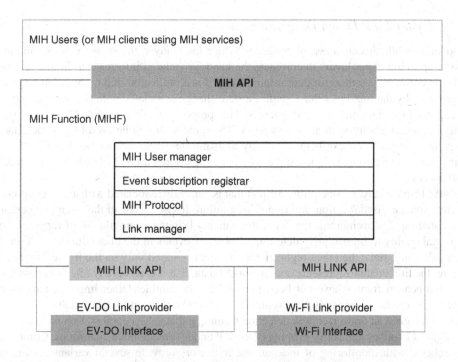

Figure 6.51 MIHF implementation stack

Link Provider component is defined for each network interface type. The Link Providers are considered as adapters to the network interfaces and can be implemented either inside or outside network interface drivers. The current Link Providers are implemented outside the network interface drivers and support MICS and MIES for IEEE 802.11 and CDMA2000 EV-DO interfaces in a Linux environment. The Link Providers are implemented in Java with Java Native Interface to utilize device-specific C calls, since most device drivers have C APIs rather than Java APIs.

A Link Provider implements a Link_Parameter_Report event notification, which generates event notifications when the related interface has crossed a configured threshold level. In order to avoid flooding with event notifications because of a frequently changing signal strength, the proposed Link Provider implements a function to average out the actual signal strength before reporting to the MIHF. On the other hand, this may delay the reaction time when the threshold is actually crossed. The Event Subscription Registrar component manages local and remote event subscriptions for the link-layer events monitored by the MIHF. It also aggregates multiple event subscriptions by multiple users of the same MIHF into a single event subscription and delivers notifications to the subscribed MIH users when event notifications are received.

The IEEE 802.21 Information Server (IS) is implemented as an MIH user that responds to MIIS queries through interaction with the MIH Protocol component. At initialization, the IS registers with its local MIHF to receive IS queries carried in MIH_Get_Information request messages. After registration, it is ready to respond to queries sent by other MIH users. The current implementation supports IS queries for RDF data using the SPARQL query language. The IS uses an Oracle 11g database to query RDF data.

We present the experimental results associated with the 802.21-based cross-layer triggers in Chapter 11.

6.11.4   Faster Link-Down Detection Scheme

The sooner a mobile detects a loss of connectivity in a lower layer, the sooner this information can be passed up to the upper layers to complete the handover-related information. In this section, we describe an optimized method for determining a *link down* indication in a mobile. *Link down* is an event provided by the link layer that signifies a state change associated with the interface no longer being capable of communicating data packets. This proposed method uses MAC layer operations to verify communicability with an access point. These methods can be used to provide a fast *link down* event indication and can help in quickly assisting the upper-layer protocols to take action. This link-down detection technique can be used in conjunction with 802.21 triggers to expedite the handoff process.

In an 802.11-based layer 2 operation, a client that is currently associated with an access point may experience sudden disconnection due to a device failure in the AP or in the client, or perhaps an unanticipated rapid movement of the client that quickly brings the mobile out of range of the AP. Using signal quality to immediately determine *link down* events in the client during such scenarios can be misleading, since the client registers the link quality based only on the frames last received. Therefore, the link quality represents only historical data, and it is reset only after a certain number of expected beacon frames have not been received by the mobile. Other implementations verify connectivity using failed transmission events (RTS/CTS) (Velayos and Karlsson, 2004).

We have designed an optimized link detection technique that uses a combination of passive monitoring of 802.11 frames and active probing of the AP under some defined conditions. Combinations of this scheme with monitoring of independent indicators provide several variants of link-down detection that are applicable in different scenarios. By such methods, the mobile can rapidly determine a sudden disconnection event and quickly propagate *link down* indications to upper-layer protocols, such as the Mobile IP stack in the mobile. Event triggers such as *link down* and *link going down* are useful for optimizing the overall sequence of handoff operations.

We now describe the details of the fast detection algorithm. The purpose of the fast detection algorithm is to provide a definitive measure that indicates that complete link loss has occurred within a relatively short period of time. Normally, definitive indicators can be ambiguous for 802.11 networks when there are very tight time constraints (on the order of milliseconds), because of the following factors:

1. The signal strength can vary by several dB when one moves just a few meters. The low points of these fluctuations can cause a "false" indication of link loss for a short period of time.
2. An unstable packet loss rate can also cause "false" indications. Most 802.11 links have a stable loss rate, although it is expected that there will be periods of bursty loss rate, perhaps due to an increase in traffic demand or to bursty traffic.
3. There can be propagation interference from an adjacent AP within or near the non-AP STA transmit channels. This relates to the previous factor, where the packet loss rate is at a high stable level. This is aggravated by the DCF algorithm of 802.11 MAC (del Prado Pavon and Choi, 2003), which may cause delay for any frame that requires an ACK (acknowledgement).
4. Multipath fading can occur. However, although this can contribute to the previous factor, it is less likely to occur in more recent DSP/chipset implementations, which can compensate for this effect.

In the following, we describe three basic ways in which link-down detection can be carried out in a IEEE 802.11 network.

6.11.4.1   Passive Scanning

During passive scanning, the access point broadcasts beacons at regular intervals. A mobile counts the number of consecutive beacon losses. A beacon is considered as lost if the mobile does not receive a beacon for a period of time T_B since the receipt or loss of the last beacon. A passive scan is said to have failed if the number of consecutive beacon losses reaches a threshold N_B. The link is considered to be down when a passive scan fails. Thus, the detection time is $T_B \times N_B$. However, this scheme suffers from a few drawbacks. The detection speed is slow if T_B is large; on the other hand, if T_B is small, there are too many beacons. There is also an increased probability of false link-down detection if N_B is small.

6.11.4.2   Active Scanning

A mobile unicasts a probe request every interval of time T_P. An active scan is said to have failed if the number of transmissions reaches a threshold N_P and no probe response has been received. If at least one probe response has been received before N_P is reached, then the active scan has succeeded. The link is considered to be down when an active scan fails. Thus, the detection time is $T_P \times N_P$. However, there are many issues with active scanning. There is too much probe traffic if T_P is small. Also, there is an increased probability of false link-down detection if N_P is small. However, the detection speed is slow if T_P is large.

6.11.4.3   Hybrid Scanning

Hybrid scanning is a combination of passive and active scanning. Here, the mobile normally performs passive scanning. When a passive scan reports a failure, the mobile starts active scanning instead of immediately considering the link to be down. The link is considered to be down if the active scan also fails. If the active scan succeeds, the mobile switches back to passive scanning. Thus, the detection time can be considered as equivalent to $T_B \times N_B + T_P \times N_P$. This combination of schemes lessens the chance of false detection compared with either passive or active scanning by itself, but under heavily loaded conditions, the chance of false detection is greater than under lightly loaded conditions for the same pair of N_B and N_P values.

6.11.4.4   Independent Modifiers

Traffic flows can be considered by using independent modifiers, since these modifiers directly affect the basic set of algorithms. The addition of these modifiers to the basic algorithms produces variants that use sent and received frames as replacement indicators for beacon and probe responses. The following are two of these modifiers:

1. *Received frame.* The receipt of any frame can be considered as the receipt of a beacon frame or probe response. Therefore, a passive or active scan is considered successful if any frame is received from the AP. This takes advantage of heavily loaded conditions, where beacon or probe responses can be lost and trigger a false link-down event.
2. *Transmission failure.* Failure to transmit a data frame can be used as an indication of link failure. Under 802.11 MAC, each data frame sent by a mobile requires an ACK from the AP. The mobile

will retransmit the data frame if an ACK is not received within a certain amount of time (normally implementation-specific). The link is considered down if the number of retries exceeds a configured threshold (also implementation-specific).

The independent modifiers are not considered as independent solutions, since they rely on the applications generating the data traffic. However, they can be combined with the basic algorithms, namely active scanning, passive scanning, and hybrid scanning, to produce several variants. One or more of these variants can be used as actual solutions for fast link-down detection based on preference or on scalability or implementation considerations. Using these variants helps reduce false link-down detection caused by heavy traffic conditions, since it takes advantage of the ongoing data exchanges. The two modifiers can be combined with the basic algorithms in various ways. The following are some of the proposed fast detection techniques that use a combination of scanning and modifiers.

1. *Passive scanning combined with modifiers*. Passive scanning combined with the two modifiers can result in the following variants:
   (a) *Received frame*. A passive scan is successful if a frame is received from the AP within $T_B \times N_B$; otherwise, the passive scan fails if the threshold N_B is reached without receipt of a frame. Any received frame becomes a substitute for an expected beacon frame.
   (b) *Transmission failure*. A passive scan fails if data transmission fails, even if N_B has not yet been reached. Likewise, a passive scan succeeds if data transmission succeeds, even if N_B has not yet been reached. If no non-AP STA application is generating data, the passive scan proceeds as normal.
   (c) *Received frame and transmission failure*. A passive scan fails if (a) or (b) in this list fails. Likewise, a passive scan succeeds if (a) or (b) succeeds. The detection time is determined by which failure occurs first ((a) or (b)).
2. *Active scanning combined with modifiers*. Active scanning combined with the two modifiers can result in the following variants:
   (a) *Received frame*. An active scan is successful if a frame is received from the AP, even before N_P is reached; otherwise, the active scan fails if the threshold N_P is reached without receipt of a frame. Any received frame becomes a substitute for an expected probe response.
   (b) *Transmission failure*. An active scan fails if data transmission fails, even if N_P has not yet been reached. Likewise, an active scan succeeds if data transmission succeeds, even if N_P has not yet been reached. If no non-AP STA application is generating data, the active scan proceeds as normal.
   (c) *Received frame and transmission failure*. An active scan fails if (a) or (b) in this list fails. Likewise, an active scan succeeds if (a) or (b) succeeds. The detection time is determined by which failure occurs first ((a) or (b)).
3. *Hybrid scanning combined with modifiers*. Hybrid scanning combined with the two modifiers can result in the following variants:
   (a) *Received frame*. A hybrid scan is successful if a frame is received from the AP, even before the passive and active scan thresholds have been reached. The receipt of any frame constitutes receipt of an expected beacon or probe response. The hybrid scan fails when the passive and subsequent active scans fail without receipt of any frame.
   (b) *Transmission failure*. A hybrid scan fails if data transmission fails, even before the passive and active scan thresholds have been reached. Likewise, a hybrid scan succeeds if data transmission succeeds, even before the passive and active scan thresholds have been reached. If no non-AP STA application is generating data, the hybrid scan proceeds as normal.

(c) *Received frame and transmission failure.* A hybrid scan fails if (a) or (b) in this list fails. Likewise, a hybrid scan succeeds if (a) or (b) succeeds. The detection time is determined by which failure occurs first ((a) or (b)).

We have used 802.21-based cross-layer triggers in conjunction with the faster link-down detection scheme described in this section to reduce the handover delay and packet loss during movement between an 802.11 access network and a CDMA network. In this specific scenario, the mobile was connected to the 802.11 network and there was a sudden power failure at the 802.11 access point. Using this faster link-down detection mechanism and the 802.21 event notification triggers, the hand-off delay was reduced. In Chapter 11, we describe how such faster link-down detection techniques are useful for optimizing handovers involving CDMA and 802.11 networks.

6.12   Concluding Remarks

The proposed optimization techniques for the various handoff components (e.g., discovery, authentication, configuration, security association, binding update, and media rerouting) can be categorized primarily into three types, namely *reactive*, *proactive*, and *cross-layer*.

The proposed reactive mechanisms are applicable after handover of a mobile to a new network. These techniques use additional elements in the network such as an anchor agent, a mobility proxy, or an external home agent to mitigate the effects of IP address acquisition delay and longer binding update delays, to avoid triangle routing, and to reduce the delays due to security association.

The proposed proactive mechanisms are applied prior to handover and offer better handoff performance than the reactive techniques at the cost of additional resources in the network such as proactive tunnels with the candidate target networks, multiple media streams due to proactive binding update, caching of IP addresses from neighboring networks, and buffering at the edge routers.

The proposed cross-layer mechanisms can expedite the detection of a new point of attachment or of the sudden loss of a network. Cross-layer triggers are most useful when the mobile performs a handover across heterogeneous access networks (e.g., CDMA and 802.11), as the mobile can prepare for many of the upper-layer handoff-related operations based on layer 2 triggers. Cross-layer triggers can be applied both *proactively* and *reactively*. For example, proactive cross-layer triggers can help initiate an application layer discovery process before handoff, whereas reactive cross-layer triggers can trigger a handoff by passing lower-layer information such as the signal-to-noise ratio to initiate the layer 3 handover operations.

Depending upon the type of network environment, performance requirements, and resource availability, either reactive, proactive, or cross-layer techniques can be applied to optimize specific handoff components.

7

Optimization with Multilayer Mobility Protocols

In this chapter, we describe our proposed multilayer mobility optimization techniques that use triggers from the data link layer (e.g., layer 2) and the application layer and optimize several handoff operations, namely address configuration, layer 3 binding update, and media traversal. During the discovery process, the data link layer trigger specifies the mobile's movement type based on the layer 2 beacon id and the layer 3 subnet prefix or domain id. Similarly, the application layer trigger specifies the type of application, such as a TCP-based (e.g., ftp) or RTP-based (e.g., VoIP) application, on the mobile to trigger the right type of mobility protocol.

The proposed mechanism optimizes handoff delay and reduces packet loss by way of limiting the number of signaling update messages during the mobile's movement within a domain and reduces the end-to-end transport delay for the media by use of a direct binding update for real-time applications. This mechanism uses a policy-based approach based on the mobile's movement pattern and the application, and decides on the mobility protocol that is most appropriate.

7.1 Summary of Key Contributions and Indicative Results

The network layer mobility protocol, application layer mobility protocol, and local mobility protocols can operate independently without interacting with each other. However, each of these mobility protocols has its own pros and cons. For example, a network layer mobility protocol such as MIP (Perkins, 2002b) needs an additional networking element (e.g., a home agent) in the home network to support terminal mobility and is thus not optimized; an application layer mobility protocol such as SIP (Schulzrinne and Wedlund, 2000a) is best optimized for real-time applications (e.g., VoIP) but cannot support mobility for TCP-based traffic in its current form. A local mobility protocol such as Cellular IP (Campbell et al., 2000) cannot support mobility across subnets or interdomain mobility. An integrated mobility management scheme in which a mobile can use any of these mobility protocols based on some policy (e.g., based on the type of movement or the application) will enable the mobile to use the best features of each of these protocols and thus will offer optimized handoff performance.

Mobility Protocols and Handover Optimization: Design, Evaluation and Application, First Edition.
Ashutosh Dutta and Henning Schulzrinne.
© 2014 John Wiley & Sons, Ltd. Published 2014 by John Wiley & Sons, Ltd.

We have developed a multilayer mobility management scheme that uses cross-layer triggers from data link layers and application layers and optimizes several handoff operations, namely address configuration, layer 3 binding update, and media rerouting. The proposed mechanism uses a policy-based approach based on the mobile's movement pattern and type of application and executes the mobility protocol that is most appropriate for use in any specific network and application environment. This mechanism uses SIP-based application layer mobility to support real-time traffic and MIP-LR-based mobility to support non-real-time traffic during interdomain movement, while it uses a local mobility management protocol, MMP (Wong et al., 2002), to support real-time and non-real-time traffic during intradomain and intrasubnet movement.

The proposed multilayer mobility mechanism has the following key advantages:

- The proposed mechanism increases the data throughput by 50% in a high-mobility scenario by reducing the binding update traversal during intradomain mobility, and uses lower-layer triggers such as information from the layer 2 beacon id to determine intradomain and interdomain mobility based on the gateway's identifier.
- The proposed mechanism expedites the discovery operation by discovering the layer 3 and layer 2 points of attachment using optimization by way of parallelism, and reduces the packet loss during handover.
- Using application layer triggers, the proposed mechanism uses a mobility protocol that is optimized for a specific type of application (e.g., SIP for RTP-based traffic and MIP for TCP-based traffic).

After we developed the proposed policy-based mobility management scheme back in 2001, a few other integrated mobility management schemes were developed by Politis et al. (2004) and Lee et al. (2003), which use SIP for personal mobility and MIP for terminal mobility and carry SIP registration information as part of the MIP binding update. However, none of these existing approaches use any cross-layer triggers to optimize the handoff performance, nor do they provide a throughput increase comparable to the experimental results obtained from the mechanisms proposed here.

In the rest of the chapter, we describe the details of the proposed mechanisms; present a few integrated mobility management schemes that use a combination of SIP, MIP-LR, and the micro-mobility protocol MMP; cite some related work; and present results from the experimental systems that we built.

7.2 Introduction

The proposed multilayer integrated mobility management scheme was designed keeping in mind the requirements for real-time and non-real-time traffic. Currently, there is no framework that handles global macromobility as well as micromobility, but both are important and necessary. We have designed a micromobility management protocol (MMP) to handle mobility in layer 2 for the proposed integrated mobility management scheme described in this chapter. For macromobility involving subnet and domain movement, we used SIP (Rosenberg et al., 2002) to handle mobility for real-time traffic, and MIP with Location Registers (MIP-LR) to handle mobility for non-real-time traffic. In either case, MMP handles micromobility to support layer 2 movement. This multilayer mobility management architecture introduces several novel features as follows:

1. A mechanism that introduces a policy to use SIP to support macromobility[1] for real-time traffic and MIP-LR to support non-real-time traffic.
2. Use of SIP for macromobility together with MMP for micromobility for real-time traffic.
3. Use of MIP-LR for macromobility together with MMP for micromobility for non-real-time traffic.

7.3 Key Principles

The following are the key principles that were used to design the multilayered mobility optimization scheme:

1. Based on the type of movement in a specific layer such as layer 2 or layer 3, binding updates can be confined to a domain by using an anchor agent to optimize the handoff delay.
2. Based on the transport protocol that an application uses (e.g., RTP or TCP), a policy can be applied to invoke either an application layer or a network layer mobility protocol.
3. Layer 2 and application layer triggers help to determine the type of mobility protocol that is needed to optimize the handoff.
4. The mobile needs to be able to execute a specific type of mobility protocol based on the policy in item 2.

7.4 Related Work

At the time when this specific work was done and published (Dutta et al., 2002c; Wong et al., 2003b), to the best of our knowledge there was no other prior work that proposed the use of a multilayer mobility scheme based on policy governed by layer 2 or application layer triggers. Soon after this work or at around the same time, a few other papers proposed integration of mobility protocols in multiple layers. Politis et al. (2004) described a multilayer mobility management scheme involving SIP and MIP along with an enhancement of the AAA architecture. Carli et al. (2001) described how Mobile IP and Cellular IP could provide an integrated mobility solution.

Lee et al. (2003) described a mobility management scheme that was based on the integration of Mobile IP and SIP. In this scheme, the client does not register with the SIP registrar even if its IP address changes. Thus, all the data still flows through the home agent. SIP registration is invoked only when personal or service mobility is needed. Thus, the advantage of SIP-based terminal mobility is not realized in this specific scheme. Zeadally et al. (2004) described an architecture that integrates SIP and Mobile IP to support seamless mobility. That proposal used MIP to support terminal mobility and SIP to support personal mobility.

Kim et al. (2004) described how to route SIP messages as part of the Mobile IPv6 binding update message. However, in this mechanism, a home agent (HA) on the home subnet acts as a redirect server and a registrar for SIP as well as a home router for Mobile IPv6. Thus, the binding cache in the HA contains the location information for SIP as well as home registration entries for Mobile IPv6. The method requires that the MIPv6 binding update is modified so that it can carry the SIP registrations.

---

[1] Macromobility and micromobility are defined as in Appendix B.

Wong (2002) provided some architectural alternatives for integrating Mobile IP and Cellular IP. These alternatives also highlight some of the drawbacks of using Cellular IP with MIP in co-located mode. Thus, their analysis suggested that other global mobility protocols should be used when the mobile uses a co-located care-of address. More recently, the IETF has been considering a network-controlled localized mobility protocol (Kempf, 2007) that adopts an optimization technique similar to Cellular IP and helps to reduce handoff delay during a mobile's movement within a domain. This introduces a new network element, the LMA (local mobility agent), that acts as an anchor agent and identifies a specific mobility domain. Based on its movement, the mobile limits the binding update to the domain of the LMA.

In contrast to all of these proposals, our proposed mechanism does not need changes in the SIP or MIP specification, and uses layer 2 and application layer triggers to determine the right mobility protocol based on the mobile's movement pattern. This mechanism is the first to be developed that uses a policy-based approach and takes advantage of global and local mobility protocols to optimize the handoff performance.

7.5 Multilayer Mobility Approach

The mobility protocols in each layer are best suited to working with a specific type of application that uses either TCP or RTP as the transport protocol, and a specific movement pattern of the mobile such as interdomain or intradomain mobility. For example, an application layer mobility protocol may be suited to working with interactive traffic such as VoIP, and a network layer mobility protocol works well to support TCP-based applications. Cross-layer optimization techniques help the mobile to reduce the handoff delay by avoiding layer 3 reconfiguration, limiting the binding update, and reducing the media traversal by choosing the appropriate mobility protocol. Thus, during intradomain movement, a local mobility protocol takes care of redirecting the traffic to the new point of attachment of the mobile and confines the binding update to the domain itself.

We have designed, prototyped, and validated our proposed cross-layer mobility optimization scheme (Dutta et al., 2002b) using three mobility protocols in different layers, namely an application layer SIP-based protocol (Wedlund and Schulzrinne, 1999), the network layer protocol MIP-LR (Dutta et al., 2005a), and the layer-2-based protocol MMP (Micro Mobility Protocol) (Wong et al., 2002), that operate in collaboration with each other. These mobility protocols are triggered by cross-layer information during the handover.

MMP is a modified version of Cellular IP (Valkó, 1999) that provides additional survivable features in an ad hoc network by adding multiple gateways for each domain. A SIP-based mobility protocol is used for real-time traffic, and MIP-LR is used for non-real-time traffic during a node's movement between two different domains, while MMP takes care of the movement within a domain. MMP is designed as a micromobility protocol to handle intradomain mobility and works in conjunction with SIP and MIP-LR. To support real-time communication during the mobile's movement between domains, the mobile sends a SIP re-INVITE to the correspondent host (CH) to keep the session active. Similarly, a MIP-LR UPDATE message is sent to the CH for the TCP/IP traffic. However, for any subsequent move within the new domain, re-INVITE and update messages are not sent, since MMP takes care of routing the packets properly within that domain. This helps to limit the extent to which the binding update has to traverse a long distance.

7.5.1   Policy-Based Mobility Protocols: SIP and MIP-LR

For macromobility, we used both SIP and MIP-LR. Although MIP-LR alone can handle macromobility for both real-time and non-real-time traffic, we used MIP-LR for non-real-time traffic and SIP for macromobility for real-time-traffic because:

1. SIP is already used for session control signaling for real-time applications, and mobility for these applications can be handled using the same signaling mechanisms.
2. SIP-based terminal mobility integrates well with SIP personal mobility, by employing a unique URI for the user and obtaining the assistance of SIP proxies.
3. A SIP-based solution exists for smooth handoffs of real-time traffic streams.

In order to use both SIP and MIP-LR for macromobility management, we used a policy table. Between the IP level and link layer processing, there is an entity that examines each IP packet and dispatches it to the appropriate handler. The decision is based on the policy table. For example, the MIP-LR software module can capture every IP packet and process every packet that is not related to real-time traffic (i.e., RTP packets or SIP signaling). The real-time traffic passes through untouched, and is redirected by the SIP application when the IP address changes. Figure 7.1 shows how SIP

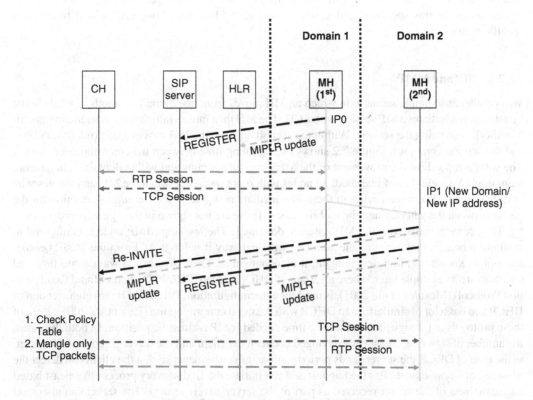

Figure 7.1   Integration of SIP and MIP-LR

and MIP-LR can both manage mobility at the same time for RTP and TCP packets, respectively. Suppose a voice or video session (carried by RTP) and a file transfer (e.g., using ftp over TCP) are in progress at the same time. The mobile host (MH) starts in domain 1, labeled "MH (1st)," referring to the first phase of its movement. The MH then moves to domain 2, labeled "MH (2nd)," referring to the second phase of its movement. The solid arrow shows the movement of the MH between domains. When the MH detects that it is in a new domain (after arriving in domain 2), it performs autoconfiguration. MIP-LR then updates the CH and the home location registers (HLRs) with this new address, so the CH can update the destination IP address of the TCP packets. At the same time, the SIP client (on the MH) issues a re-INVITE request and also updates the SIP registrar for location management. The SIP UA (user agent) on the CH then informs the real-time applications that the address of the MH has changed. Additionally, for real-time traffic, a fast-handoff scheme could be deployed without affecting the MIP-LR-based mobility management.

7.5.2   Integration of SIP and MIP-LR with MMP

The signaling time for the global update in SIP, as in MIP, can result in significant handoff latency. It has previously been suggested at a high level (Wedlund and Schulzrinne, 1999) that micromobility schemes could be used together with SIP to improve the mobile's performance in micromobility situations. In this section, we describe the details of how these two can coexist based on a specific policy.

7.5.2.1   SIP and MMP

We consider an example scenario in which an MH moves from one domain to another. While in the first domain, it initiates a SIP session with a CH. The MH then moves into the second domain (macromobility), continuing the session. Within the second domain, the MH moves again (micromobility), and the session continues. Figure 7.2 shows the signaling flow between the communicating nodes. The solid arrows show the movement of the MH between domains and within domain 2. In general, there might be a number of intermediate nodes with route caches between the MH and the gateway in each domain. The route caches in these intermediate nodes store the routing information for the traffic between the gateway and the mobile nodes. These are not shown in the figure, to reduce clutter. The scenario starts with the MH entering domain 1. The last-hop MMP node is configured to send out a beacon that contains the address of the gateway it belongs to. From the MMP beacon, the mobile knows it is in a new domain; it autoconfigures. There are several ways to do this, and we illustrate an example later, where a variant of DHCP, namely DRCP (Dynamic Rapid Configuration Protocol) (Mcauley et al., 2001), is used for autoconfiguration. DRCP is a lightweight version of DHCP and has a lot of similarities to DHCP with the rapid commit option (Park et al., 2005). Both of these protocols are designed to reduce the time needed for IP address acquisition. In both the cases, the number of signaling messages exchanged between the client and the server is reduced. However, in the case of DRCP, the server sends periodic server advertisements so that the client can detect the presence of a new subnet after the handoff and can initiate the IP discovery process in unicast based on the address of the server received as part of the server advertisement. The server can also send a unicast-based offer to the client's address directly instead of sending it on a broadcast address. In both of these protocols, the client does not perform a duplicate address detection; instead, the server performs an IP address check before it offers the address to the client.

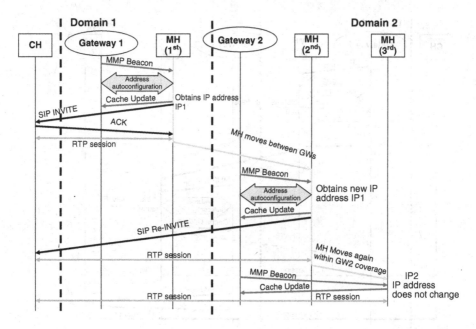

Figure 7.2   Integration of SIP and MMP

Having obtained a local address from the DRCP server in domain 1, IP1, the client updates the MMP gateway. It should then send one or more SIP REGISTER messages to the appropriate SIP servers (not shown in the figure, to reduce clutter). Some time later, it initiates a SIP session with the CH. After a subsequent move into the new domain (domain 2), the MH listens to the gateway beacon and realizes that it is in a new domain. It autoconfigures and sets itself up for micromobility management with its new local address. It then sends a SIP re-INVITE to the CH with its new address so the SIP handoff can be completed, with the CH changing the destination address of the packets it sends to the MH. The MH also sends one or more SIP REGISTER messages to appropriate SIP servers, which are not shown for brevity. When the MH moves again, its movement is confined to domain 2. Hence, it listens to the MMP beacon and knows that the move is only a local move. Therefore, it only updates the MMP gateway. SIP is completely uninvolved in the process because the IP address is unchanged. Compared with an interdomain handoff, this intradomain handoff incurs very low handoff delay.

7.5.2.2   MIP-LR and MMP

We illustrate the integration of MIP-LR and MMP in Figure 7.3. While in the first domain, the MH initiates a TCP session (e.g., a file transfer) with a CH. In domain 1, the MH sends MIP-LR update messages to the appropriate HLRs (not shown for brevity). Then it initiates a file transfer session. After moving into domain 2, the MH hears the gateway beacon, autoconfigures, and performs micromobility setup signaling. It then sends a MIP-LR UPDATE to the CH with its new address. The MH should also send MIP-LR UPDATE messages to the appropriate HLRs.

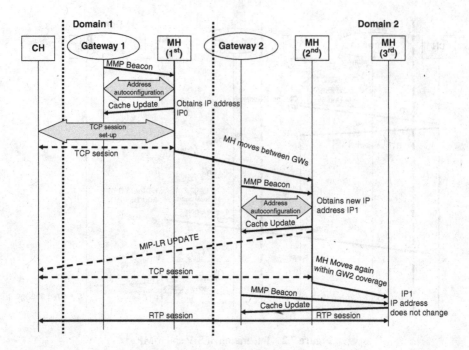

Figure 7.3   MIP-LR–MMP flow

7.5.3   Integration of Global Mobility Protocol with Micromobility Protocol

Figure 7.4 illustrates a policy-based mobility management scenario where macromobility and micromobility protocols work together based on layer 2, layer 3, and application layer triggers. The figure shows how, based on the type of application on the mobile and the movement pattern of the mobile (e.g., movement between subnets or between cells within a subnet), a specific mobility protocol is used. The results shown in Figure 7.5 demonstrate how the mobile can obtain higher data throughput when a micromobility protocol is used for intradomain movement compared with a macromobility protocol because this limits the number of signaling message updates and the traversal of updates. We have described the complete implementation of the cross-layer mobility management scheme and experimental results in Dutta et al. (2002c). Simulations and experimental results for MMP are presented in Wong et al. (2002). We have described the architectural details of this scheme in Wong et al. (2003a).

7.5.4   Implementation of Multilayer Mobility Protocols

Figure 7.6 shows the setup of our Linux-based laboratory prototype using an 802.11 wireless LAN (WLAN) for the wireless links. IP address management (including autoconfiguration) is provided by DRCP and DCDP (Dynamic Configuration Distribution Protocol) servers. DRCP is a version of the Dynamic Host Configuration Protocol (DHCP) optimized for wireless environments by reducing the

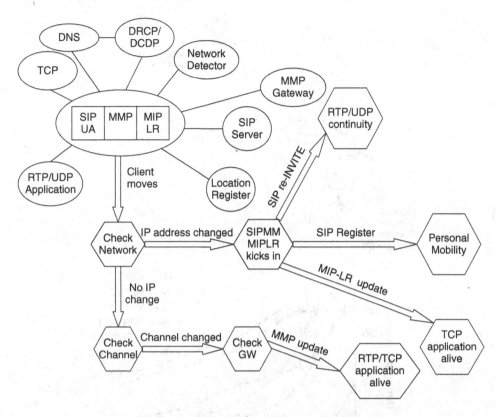

Figure 7.4   Policy-based mobility management

size of the protocol headers and the number of messages between the DHCP client and server. The DRCP server configures a node's interface with an IP address, and provides the addresses of the DNS server, SIP server, and other application servers. DCDP works in conjunction with DRCP, which distributes pools of IP addresses to the nodes in a quasi-static network so that these nodes become DRCP servers and dispense IP addresses to the clients. The current implementation of MIP-LR eliminates the tunneling function (and its encapsulation overhead) by using Linux's new *libipq* and *iptables* utilities to modify the packets (change the IP header fields) appropriately at the end points.

The MH obtains a new IP address once it has moved to a new domain, and keeps this IP address as long as it remains within that domain. This is handled automatically by DRCP. As shown in Figure 7.6, as the mobile node moves between the domains it uses SIP or MIP-LR, depending on the type of application being supported. But, while it is moving within a domain, mobility management is handled by MMP, where the gateway acts as a DRCP/DCDP server, and one of the MMP nodes acts as a DRCP server. For convenience, in our test bed, all access points within a domain used the same WLAN frequency, whereas access points in different domains used different frequencies. It is reasonable for all access points within a domain to use the same frequency, and micromobility handoff is optimized in this manner. However, using the same frequency in an adjacent channel may cause interference, leading to lower capacity.

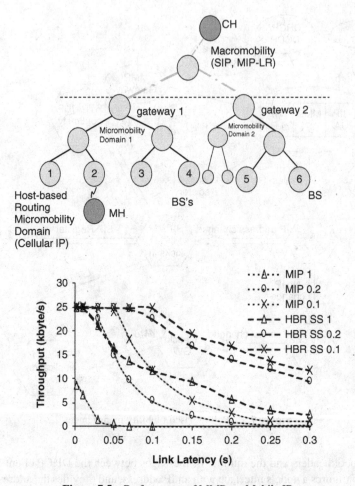

Figure 7.5   Performance of MMP vs. Mobile IP

7.5.5   *Implementation and Performance Issues*

SIP, MIP-LR, and MIP all provide a binding update mechanism that updates the mapping between a permanent address and a temporary one. With SIP, this is done with REGISTER (for presession mobility) and re-INVITE (for mid-session mobility). With MIP and MIP-LR, this is done with registration (with HAs and HLRs, respectively). MIP (with route optimization), SIP, and MIP-LR all allow binding updates for CHs to route packets directly to the MHs after mid-session mobility events. SIP servers and MIP-LR location registers can be replicated for survivability.

How well does the new mobility management scheme meet the requirements stipulated earlier? By virtue of the use of macromobility protocols such as SIP and MIP-LR, the triangular-routing problem of MIP is eliminated. We have found that this significantly increases routing efficiency when the home network of the MH is far from the visited network and the CH is closer to the MH. This scheme has much less overhead than MIP because encapsulation is not used by any of the component protocols, and the use of micromobility significantly reduces the global signaling

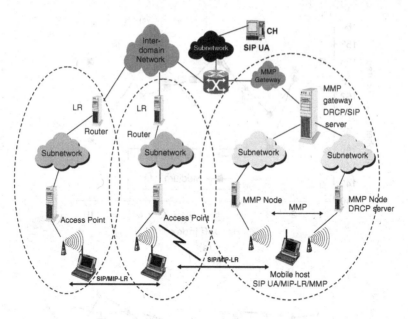

Figure 7.6   Integrated mobility management

overhead. The avoidance of triangular routing and the absence of encapsulation contribute to low latency for both real-time and non-real-time communication. The scheme can be made survivable by having SIP proxies and multiple HLRs that act like dynamic HAs. In general, the MH maintains a current list of SIP proxies or HLRs that can be contacted prior to a session or during communication.

We investigated the performance of the multilayer mobility management scheme using the laboratory test bed shown in Figure 7.6. We used SIP to initiate a video session between the MH and CH. During the movement of the MH, both micromobility and macromobility handoffs occurred. For micromobility handoffs (within a domain), since the two access points were on the same frequency, the handoff did not require binding to a new frequency. The IP address also remained unchanged. The only difference (when the MH was transmitting) was that the default gateway and the destination medium access control (MAC) address of the outgoing packets were changed to the new access point. This resulted in practically no disruption to the outgoing packets. For incoming packets, since the mobile could receive packets from both access points (on the same frequency), no dropped packets were observed. However, there was a short time during the handoff when the same packets were transmitted through both access points, resulting in duplicate packets. Figure 7.7(a) shows the number of duplicate packets measured in different handoffs. By "handoff index," we simply mean an index number denoting a specific handoff among all the handoffs considered. The variation in the number of dropped packets was low (less than 5%), and this number did not change significantly when the video bit rate was doubled from 10 kb/s (low rate) to 20 kb/s (medium rate); this was because when the packet size changed, the packet rate remained roughly the same. Duplicate RTP packets should not pose a problem for most streaming video receivers. However, duplicates could be eliminated by performing a hard handover in the MMP gateway between sending the packets to the old and new access points, although this may lead to a few dropped packets.

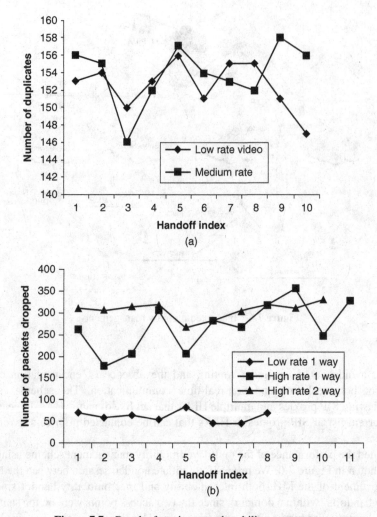

Figure 7.7   Results from integrated mobility management

Figure 7.7(b) shows the handoff behavior when the same MH moves across domains. The MH acquires a new IP address using DRCP, which triggers SIP re-INVITEs. However, it takes time to change the frequencies of the access networks and resume the physical-layer connectivity, and then to autoconfigure with a new IP address. Furthermore, more packets are lost owing to the longer traversal path of the redirected traffic. We used high-rate video traffic at a rate of 200 kb/s, either one-way ("high 1 way" in Figure 7.7) or in both directions ("high 2 way"), and low-rate one-way traffic ("low 1 way") at 10 kb/s. The rate of dropped packets increased slowly with the data rate. However, a SIP-based fast-handoff mechanism, as discussed in Chapter 6, can be used here to reduce the packet loss due to the longer binding update.

7.6 Concluding Remarks

In the course of developing and designing the prototype test bed, the following observations were made:

- Care must be taken to be consistent regarding the IP address that the MH uses to identify itself in micromobility zones. The IP address that the MH uses to identify itself is the address that is stored in the route caches of the MMP nodes. When this was the mobile's home address, we found that the technique worked best with FAs co-located with the micromobility gateway (e.g., MIP-LR can be used with FAs). Otherwise, packets would arrive for the MH addressed to its autoconfigured foreign network address, and the route caches would need to associate the two addresses (this can be handled by an MMP extension, but is less elegant). Conversely, when the MH identified itself by its foreign-network autoconfigured address, the technique worked best without FAs, since the route caches would be set up to forward to the foreign-network address in this case.
- Separation of real-time and non-real-time traffic is becoming more reasonable in practice. With standard tools such as iptables for Linux 2.4.7-10 and above, it is easy to set policy-based handling of different types of traffic, for instance to do MIP-LR processing only for non-RTP packets, bypassing SIP signaling packets and RTP packets based on the port numbers.
- There are other significant contributors to macromobility handoff latency besides MIP signaling latency. We found that the complete autoconfiguration process of IP address distribution using DCDP and IP address configuration using DRCP could take a few seconds, including reconfiguration of the wireless interface. In fact, our test bed typically did not have high network latency, but the macromobility handoff latency was still significantly higher than that for micromobility handoff.
- Changing the IP address as a result of mobility may require slight application-level changes. For MIP-LR-based macromobility, applications are unaware of IP address changes with mobility. However, for SIP macromobility, we had to modify our video and audio applications (VIC and RAT, respectively, both available as freeware on Linux), and to add modifications for interprocess communication with SIP UAs. In general, a mobility-aware RTP stack should be built to adapt itself to IP address changes. Some recently built RTP stacks (www.vovida.org) are in fact mobility-aware and can adapt themselves to changes during mobility.

8

Optimizations for Simultaneous Mobility

In this chapter, we analyze the problem due to nonreceipt of binding updates that results when two mobile nodes move simultaneously, and propose optimization techniques that increase the probability of successful handover in the simultaneous-mobility scenario. These optimization techniques could be applied to mobility protocols in several layers – network layer mobility protocols such as MIPv6 (Johnson et al., 2004) and MIP-LR (Jain et al., 1999), and application layer mobility protocols such as SIP-based mobility (Schulzrinne and Wedlund, 2000a).

8.1 Summary of Key Contributions and Indicative Results

Without a thorough analysis of the simultaneous mobility problem that arises from nonreceipt of binding updates when two hosts move when they are in communication with each other, it is difficult to predict the parameters that affect the simultaneous mobility problem and to propose solutions to mitigate the problems. Prior to the present study, there was no comprehensive study that analyzed the simultaneous mobility problem, nor was there any existing solution to mitigate the problems in an infrastructure-based mobility environment.

Here, we analyze the simultaneous mobility problem and develop an analytical framework to study the effect of the interhandoff rate of the mobiles and the binding update latency on the probability of occurrence of the simultaneous mobility problem. We have proposed timer-based retransmission, forwarding, and redirecting mechanisms using binding-update and location update proxies and the use of simultaneous bindings by the mobiles to eliminate the vulnerability of the binding update to simultaneous mobility. We have applied these solution mechanisms to several application layer and network layer mobility protocols, namely SIP-based mobility, MIPv6, and MIP-LR.

The proposed analytical framework for simultaneous mobility can predict the probability of simultaneous mobility based on the mobiles' interhandoff time and binding update latency. Each of the proposed techniques can be applied on either the sender side or the receiver side and work for both network layer and application layer mobility protocols, unlike the protocol-specific mechanisms proposed by Tilak and Abu-Ghazaleh (2001) and Dreibholz et al. (2003), which use TCP migrate and

Mobility Protocols and Handover Optimization: Design, Evaluation and Application, First Edition.
Ashutosh Dutta and Henning Schulzrinne.
© 2014 John Wiley & Sons, Ltd. Published 2014 by John Wiley & Sons, Ltd.

SCTP extensions. Each of our proposed solution mechanisms reduces the interval of vulnerability to simultaneous binding updates.

In the rest of the chapter, we introduce the simultaneous mobility problem and illustrate this problem for different mobility protocols, develop the analytical framework, prove lemmas covering two cases of simultaneous mobility, and propose solution mechanisms that can be applied to several network layer and application layer mobility protocols, namely MIP-LR, MIPv6, and SIP.

8.2 Introduction

Zhuang et al. (2005) proposed seven properties that are needed to fully realize the promise of ubiquitous mobility. These properties also include simultaneous mobility. It is expected that, in most scenarios, nonsimultaneous mobility will occur more frequently than simultaneous mobility. Nonsimultaneous mobility refers to mobility of one end host while the other remains stationary. Nevertheless, simultaneous mobility will happen once in a while and must be handled properly by the mobility protocols.

The simultaneous mobility problem occurs when two mobile nodes are part of a communication session in the normal state, and they both move such that the binding updates that they send to each other are both lost through belated arrival of a binding update, and such that the communication session never returns from the interrupted state to the normal state. More precisely, the simultaneous mobility problem can be defined as the problem of losing a binding update from one mobile node because it has been sent to a previous address of the other mobile node that is also moving at around the same time. The disruption caused by simultaneous mobility may far exceed the disruption caused by nonsimultaneous mobility.

Thus, the optimization techniques related to simultaneous mobility are a special type of the optimization for binding update discussed in Chapter 6. Any solution for simultaneous mobility should ensure that the end hosts should be able to move simultaneously without breaking an ongoing session between them owing to a delayed binding update.

8.2.1 Analysis of Simultaneous Mobility

In this section, we analyze the simultaneous mobility event and describe several concepts associated with simultaneous mobility.

We limit the analysis of simultaneous mobility primarily to layer 3 handoff, that is, where the IP addresses of the mobile nodes change. A binding update carries information about the location of the sending mobile host, including its new IP address. A binding update is lost if it does not arrive at its intended recipient mobile host. It makes a belated arrival if it arrives at a network where the destination address used to be valid for the intended recipient mobile host but is no longer valid at the moment of arrival. Binding updates do not contain information about future moves of the sending mobile host. While two mobile hosts are in a communication session, they get information about the location of the other mobile host only from binding updates. In other words, they do not actively seek the location of the other mobile host, but only passively accept binding updates. Binding updates are sent directly to the most current known address (known by the sender) of the intended recipient mobile host. In general, the latency associated with binding updates is assumed to be much smaller than the average interhandoff time; therefore, it is extremely unlikely that a binding update will be sent and the recipient mobile host will move twice before the binding update arrives in the previous network of the recipient.

The most basic version of the simultaneous mobility problem is shown in Figure 8.1. There are two nodes, A and B. Time is in the vertical direction (and flows downward), whereas the spatial location is in the horizontal direction. Node A moves from domain A1 to A2 while Node B moves from domain B1 to domain B2. After their respective moves, these two nodes send binding updates to each other, and both binding updates are lost. Additionally, there are proxies and servers in the network as well.

Standard mobility protocols such as the original Mobile IP (MIP) handle simultaneous mobility adequately, because of the nonmobile home agents. The home agent of a mobile host functions as an anchor point for the mobile host. No matter where the mobile host moves, packets for it always go first to its home network for interception and are tunneled by its home agent. If it turns out that the correspondent host is also mobile, it will also have a home agent, and packets from the mobile host will similarly be intercepted and tunneled to the appropriate network by its home agent. Since both home agents are stationary and can always be reached through IP routing, simultaneous mobility does not present a problem in MIPv4.

However, the simultaneous mobility problem occurs in scenarios where the end hosts can send binding updates directly to each other. Therefore, we have analyzed the simultaneous mobility problem for network layer mobility protocols such as MIPv6 and MIP-LR and application layer mobility protocols such as SIP, and proposed a common framework for solution of the problem. It is important to note that the problem of simultaneous mobility is very similar in these protocols because these protocols allow binding updates to be sent to the communicating hosts directly. The proposed solutions are designed to impose minimal changes on the existing protocols while efficiently dealing with the simultaneous mobility problem. We focus on situations where the handoff rate of a mobile node is such that consecutive handoffs of the same mobile node do not overlap. We do not focus on the situation of overlapping consecutive handoffs of the same mobile node, where one handoff has not completely finished before the next one begins, for example when there has not been enough time after the acquisition of an IP address for binding updates to reach their destination networks.

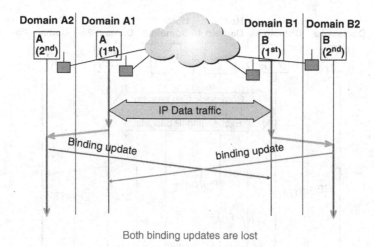

Both binding updates are lost

Figure 8.1   Simultaneous-mobility scenario

The following are the reasons for the assumptions that we have made:

- The problems encountered with overlapping consecutive handoffs are not so much a problem of simultaneous mobility as one of excessive handoff rate. There will be severe problems, leading to complete deadlock situations, when the mobile node changes its IP address before the binding updates for its previous IP addresses have even arrived at their destinations.
- For the foreseeable future, the extreme case of handoff rates high enough for overlapping consecutive handoffs is highly improbable. Hence, we assume that consecutive handoffs of the same mobile node are nonoverlapping, and we focus on the overlap of handoffs of different mobile nodes, that is, simultaneous mobility.

8.3   Illustration of the Simultaneous Mobility Problem

In this section, we illustrate how the simultaneous mobility problem is encountered in SIP-based mobility, MIPv6, and MIP-LR.

Figure 8.1 is easily adapted to illustrate the simultaneous mobility problem with SIP, as shown in Figure 8.2. In Figure 8.2, the binding updates are denoted by re-INVITE messages. The main difference is that there are two additional SIP servers with proxy and redirect functionality in each network, one for each mobile node in its home network. In general, when the re-INVITE messages are lost, the servers do not intervene, but we have proposed a solution that could use these servers and solve the simultaneous mobility problem in SIP-based mobility. These proposed solutions are described in Section 8.6. Furthermore, Figure 8.2 can also be used to illustrate MIP-LR's simultaneous mobility problem. In the case of MIP-LR, the SIP servers are replaced with MIP-LR home location registers (HLRs), and the re-INVITE messages are replaced with MIP-LR binding updates.

Similarly, MIPv6 is vulnerable to the simultaneous mobility problem because of the direct binding updates and associated return routablity procedures. The direct binding updates from the mobile node to the correspondent nodes pose a security problem. Thus, the return routability procedure allows the mobile node and correspondent node to set up a shared key in a "reasonably" secure manner.

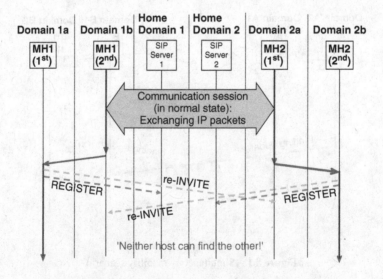

Figure 8.2   Simultaneous mobility in SIP

In the return routability procedure, the mobile node sends two messages to the correspondent node, namely home test init (HTI) and care-of test init (CTI) messages. These messages are sent through the home agent (reverse tunneled to the home agent from the mobile node, and then forwarded to the correspondent node) and directly to the correspondent node, respectively. The correspondent node replies by sending two tokens to the mobile node, one directly to the mobile node addressed to its care-of address (the care-of test message), and the other to the home address of the mobile node (the home test message). The mobile node needs both of these tokens to be able to generate the shared key. Thus, the return routability procedure ensures that the mobile node is what it claims to be by testing that it is reachable both by the direct path and through its home address. Subsequently, the correspondent node can accept binding updates directly from the mobile node.

However, the additional message exchange due to the return routability procedure adds to the existing simultaneous mobility problem. Figure 8.3 illustrates the simultaneous mobility problem in Mobile IPv6. The following are three different possible scenarios that could result in simultaneous mobility in MIPv6:

1. Both sides' CTI and HTI messages are lost because of simultaneous mobility. This would look like Figure 8.2 except that CTI and HTI messages are lost instead of re-INVITE messages (and home agents are used instead of SIP servers).
2. One side actually completes return routability, but then its binding update is lost because the other side moves. This interesting asymmetric scenario is illustrated in Figure 8.3.
3. Both sides complete the return routability checks, but then their binding updates are lost owing to simultaneous mobility.

We propose solutions to deal with the simultaneous mobility problem for SIP-based mobility, MIP-LR, and MIPv6 in Section 8.6.

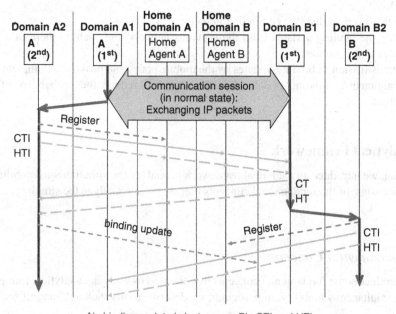

A's binding update is lost, as are B's CTI and HTI

Figure 8.3 Simultaneous mobility in MIPv6

8.4   Related Work

There have been only a few papers that discuss simultaneous mobility. Tilak and Abu-Ghazaleh (2001) extended the TCP migration mobility protocol (Snoeren and Balakrishnan, 2000) to handle simultaneous mobility, but there are significant differences between the TCP migration schemes (where mobility is handled in the transport layer) and MIP-related protocols or SIP-based mobility protocols. Dreibholz et al. (2003) proposed a scheme that handles simultaneous mobility in a layer between the transport and application layers. In that scheme, mobility is handled using Stream Control Transmission Protocol (SCTP) extensions. However, no analytical framework or theorems and proofs related to the simultaneous mobility problem for SIP, MIPv6, and MIP-LR have been proposed before.

   We have analyzed the simultaneous mobility problem for MIP-LR, SIP, and MIPv6 in Wong et al. (2003b) and Wong and Dutta (2005). We have also developed some common approaches that could be applied to provide solutions for mobility protocols such as MIPv6, MIP-LR, and SIP-based mobility. We have described these results in Daniel Wong et al. (2007).

8.5   Key Optimization Techniques

The following are some of the key fundamental techniques that should be considered for optimizing the handoff event during simultaneous mobility of the communicating hosts:

1. Many of the principles related to optimization of binding update are applicable to the simultaneous-mobility scenario. However, the handoff rate of the mobile nodes will determine if any of those techniques can be applied to the simultaneous-mobility scenario.
2. Reducing the effect of delayed direct binding updates by introducing an anchor point closer to the mobile.
3. Limiting the traversal distance of binding updates.
4. Forwarding the binding updates from the previous network and caching them in a forwarding agent closer to the mobile.
5. Using retransmission of binding updates by the mobile nodes and proxies to complete the update.
6. Using simultaneous binding updates by the mobile to reduce the probability of failure of reconnection.

8.6   Analytical Framework

In this section, we introduce an analytical framework to analyze the simultaneous mobility problem. We first define some of the fundamental concepts that are used to analyze the simultaneous-mobility framework.

8.6.1   Fundamental Concepts

Here, we introduce some fundamental concepts that are used to study the analytical framework associated with simultaneous mobility. In particular, we describe terms such as "handoff sequence" and "binding update."

Definition

Two mobile nodes are in a communication session if they are actively exchanging data. A communication session may be in a normal state or an interrupted state. The session is in a normal state when data from one node is arriving at the right location for the other node, and vice versa. It is in an interrupted state otherwise.

Example

A communication session is typically in an interrupted state from the moment a handoff occurs until data starts arriving again at the new attachment point (e.g., after a binding update has been received at the other node). This alternation between the normal state and the interrupted state is shown in Figure 8.4. We explain this figure in more detail later.

8.6.2 Handoff Sequences

As defined in Appendix B, a handoff is a movement of a mobile node from a previous attachment point to a new attachment point. During simultaneous mobility, the handoff time of a handoff instance is the moment in time when it changes from being reachable at the previous attachment point to not being reachable at that attachment point. Let the handoff time (of a particular handoff instance) be T. Then the node needs time for network configuration, so it becomes reachable (with a valid IP address in the new network) at a time $T + \gamma$. If there is a correspondent node, then at some time later, $T + \gamma + \zeta$, it sends a binding update to the correspondent node. The binding update arrives at time $T + \gamma + \zeta + \Delta$. We shall use these symbols to represent these differential times after a handoff. For convenience, we can write $X(i) = \gamma(i) + \zeta(i) + \Delta(i)$, as shown in Figure 8.4. So, $T(i) + X(i)$ denotes the time when the binding update arrives at the other node. Given that time is continuous, it is assumed that only one handoff can occur at any given moment in time; that is, handoff times are unique. It should be noted that the definition of handoff and handoff time may not be applicable to certain types of IP-layer soft handoff or physical-layer soft handoff, as in CDMA systems and bicasting or multicasting schemes.

Figure 8.5 shows a scenario where A and B are two mobile nodes that are in a communication session with each other, during which each node performs zero, one, or more handoffs.

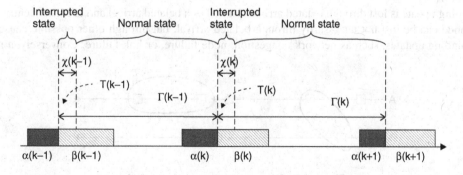

Figure 8.4 Notation for simultaneous-mobility framework

Definition

The handoff sequence of A is the ordered set

$$H_A = \{T_A(0), T_A(1), \ldots, T_A(N_A - 1)\} \tag{8.1}$$

and the handoff sequence of B is the ordered set

$$H_B = \{T_B(0), T_B(1), \ldots, T_B(N_B - 1)\}, \tag{8.2}$$

where $T_A(i)$ is the handoff time of the ith handoff of A, so that $T_A(i) < T_A(j) \; \forall \; i,j$ such that $0 < i < j < N_A - 1$, and the same holds good for B. The function arguments i,j are the handoff index numbers. In general, when necessary, we shall use subscripts to indicate the mobile node and show the handoff index number in function arguments.

Two handoffs are consecutive (with respect to a pair of mobile nodes) if neither of the mobile nodes performs another handoff in between the two handoffs. For example, if the two handoffs are at A and B, at times $T_A(i_0)$ and $T_B(j_0)$, and if A's handoff is earlier, then saying they are consecutive is equivalent to saying $T_A(i) \in H_A : T_A(i_0) < T_A(i) < T_B(j_0) = \emptyset$ and $T_B(j) \in H_B : T_A(i_0) < T_B(j) < T_B(j_0) = \emptyset$. As defined, then, consecutive handoffs could be at the same mobile node or at two different mobile nodes. Figure 8.5 shows two examples of consecutive handoffs, one in which the two handoffs are at different mobile nodes and one in which they are at the same mobile node.

8.6.3 Binding Updates

Definition

A binding update is lost if it does not arrive at its intended recipient node.

Definition

A binding update makes a belated arrival if it arrives in a network where the destination address used to be valid for the intended recipient node but is no longer valid (for the intended recipient) at the moment of arrival. For example, if A is the sender and B is the intended recipient, and we are considering the binding update for A's ith handoff, then if B's next handoff is its jth, then the binding update makes a belated arrival if and only if $T_A(i) + \gamma_A(i) + \zeta_A(i) + \delta_{A \rightarrow B}(i,j) > T_B(j)$.

Definition

A binding update is lost through belated arrival if it makes a belated arrival and is consequently lost.

A node can be lost not necessarily through belated arrival, but through other possible causes of lost binding updates, such as network congestion, node failure, or link failure. Conversely, a node

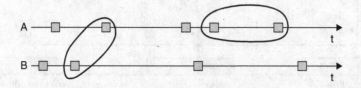

Figure 8.5 Examples of consecutive handoffs

can make a belated arrival and not be lost, for example if there is an agent in the network that can forward the binding update to the current location of the intended recipient.

Furthermore, the following assumptions are made about binding updates for simultaneous mobility:

1. Binding updates cannot and do not contain information about future moves of the sending node.
2. While two nodes are in a communication session, they get information about the location of the other node only from binding updates; that is, they do not actively seek the location of the other node, but only passively accept binding updates.
3. Unless otherwise stated, a binding update is sent directly to the most current known address (i.e., known by the sender) of the intended recipient.
4. Regarding the relative timings of binding update latencies and consecutive handoffs of a receiving mobile node, the timescale of the latencies for binding updates is assumed to be much smaller than the average interhandoff time. In other words, $\delta \ll ET(i+1) - T(i)$, where $E(.)$ denotes the expectation.
5. It is extremely unlikely that a binding update will be sent and the recipient moves twice before the binding update arrives at the previous network of the recipient.
6. It is also assumed that if there is a forwarding location proxy (defined in Section 8.6.4) in the previous network of the recipient, it will correctly forward the binding update to the recipient, which would only have moved once from the previous network.

8.6.4   Location Proxies and Binding Update Proxies

Here we introduce two kinds of stationary proxy for mobility signaling. These proxies, if used carefully, can help prevent the simultaneous mobility problem. These proxies are abstract proxies –the definitions are more about network functionality than specific implementations as network elements.

Thus, it can be seen how familiar network elements such as home agents can be described as having certain proxy functions, or can be enhanced for such purposes. The abstraction of these proxies will allow general problems and solutions (related to simultaneous mobility) to be discussed without us being unnecessarily bogged down in details of specific mobility protocols. It is also assumed that these proxies are stationary, not mobile.

8.6.4.1   Location Proxy

A location proxy (of a mobile node) is a network function that is used to locate the mobile node. There can be three kinds of location proxy. A *forwarding location proxy* forwards messages (including binding updates) to the most recent location that it knows for the mobile node. A *redirecting location proxy* redirects messages (e.g., by responding to a query with the latest address) to the most recent location that it knows for the mobile node. An *intercepting location proxy* intercepts, and may act on (forward or redirect), messages in packets not addressed to it. A nonintercepting location proxy only acts on messages in packets addressed to it. The fundamental differences between the types of proxy are shown in Figure 8.6. It is important to note that whereas a forwarding location proxy will pass messages along towards the final destination, a redirecting location proxy will not do this, but just return location information that can be used to send the message towards the final destination.

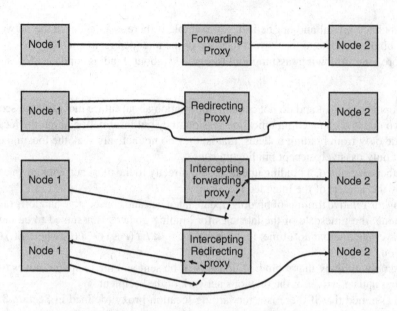

Figure 8.6　Abstract functions of the location proxies

A location proxy is up to date with respect to a particular mobile node if that mobile node continually updates the proxy with its latest address after each move.

8.6.4.2　Proactive Location Proxy

A *proactive location proxy* keeps a copy of mobility-related signaling messages (typically binding updates, but possibly other messages such as care-of test init messages, when procedures such as return routability are used before the binding update is sent). It keeps the messages for a short time, X, after receiving and acting on them (i.e., after redirecting and/or forwarding the message). The messages are kept in the location proxy cache and indexed by the destination node. The messages are discarded after a time X has elapsed. If, during this period of time, the proactive location proxy receives a binding update from any one of the destination nodes in its location proxy cache, it either (a) redirects to the new address (if it is a redirecting proactive location proxy) or (b) forwards the corresponding saved message(s) to it (if it is a forwarding proactive location proxy).

Here are some examples of how some of the mobility components in different mobility protocols behave as different kinds of proxies. The Mobile IP home agent is a forwarding location proxy (of the intercepting kind). DNS servers are nonintercepting redirecting location proxies. MIP-LR home location registers are nonintercepting redirecting location proxies. SIP proxy servers are nonintercepting proxies that can be either forwarding location proxies, known as proxy servers in SIP terminology, or redirecting location proxies, known as redirect servers in SIP terminology. Except for DNS servers, the examples given here are typically used in mobility schemes as up-to-date location proxies. In TCP migration, though, DNS servers are part of the mobility scheme, and so they are up-to-date location proxies in that scheme. Most existing location proxies are not proactive location proxies. However, proactive location proxies (defined in Section 8.6.4) may be useful for providing solutions to the simultaneous mobility problem. The current solutions take only signaling into account. It is assumed that if the mobility signaling gets to its intended recipient, the mobility scheme will, and must, take care of the data traffic correctly. However, in some cases, for example with Mobile IP, the home agent forwards both signaling and data, whereas in other cases, for example MIP-LR and SIPMM (SIP-based mobility), the location registers or SIP servers are involved only in the signaling.

8.6.4.3 Binding Update Proxy

A binding update proxy acts on behalf of a mobile node to send its binding updates to its correspondent node's latest addresses. It typically engages the services of a location proxy of each correspondent node either for redirection to the correspondent node's latest address or for forwarding the relevant binding update. At the same time, it also forwards a copy of the message to the latest address that it knows for the correspondent node. A mobile node on whose behalf a binding update proxy acts may be referred to as a master of that binding update proxy.

8.6.4.4 Proactive Binding Update Proxy

A proactive binding update proxy not only queries for the latest addresses of the correspondent nodes of its master(s), but also keeps the binding updates for a short time α after receiving and forwarding them. The messages are kept in the binding update proxy cache and indexed by destination node. The messages are discarded after a time α has elapsed. If, during this period of time, the proactive binding proxy receives a redirection regarding any one of the destination nodes in its binding update proxy cache, it forwards the corresponding saved binding update(s) to it.

8.7 Analyzing the Simultaneous Mobility Problem

We now prove four lemmas that cover the two cases of what happens when there is a pair of handoffs at two mobile nodes, and the binding update from the earlier one arrives (a) later than the time the other node moves (Lemmas 8.1 and 8.2) or (b) earlier than the time the other node moves (Lemmas 8.3 and 8.4).

Lemma 8.1 *Given a pair of consecutive handoffs, one for each of the two mobile nodes in a communication session in the normal state, in the absence of location proxies for either mobile node, any binding update sent by the earlier-moving mobile node will be lost through belated arrival if and only if the binding update does not arrive at the other mobile node before it moves.*

Proof. Suppose, without loss of generality, that node A moves before node B. Let the handoff times be $T_A(i_0)$ and $T_B(j_0)$, so $T_A(i_0) < T_B(j_0)$. Since node A and node B are in a communication session in the normal state up until $T_A(i_0)$, then up until $T_A(i_0)$, anything sent by A arrives at B and vice versa. Since the two handoffs are consecutive, then by definition there is no other handoff in the time interval $[T_A(i_0), T_B(j_0)]$. By our third assumption about binding updates, A's binding update would be addressed to the latest address that it has for B. So, for the time interval $[T_A(i_0), T_B(j_0)]$, anything sent by A will still be addressed to B's prehandoff address, and still arrive at B, including A's binding update. However, as soon as $t \geq T_B(j_0)$, B is no longer be reachable at its prehandoff address. In some scenarios, a location proxy for B would be able to prevent the binding update from being lost. However, in the absence of a location proxy, the binding update would just go to B's previous address and disappear there. Hence, it would be lost through belated arrival.

Conversely, suppose A's binding update is lost through belated arrival. As shown, for the time interval $[T_A(i_0), T_B(j_0)]$, anything sent by A will still be addressed to B's prehandoff address and still arrive at B, including A's binding update. So, if it arrives before $T_B(j_0)$, it will not be lost through belated arrival. Thus, A's binding update cannot arrive before $T_B(j_0)$. Therefore it arrives after B has moved.

This lemma and the next make assertions about cases where the binding update from the earlier-moving mobile node arrives after the later-moving mobile node has moved. This is shown in Figure 8.7.

Lemma 8.2 *Given a pair of consecutive handoffs, one for each of two mobile nodes in a commu-nication session in the normal state (up until the first handoff), in the absence of location proxies for either mobile node (or there might be location proxies but they are not used or not involved), the simultaneous mobility problem will occur if and only if the binding update sent by the earlier-moving mobile node does not arrive at the other mobile node before it moves.*

Proof. We use A and B as the first and second mobile nodes again. If the binding update from A does not arrive at the other node before it moves, then by Lemma 8.1, it is lost through belated arrival. Then, at time $T_B(j)$, B does not have A's new address. Since A's binding update is lost, by the time B sends its binding update (i.e., $T_B(j) + \lambda_B(j) + \zeta_B(j)$), it will send it to A's previous address. Thus, in the absence of location proxies, B's binding update will also be lost through belated arrival. So both A's and B's binding updates are lost through belated arrival. By definition, the simultaneous mobility problem has occurred.

Conversely, if the simultaneous mobility problem has occurred, then if both of the binding updates have been lost through belated arrival, then A's binding update is lost through belated arrival.

From the above proof, the following corollary emerges.

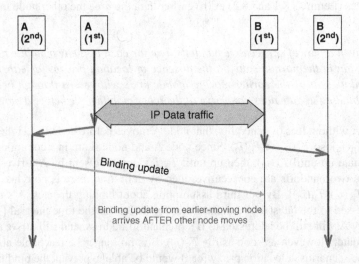

Figure 8.7   Lemmas 8.1 and 8.2

Corollary 8.1 *Given a pair of consecutive handoffs, one for each of two mobile nodes in a communication session in the normal state (up until the first handoff), in the absence of location proxies for either mobile node (or there might be location proxies but they are not used or not involved), if the binding update from the node that moved first is lost through belated arrival, the binding update from the node that moved second will also be lost through belated arrival.*

Lemma 8.3 *Given a pair of consecutive handoffs, one for each of two mobile nodes in a communication session in the normal state (up until the first handoff), the simultaneous mobility problem does not occur if the binding update from the node that moved earlier reaches the other node before that node moves.*

Proof. As in the proof of Lemma 8.1, one can argue that for the time interval $[T_A(i_0), T_B(j_0)]$, anything sent by A still arrives at B, including A's binding update. Thus, B can then send its binding update correctly to A's new address after it moves at $T_B(j_0)$. Therefore, the simultaneous mobility problem does not occur.

This lemma and the next make assertions about cases where the binding update from the earlier-moving mobile node arrives before the later-moving mobile node has moved. This is shown in Figure 8.8. NB: the converse of Lemma 8.1 is not necessarily true; that is, one cannot say that if the simultaneous mobility problem does not occur, then the binding update from the node that moved earlier reaches the other node before that node moves. The reason this is not necessarily true is that location proxies could be used, as we will demonstrate later. However, we first extend Lemma 8.3 to the case in which location proxies are excluded, where we can make a stronger statement.

Lemma 8.4 *Given a pair of consecutive handoffs, one for each of two mobile nodes in a communication session in the normal state (up until the first handoff), in the absence of location proxies for either mobile node (or there might be location proxies but they are not used or not involved), the simultaneous mobility problem does not occur if and only if the binding update from the node that moved earlier reaches the other node before that node moves.*

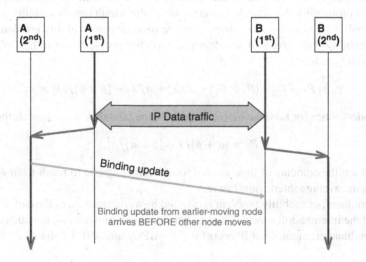

Figure 8.8 Lemmas 8.3 and 8.4

Proof. This has been partially proved in the proof of Lemma 8.3. What remains is to prove that if the simultaneous mobility problem does not occur, the binding update from the node that moved earlier reaches the other node before that node moves. Supposing the simultaneous mobility problem does not occur, this means that both binding updates arrive at the other node. B's binding update therefore cannot be lost through belated arrival, so B must have successfully received A's binding update. By the third assumption about binding updates, A's binding update could not have been addressed to B's new location, since A moved first. Given that location proxies are not used, there is no way that B could successfully receive A's binding update after $T_B(j_0)$. Therefore, A's binding update must have reached B before $T_B(j_0)$, that is, before B moved.

Remark 8.1 *What if A's binding update successfully reaches B before it moves, but B's binding update does not reach A because A's next handoff happens before the arrival of B's binding update? Does Lemma 8.3 break down? No, in that case the lemma, applied to $H_A(i_0)$ and $H_B(j_0)$, would correctly show that the problem is not between those two handoffs, but, applied to $H_B(j_0)$ and $H_A(i_0 + 1)$, it would correctly show that the simultaneous mobility problem occurs then, with B as the earlier-moving node.*

8.8 Probability of Simultaneous Mobility

In this section, we analyze the probability of simultaneous mobility. In Wong et al. (2003b), we introduced a simple mathematical model for estimating the probability of occurrence of simultaneous mobility. We describe this model briefly in this section.

The probability that any particular handoff (either separately or both at the same time) suffers from the simultaneous mobility problem is P_0, and the probability that at least one out of N handoffs in a given session suffers from the simultaneous mobility problem is P_N. Thus $P_N = 1 - (1 - P_0)^N$. The system goes into an interrupted state if either or both of the mobiles are subject to the simultaneous mobility problem. The interhandoff times for mobile 1 (A) and mobile 2 (B) are λ_1 and λ_2, respectively.

α is the time taken for A's binding update to reach B. β is the time taken for B's binding update to reach A. The probability that mobile 1 contributes to the simultaneous mobility problem is $P_1 = \beta/\lambda_1$. The probability that mobile 2 contributes to the simultaneous mobility problem is $P_2 = \alpha/\lambda_2$. Thus, the probability that there is a simultaneous mobility problem due to handoff by mobile 1, mobile 2, or both is as follows:

$$P_0 = P_1 + P_2 - (P_1 \times P_2) = \beta/\lambda_1 + \alpha/\lambda_2 - [\beta \times \alpha]/[\lambda_1 \times \lambda_2].$$

If the interhandoff times for both the mobiles are the same, and thus $\lambda_1 = \lambda_2 = \lambda$, then

$$P_0 = [\alpha + \beta]/\lambda - [\beta \times \alpha]/\lambda^2,$$

where α and β are the amounts of time needed for a binding update to reach from A to B and vice versa, and λ is the average interhandoff time.

Thus, the simultaneous mobility problem is affected by a combination of the end-to-end latency of the packet and the interhandoff time. As part of our initial experiments, we conducted a preliminary analysis of simultaneous mobility of IP hosts for SIP, MIPv6, and MIP-LR-based mobility protocols.

Consider two consecutive handoffs, one each at mobile nodes A and B. According to Lemma 8.2, the simultaneous mobility problem occurs if and only if the binding update from the earlier-moving node arrives after the other node has moved. Mathematically, this is written as $T_A + \gamma + \zeta + \delta_{A \to B} > T_B$ (if A is the earlier-moving node) or $T_B + \gamma + \zeta + \delta_{B \to A} > T_A$ (if B is the earlier-moving node). Putting the two inequalities together, the following equation is obtained:

$$T_A - \alpha < T_B < T_A + \beta, \tag{8.3}$$

where $\alpha = \gamma + \zeta + \Delta_{B \to A}$ and $\beta = \gamma + \zeta + \Delta_{A \to B}$ are convenient short forms.

We now define the concept of the "vulnerability interval" $\beta + \alpha$, which is the time around a handoff during which the two mobile nodes are vulnerable to the simultaneous mobility problem if another handoff occurs at the other mobile node.

It is reasonable to model the handoff times for A and B as independent Poisson processes. In this model, the intervals between consecutive handoffs at A, $\Gamma_A(k-1) = T_A(k)A - T_A(k-1)$, $\Gamma_A(k) = T_A(k+1)A - T_A k$, and so on, are independent exponentially distributed random values, and similarly for the corresponding intervals between consecutive handoffs at B. It is then easy to argue that the probability of the simultaneous mobility problem occurring can be estimated by the following equation:

$$P_0 = \frac{E(\alpha + \beta)}{E(\Gamma)} - \frac{E(\alpha \times \beta)}{E(\Gamma^2)}. \tag{8.4}$$

If there are N handoffs occurring at each of the two mobile nodes, then the probability of the simultaneous mobility problem occurring can be estimated as

$$P_N = 1 - (1 - P_0)^N. \tag{8.5}$$

Based on experimental measurements, $E[\alpha + \beta]$ ranges from 50 to 500 ms, while λ may range from 5 s (movement at vehicular speeds across picocells a few hundred meters in diameter) to 500 s or more (larger cells, slower speeds, and nonlinear movement pattern). $E(\Gamma)$ is the average value of the interhandoff time, which can be equated to λ. In Figure 8.9, we plot P_0 for approximately this range of $E[\alpha + \beta]$ and λ. Figure 8.9 shows how the probability of simultaneous mobility P_0 is affected by the binding update latency and mean handoff time of the mobile based on Equation 8.4 . Figure 8.9(a) shows that for a given interhandoff time (500 s), the probability of failure increases as the one-way latency of the binding update increases, while Figure 8.9(b) shows that for a given one-way-delay (50 ms) the probability of failure increases as the interhandoff time decreases. Figure 8.10 shows the probability of simultaneous mobility P_3 when the total number of handoffs is 3, based on Equation 8.5 . For the same values of interhandoff time and one-way latency, the probability of failure due to simultaneous mobility increases when the number of handoffs is increased to 3. As expected, as shown in Figure 8.9(a), the highest probability of simultaneous mobility occurs when the one-way packet latency is largest and the average interhandoff time is smallest. Thus, the effect of the simultaneous mobility problem could be quite significant. If the problem was not fixed, the binding updates of the two mobile hosts would never reach the other host, and so the connection would be lost. It is important to note that this analysis is optimistic, as it is assumed that the binding update from A to B would not be lost. Since there is a small chance that this binding update might also be lost, the values computed in this analysis could be viewed as merely providing a lower bound on the likelihood of simultaneous mobility occurring. In the case of simultaneous

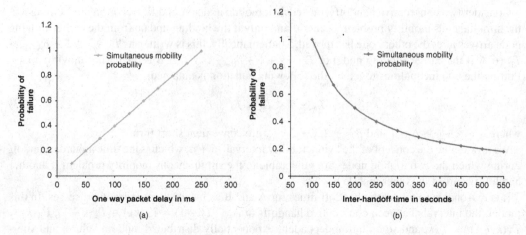

Figure 8.9   Plots of P_0 against latency and mean handoff time

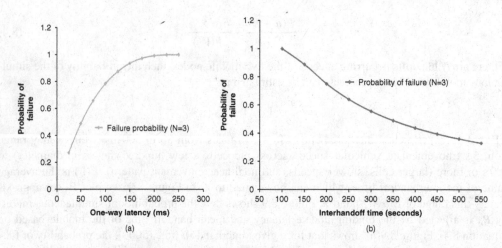

Figure 8.10   Plots of P_N against latency and mean handoff time

mobility during session initiation signaling, the probability of failure also depends upon the mobility rate of the mobiles. From our laboratory measurements, it takes about 200–300 ms to complete the whole session initiation signaling sequence. A complete registration takes about 150 ms. Hence, the probability of simultaneous mobility occurring during session initiation signaling is nontrivial.

8.9   Solutions

Kravets et al. (2001) have hinted at the benefits of using some kind of proxies in fixed locations to enable communication to continue even when both end hosts move simultaneously. However, as far as we know, previous work has not analyzed the problem to the level we have described in Section 8.7, nor has a systematic analysis of solutions applied to a range of mobility protocols been previously provided. We describe some solution mechanisms in this section. These solution mechanisms are specific mechanisms and functions that could be used (typically in conjunction with other mechanisms and functions) to provide solutions to the simultaneous mobility problem for a given mobility protocol.

The proposed techniques can be broadly classified into three types: *soft-handoff*, *sender-based*, and *receiver-based* mechanisms.

8.9.1 Soft Handoff

Suppose a mobile node can have more than one valid IP address. In such cases, the mobile can have two bindings associated with its home address. These are sometimes referred to as simultaneous mobility bindings, and should not be confused with the simultaneous mobility problem. We call this the soft-handoff approach, since it is similar to the soft handoffs in CDMA mobile systems (Wong and Lim, 1997). The idea is that if the previous IP address and the new IP address can both be used to reach the mobile node during the handoff process, that can solve the simultaneous mobility problem. Binding updates sent to the previous IP address would arrive correctly. However, although most of the current operating systems can support multiple concurrent IP addresses for wireless interfaces, this is not a universally applicable solution, for the following reasons:

- The operating system needs to be able to support multiple concurrent IP addresses for the wireless interface(s).
- The network interface of the mobile needs to be able to connect simultaneously to multiple base stations, which may belong to two different subnets. CDMA technology provides this ability, whereby the network interface can connect to two network access points simultaneously. However, this is limited to the CDMA access technology only.
- Resource utilization is not efficient, because of redundant allocation of bandwidth resources (on both communication paths) during the period of simultaneous mobility bindings.
- It is also important to make sure that the simultaneous mobility bindings are active to ensure that no problems will occur during simultaneous mobility. The longer it waits, the more this specific scheme uses valuable network resources redundantly. Since this solution must work for any radio network technology, the use of simultaneous bindings is not a satisfactory solution. Hence, mechanisms related to the soft-handoff approach will not be described further in this chapter.

8.9.2 Receiver-Side Mechanisms

Receiver-side mechanisms can typically be deployed in the previous network or home network of the receiver and act on behalf of the receiver to help it to be located. Retransmission, forwarding, redirecting, proactive forwarding, and proactive redirecting are some of the mechanisms that have been analyzed for the two mobility protocols SIP and MIP-LR. Details of the proposed mechanisms and results can be found in Wong et al. (2003b) and Daniel Wong et al. (2007). The following is a list of receiver-side mechanisms.

8.9.2.1   Timer-Based Retransmissions

One can imagine a forwarding location proxy automatically retransmitting a binding update if it has not received confirmation that the binding update was successfully received by the intended receiver. This location proxy could be located in the receiver's home network or in a visited network (e.g., the previous network or the latest network). Location proxies that retransmit based on timeouts are similar to proactive location proxies in that both need to store the message briefly, to be retransmitted if necessary. The difference is in the conditions for retransmission. A proactive proxy retransmits as soon as a new address is obtained, whereas a timer-based retransmission may be too slow. In the existing implementations, a stateful SIP proxy can retransmit binding updates (re-INVITE) after the expiration of a timer. This can be located in the home network or visited network of the receiver. In order to ensure that the re-INVITE message (and other signaling) goes through this server, the Record-Route option could be used in the initial INVITE message to add the server to the signaling path.

8.9.2.2   Regular Passive Forwarding

Forwarding mechanisms on the receiver side allow binding updates to be forwarded from a location proxy in the previous network to the correct new location of the receiver. Forwarding mechanisms from a previous network may also forward data packets (since the location proxy is forwarding packets anyway, it might as well forward data packets). One could also imagine such a location proxy in the receiver's home network as well.

The following are some existing implementations where the forwarding agents are in the receiver's previous network. In MIP-RO (Mobile IP with Route Optimization), the previous foreign agent serves this role by forwarding data packets. Unfortunately, this ability is missing from MIPv6, perhaps because no foreign agents are used in MIPv6 (and so there is no natural forwarding agent present in the previous network). Thus, the problem of simultaneous mobility remains in MIPv6. Similarly, SIP-based mobility management and MIP-LR lack such functionality. In some examples, the forwarding agents can be in the receiver's home network. An example is the home agent in MIPv4 and MIPv6. A SIP server in the receiver's home network could also serve in this capacity, for example if it places itself in the signaling path using the Record-Route field in the initial INVITE message.

8.9.2.3   Proactive Forwarding

Regular passive forwarding may be insufficient to solve the simultaneous mobility problem. A proactive forwarding location proxy may help, where forwarding takes place before the handoff.

8.9.2.4   Redirecting

Redirecting mechanisms on the receiver side can help to get messages such as binding updates to the right place. There are some existing implementations where the redirecting agents are placed in the receiver's previous network and home network. In MIP-RO, the previous foreign agent serves this role. In MIP-LR, the HLR does this, but only before a media session begins. Thereafter, it is not involved in control signaling during the communications session. Thus, it does not count as a proper implementation of a solution mechanism for the simultaneous mobility problem.

8.9.2.5 Proactive Redirecting

Regular redirecting may be insufficient to solve the simultaneous mobility problem. A proactive redirecting location proxy may help in some cases, where there is a probability of handover to a number of target networks.

8.9.3 Sender-Side Mechanisms

Sender-side mechanisms typically can be deployed in the home network of the sender or in the sender itself, and act on behalf of the sender to try to reach the receiver. The receiver may be moving simultaneously with the sender and may not receive the binding update if none of these mechanisms are used. The following is a list of sender-side mechanisms.

8.9.3.1 Timer-Based Retransmissions

A forwarding location proxy automatically retransmits a binding update if it has not received confirmation that the binding update was successfully received by the intended receiver. This location proxy can be located in the sender's home network or even in the sender itself (for end-to-end retransmission).

There are several existing implementations. A stateful SIP server can retransmit binding updates (re-INVITE) after the expiration of a timer.

8.9.3.2 Forwarding (Regular Passive Type)

Forwarding mechanisms in the sender's home network can help to get messages such as binding updates to the right place, but are probably less useful than mechanisms on the receiver side because of the time spent by the forwarded signals owing to the distance between the sender's home network and the receiver.

8.9.3.3 Proactive Forwarding

Regular passive forwarding may be insufficient to solve the simultaneous mobility problem. A proactive binding update proxy may help in some situations, where it attempts to find the most current location of the receiving node and retry the forwarding there.

8.9.3.4 Redirecting

Redirecting mechanisms in the sender's home network can help to get messages such as binding updates to the right place, but are probably less useful than mechanisms on the receiver side.

8.9.3.5 Proactive Redirecting

Regular redirecting may be insufficient to solve the simultaneous mobility problem. A proactive redirecting location proxy may help in some cases when the target network is not deterministic.

Table 8.1 Applicability of the various solution mechanisms

Solutions			MIP-RO	MIPv6	SIPMM	MIP-LR
Receiver-side	Prior network	Retransmission			Possible	
		Forwarding	Yes			
		Proactive forwarding				
		Redirecting	Yes			
		Proactive redirecting				
	Home network	Retransmission			Possible	
		Forwarding	Yes	Yes	Possible	
		Proactive forwarding				
		Redirecting				Yes
		Proactive redirecting				
Sender-side	Home network	Retransmission			Possible	
		Forwarding			Possible	
		Proactive forwarding				
		Redirecting				
		Proactive redirecting				
	At sender	Retransmission			Yes	

Table 8.2 Strengths and weaknesses of different solutions

Solutions	Strengths	Weaknesses
Timer-based retransmission of lost messages	Can be easily implemented with a timer	(a) Difficulty of choosing good timeout values; (b) retransmissions may also be lost
Simultaneous bindings	No significant increase in handoff latency to solve simultaneous mobility	(a) Not supported by all wireless networks; (b) redundant resource utilization; and (c) not clear how long to keep the simultaneous bindings active
Forwarding mechanisms from previous network	Effective if not just data but also signaling is forwarded	Handoff latency is slightly increased because of forwarding from the previous network; still vulnerable to simultaneous mobility, but with reduced vulnerability interval
Stationary proxies	Completely eliminates vulnerability interval	Handoff latency is increased

Table 8.1 shows the applicability of these solution mechanisms for different mobility protocols. Table 8.2 shows the strengths and weaknesses of the different solutions.

8.10 Application of Solution Mechanisms

In this section, we describe how the various solution mechanisms can be applied to different mobility protocols. We illustrate their applicability to a few mobility protocols, namely Mobile IPv6, SIPMM, and MIP-LR.

8.10.1 Mobile IPv6

We consider three different solution mechanisms to deal with the simultaneous mobility problem in MIPv6. These are described as follows.

8.10.1.1 Forwarding Proxy in Previous Network

As described earlier, for MIP-RO, foreign agents in the previous network act as forwarding proxies. However, foreign agents are not used in MIPv6. Thus, ordinary routers in the previous network need to be augmented with forwarding proxy functionality. This would involve significant challenges and modifications to MIPv6. For example, a mechanism would be needed to securely update a router with the latest IP address of a mobile node. However, getting ordinary IPv6 routers to perform this kind of forwarding might create a deployment bottleneck, and thus some kind of agent might need to be introduced.

8.10.1.2 Combination of Sender-Side and Receiver-Side Mechanisms

We proposed a combination of sender-side proactive binding update proxies and receiver-side proactive redirecting location proxies as a general solution in Wong et al. (2003b). Here, we propose that the same technique can also be applied to MIPv6. The home agents of the sender and receiver can serve as a proactive binding update proxy and proactive redirecting location proxy, respectively. Return routability has to be modified so that the CTI message goes through the sender's home agent. Another modification to MIPv6 is that the binding update must first be reverse-tunneled to the mobile node's own home agent before being forwarded to the correspondent node. The revised MIPv6 update procedure will work as follows. Let there be two mobile nodes, A and B. A sends its CTI and binding update messages to B through A's home agent, rather than directly to B's care-of address. However, A's home agent then forwards these messages to B at B's care-of address. A's home agent, acting as a proactive binding update proxy, also keeps a copy of any such message for a period τ. It then queries B's home agent (a proactive location proxy for B) to find out if B has a newer address. B's home agent responds immediately but keeps a copy of the query for a period ρ. If, before this period is over, B's home agent receives a registration for B at a new address, it proactively corrects its query. A's home agent then forwards the message to the new address. This solution is illustrated in Figure 8.11.

The selection of τ and ρ must be done carefully, based on reasonable estimates of the appropriate signaling and computational delays of the network. It is clear that $\tau > \rho$, so A's home agent can respond to any query response correction from B's home agent.

8.10.1.3 Receiver-Side Mechanism Only

The two solutions discussed so far are not good, since they require significant changes to MIPv6. A more MIPv6-centric solution is preferable. We therefore consider a solution where just the receiver's home agent is involved. A is the sender (of the CTI, HTI, or binding update). A sends all of these control messages to B using B's *home address*, thus forcing B's home agent to be involved. B's home agent acts as a proactive forwarding location proxy (a slight modification from its usual role

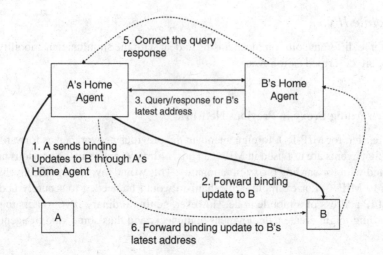

Figure 8.11   Sender- and receiver-side mechanism

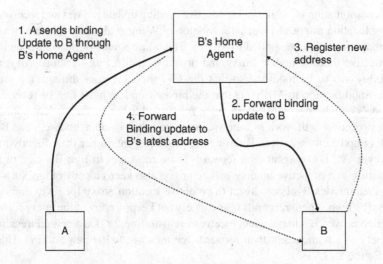

Figure 8.12   Receiver-side mechanism

as a forwarding location proxy), forwarding the control message to B as usual, but keeping a copy of it for a time τ. If it gets any binding updates from B during that time, it proactively forwards the message to B. This solution is shown in Figure 8.12. However, it requires some modifications to be made at the home agent. Home agents need to behave like a proactive forwarding location proxy in addition to behaving like a forwarding location proxy. The main modification to the mobile nodes implementing this solution is also small – to send the CTI and binding update to the home address of the correspondent node instead of directly to its care-of address.

8.10.1.4 Evaluation

It is clear that the third solution, with a receiver-side mechanism only (sending messages to the other node's home agent), is the cleanest solution, with the fewest changes to MIPv6. The addition (in the second solution) of a query and response capability to home agents is quite a drastic change for MIPv6. After the removal of foreign agents in going from MIPv4 to MIPv6, the addition of a forwarding proxy in the previous network with substantial functionality (in the first solution) is not desirable, as it increases delay.

8.10.2   MIP-LR

In this section, we describe how these solution mechanisms can be applied to deal with the simultaneous mobility problem in MIP-LR.

8.10.2.1   Forwarding Proxy in Previous Network

There are two types of MIP-LR: one with foreign agents (Jain et al., 1998) and one without (Jain et al., 1999). The version without foreign agents uses advertisement agents. The forwarding proxies use an interceptor function that intercepts the binding update and sends it to the new address of the mobile.

For the purposes of placement of the interceptor function, it does not matter whether foreign agents or advertisement agents are in use. The point is that there is some kind of agent in each of the foreign networks, and that the interceptor function can be placed here. MIP-LR needs modification so that the mobile node sends a binding update to the foreign agent (or advertisement agent) in the previous subnet as soon as it obtains its new IP address.

8.10.2.2   Sender-Side and Receiver-Side Mechanisms

In this solution, the binding update sent by a mobile node to its HLR has a list of correspondent nodes and their addresses. The HLR, which already performs the role of a redirecting location proxy, is enhanced to be a proactive redirecting location proxy. It also acts as a proactive binding update proxy, since it already obtains the current binding information as part of MIP-LR updating after each handoff. In order to do this, the HLRs must be enhanced to proactively retransmit binding updates and to query other HLRs for correspondent nodes' addresses. In order to minimize the changes to MIP-LR, the HLR-initiated binding updates are only sent when necessary, that is, when the queries return a newer address for a correspondent node than the one provided by the mobile node.

8.10.2.3   Evaluation

It is not recommended to consider a solution using only a receiver-side mechanism, as we have done for MIPv6. This is because the HLR would then have to become a proactive forwarding proxy. Such a change is too much for MIP-LR, one of whose points is that no forwarding location proxies are used (but multiple replicated HLRs are used instead).

8.10.3   SIP-Based Mobility

SIP allows much flexibility in the placement and usage of SIP servers in the signaling path between two mobile nodes. The Record-Route field is an optional field in the SIP header that allows SIP servers to remain in the signaling path between two SIP end nodes during a communication session. It is assumed that there will often be a SIP server in each mobile node's home network that serves as an up-to-date location proxy for it. An up-to-date location proxy keeps a record of the most recent location of the mobile. SIP also provides an inbuilt retransmission technique.

8.10.3.1   Timer-Based Retransmission

SIP has an inbuilt retransmission capability, where messages are retransmitted after a timeout if an acknowledgement is not received. During mid-session mobility, a re-INVITE may get lost even if it goes through the SIP server that keeps the most recent registration status of the destination. However, SIP allows automatic retransmission of INVITEs (including re-INVITEs) by SIP UAs (user agents) if a response (OK message) is not received within a specified time. Stateful SIP servers can also retransmit (re-)INVITEs.

One problem with timer-based retransmission is that significant latency could be added to the handoff when messages are lost because of simultaneous mobility. Another problem is that there is no guarantee that the retransmission will not also be lost. For example, the retransmission may be sent directly to the old address of the correspondent host, bypassing network elements (e.g., the relevant SIP servers or the home agent of the correspondent host) that know the latest address of the correspondent host. Figure 8.13 shows how a server-assisted retransmission technique can be useful for solving the simultaneous mobility problem.

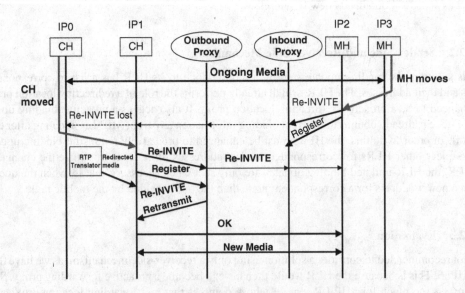

Figure 8.13   Server-assisted retransmission mechanism

8.10.3.2   Forwarding Proxy in Previous Network

As in the first proposed solution for MIPv6, we consider adding a forwarding location proxy to the previous network of the mobile node. For SIP-based mobility, the most natural choice of the forwarding proxy could be an entity similar to an RTP (Real-time Transport Protocol) translator (http://www.cs.columbia.edu/IRT/software/rtptools/), since these are already used to forward media traffic, among other things. Having a SIP server in the previous network that can forward this signaling is one solution. However, if one keeps on adding SIP servers in previous networks to the signaling path (using Record-Route), the signaling path becomes inefficient as the mobile node moves and the signaling path goes through more and more SIP servers in previous networks. Although the existing RTP translator can only forward data traffic from the previous network, a similar mechanism can be deployed that can intercept the signaling traffic and forward it to the new location of the mobile. To receive update signaling, the RTP translator can be enhanced to act as a SIP signaling translator without generally translating the RTP.

8.10.3.3   Receiver-Side Mechanism

For a SIP-based mobility scheme, the receiver-side home network SIP server already has some location proxy functionality that can be modified to act as a proactive location proxy. We first consider the case where it acts as a forwarding location proxy (it can also be a redirecting location proxy, which we will consider in the next paragraph). The SIP server immediately retransmits the re-INVITE upon receiving a REGISTER message from the destination of a pending re-INVITE. This is basically the same solution as the third one for MIPv6 (the preferred solution). Hence, Figure 8.11 also applies here, where B's home agent is replaced with B's home network SIP server, and the binding update is replaced with a re-INVITE. One difference is that, in order to get the SIP server to be in the signaling path for the re-INVITE request, the Record-Route field can be used. No modifications are needed at the mobile nodes, since SIP conveniently already has the Record-Route feature, unlike the case with MIPv6, where the mobile nodes have to be slightly modified to send control signaling to the home address of the other node rather than directly to the care-of address. The conversion of a SIP server to a proactive one is in some ways easier than the conversion of a MIPv6 home agent to a proactive forwarding proxy. This is because there is already the notion of a stateful SIP server that can retransmit messages such as a re-INVITE if no acknowledgement has been received by the time a timer expires.

8.10.3.4   Sender-Side and Receiver-Side Mechanisms

If the home network SIP server is modified to become a proactive redirecting location proxy instead of a proactive forwarding location proxy, then it needs to interact with a proactive forwarding proxy closer to the sender in the signaling path. In particular, when there is a SIP server in each mobile node's home network, there needs to be a proactive forwarding proxy in the sender's home network. This is similar to the chosen solution for MIP-LR, where the two HLRs were involved in this way. One difference is that the Record-Route feature is needed to keep both SIP servers in the signaling path.

8.10.3.5 Evaluation

It would appear that either the second or the third solution is equally simple to implement, given that SIP servers of both types (forwarding and redirecting) are available. With MIPv6, on the other hand, the clear preference was for the receiver-side solution, given that home agents are forwarding location proxies only.

8.11 Concluding Remarks

Although the original MIP did not suffer from the simultaneous mobility problem, newer mobility management protocols such as MIP-LR, SIP-based mobility, and MIPv6 do face this problem. In this chapter, we have identified the problem of simultaneous mobility, introduced a new analytical framework, and then used that framework to prove some new theorems, analyze solution mechanisms, and propose and compare solutions to the simultaneous mobility problem for MIPv6, SIP-based mobility, and MIP-LR. We also conducted a probability analysis of the likelihood of occurrence of simultaneous mobility.

The problem is further compounded by the expected rise in popularity of at least two of the three protocols we considered, namely MIPv6 and SIP-based mobility. Additionally, with the rise of smaller picocells in certain segments of the wireless market and higher mobility rates, there may be more frequent occurrences of simultaneous mobility in the future. We have explored a number of approaches to deal with the simultaneous mobility problem. In some protocols, there is existing functionality that partially helps to solve the simultaneous mobility problem or can be modified to handle simultaneous mobility. For example, with SIP-based mobility, a forwarding entity similar to an RTP translator can be used to forward signaling, including binding updates, that might have been sent to the previous network.

Most recently, we have introduced the effect of the simultaneous mobility problem in MIPv6 into the considerations of the MEXT working group in the IETF. Realizing that there is lack of solutions to the simultaneous mobility problem, a new section has been added in RFC 3775 bis (Johnson et al., 2009) to deal with the simultaneous mobility problem arising from return routability procedures. As an alternative to the solutions discussed in Section 8.9, the following modification has been added to the draft:

> In some scenarios, such as simultaneous mobility, where both the correspondent host and the mobile host move at the same time, or in the case where the correspondent node reboots and loses data, route optimization may not complete, or relevant data in the binding cache might be lost.

- Return routability signaling MUST be sent to the correspondent node's home address if it has one (i.e., not to the correspondent node's care-of address if the correspondent node is also mobile.)
- If Return routability signaling timed out after *MAX_RO_FAILURE* attempts, the mobile node MUST revert to sending packets to the correspondent node's home address through its home agent.

- The mobile node may run the bidirectional tunneling in parallel with the return routability procedure until it is successful. Exponential backoff SHOULD be used for retransmission of return routability messages.

The return routability procedure may be triggered by movement of the mobile node or by sustained loss of end-to-end communication with a correspondent node (e.g. based on indications from upper-layers) that has been using a route optimized connection to the mobile node. If such indications are received, the mobile node MAY revert to bi-directional tunneling while re-starting the return routability procedure.

9

Handoff Optimization for Multicast Streaming

In this chapter, we propose a few optimization techniques that expedite the delivery of a multicast stream during handoff in a hierarchically scoped multicast architecture. First, we propose a hierarchically scoped multicast content distribution network, describe the functional components of the architecture and their implementation, introduce optimization techniques to reduce the join and leave latencies for multicast traffic, and finally compare performance results obtained in our prototype test bed for both optimized and nonoptimized handoffs. The previous chapters have focused on fast-handoff techniques for unicast traffic. However, in this chapter, we apply some of the optimization techniques that were discussed in Chapter 6 to provide fast delivery of multicast traffic in a hierarchically scoped multicast environment.

9.1 Summary of Key Contributions and Indicative Results

Currently, multicast-based content distribution systems lack flexible features such as local and global program management and automatic advertisement insertion. These systems also do not support fast handoff when the mobile moves between subnets. Unlike unicast traffic, multicast traffic is receiver-oriented. A mobile receiving multicast traffic is subject to handoff delay and associated media interruption due to multicast join latency during its movement between layer 2 access points or layer 3 subnets. Contributions to the multicast join latency arise from periodic IGMP (Internet Group Management Protocol) router queries and the random amounts of time the client waits before it can send IGMP client reports. This join latency can be as large as 2 minutes in duration and disrupts streaming media during the mobile's movement.

We have proposed and implemented a hierarchical scope-based multicast streaming architecture that enables local and global program management and real-time advertisement insertion using RTCP (Real-time Transport Control Protocol)-based feedback control information.

In order to reduce the handoff latency for multicast join, we have proposed both proactive and reactive triggering techniques. As part of the reactive mechanism, we have developed an application layer triggering technique that sends an unsolicited RTCP *join* to join the multicast tree after the

Mobility Protocols and Handover Optimization: Design, Evaluation and Application, First Edition.
Ashutosh Dutta and Henning Schulzrinne.
© 2014 John Wiley & Sons, Ltd. Published 2014 by John Wiley & Sons, Ltd.

mobile hands off to the new network instead of the network layer technique based on IGMP. The server that receives the RTCP *join* uses, in turn, an IGMP report to join the upstream router.

As part of the proactive technique, we have proposed an application layer proxy and multicast address announcer so that the local server can join the multicast tree on behalf of the mobile when a handover of the mobile to a new network is impending. While the multicast proxy and the server join the upstream router using IGMP, the mobile triggers the multicast stream by using an RTCP *join* to the multicast proxy.

A hierarchical scope-based architecture provides the ability to manage local and global programs by using local servers in the content distribution network. By using a feedback signal such as that provided by RTCP, the proposed mechanism provides the ability to control the duration of an advertisement without relying on any additional signaling. By using an application layer triggering technique such as RTCP, the mobile does not need to depend upon the layer 3 IGMP router query interval, nor does it depend upon multicast support in the kernel. Having the ability to trigger multicast streaming during the mobile's configuration process, the mobile optimizes the operations in parallel. Compared with traditional unoptimized multicast handoff approaches, the proposed proactive optimization techniques can reduce the handover latency by a factor of 10 when the probability of the presence of a multicast group is low. The proposed proactive and parallel triggering techniques perform better by a factor of 4 compared with the proposed reactive techniques when the probability of the presence of a multicast group is low (e.g., 0.2).

In the rest of the chapter, we describe the details of the hierarchical scope-based multicast architecture, and elaborate on the proposed mechanisms that allow local and global program management and advertisement insertion. We describe the experimental test bed where we implemented the architecture and the fast-handoff mechanisms that we have proposed. We also compare the results obtained with the proposed fast-handoff mechanisms with results from nonoptimized systems.

9.2   Introduction

A CDN (content distribution network) distributes content from the origin server to replica servers that are situated closer to the end clients. The replica servers in a CDN store a very selective set of content and only the requests for that set of content are served by the CDN. This mechanism provides reduced access delay for any specific content and consumes lower bandwidth in the core of the network. There are a few commercial content distribution networks, namely Akamai (http://www.akamai.com), Digital Island, and Edgecast (http://www.edgecast.com), that distribute information from many news media, including CNN and the *New York Times*. Figure 9.1 shows a sample content distribution network and shows how the local affiliates distribute the global programs and local advertisements to the end users.

A mobile content distribution network can use multicast technology to distribute content from a single source to multiple replica servers (local affiliates) and end recipients more efficiently. Unlike unicast traffic, multicast communication is receiver-initiated. Thus, triggering techniques play an important role in ensuring efficient and timely multicast multimedia stream delivery. In order to maintain minimum loss and latency during a client's movement, it is desirable to minimize the handoff time and to enable almost instantaneous delivery of multicast streams by using optimized triggering techniques to initiate the stream delivery.

We introduced multicast mobility in Chapter 2. We reintroduce it here in the context of join latency and leave latency. Figure 9.2 shows how a mobile moves from one access point to another access point within the same router (router R1) and then moves to a new subnet connected to router R2. After the handoff to the new subnet, the mobile rejoins the same multicast group and a new multicast

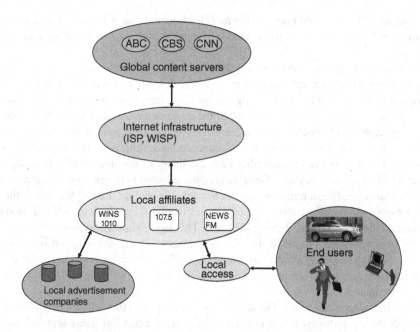

Figure 9.1 Example of a content distribution network

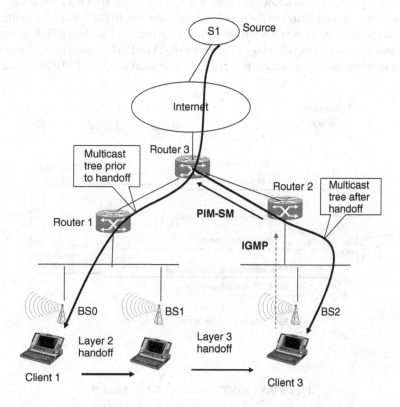

Figure 9.2 Handoff for multicast streams

tree is constructed. IGMP is used between the mobile node and router 2, and router 2 uses PIM-SM (Protocol Independent Multicast – Sparse Mode) to join the multicast tree that the mobile is part of prior to handoff.

The handoff delay during multicast stream delivery from a single source while the client moves to the next cell consists of several components, namely detection of a new cell subnet, or domain (Δ_1); address acquisition and network configuration (Δ_2); triggering of a multimedia stream to be delivered to the new subnet (Δ_3); and delivery of the multimedia stream (Δ_4). While some of these delay factors are common to both unicast and multicast traffic (e.g., cell or subnet detection, and IP parameter configuration), in this chapter we concentrate on optimization techniques that will allow faster delivery of multicast streaming traffic. Faster delivery of multicast traffic is dependent upon the join latency. The join latency is defined as the elapsed time between a host joining a multicast group and the router sending a multicast packet towards the mobile. Figure 9.3 shows the protocol interaction between the mobile and the first-hop router (router R1), which uses IGMP, and the inter-action between router 2 and router 3 using PIM-SM. This figure shows how router R2 keeps sending IGMP router query messages to all host multicast addresses (e.g., 224.0.0.1) at periodic intervals and the mobile sends a response after it has joined a specific multicast group. On receipt of an IGMP query response message, router 2 joins the new multicast group by sending a PIM-SM join message to the upstream router 3.

Fast-handoff techniques can be used in several layers to expedite the delivery of multicast streams. The delivery delay for multicast traffic depends on the layer 2 handoff delay and join latency contributed by IGMP (Fenner, 1997) in layer 3. The leave latency is defined as the time period during which the multicast traffic is still allowed to flow in the previous cell after the mobile has left the cell. Thus, the leave latency contributes to the waste of bandwidth due to the flow of the multicast stream in the previous cell after the mobile has left the cell. Several methods, such as IGMP snooping (Wang et al., 2002) and the Cisco Group Management Protocol (CGMP) (Farinacci et al., 1996/1997), take care of the handoff for multicast streams in layer 2. As discussed in RFC 3170 (Quinn and Almeroth,

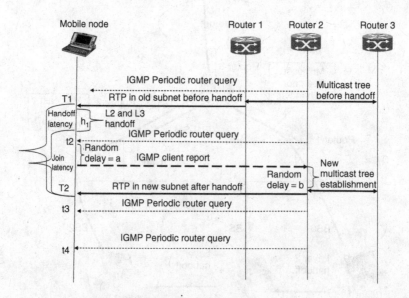

Figure 9.3 IGMP flow during subnet handoff

2001), mobile receivers within a domain can do expedient "joins" and "leaves," whereas a mobile can send to a multicast address without explicitly joining that address.

In layer 3, the triggering delay is caused by an IGMP query report (Fenner, 1997) that helps the node to be part of the new multicast tree after the mobile hands over. However, if there is currently at least one other active participant in the subnet, the mobile host can continue to receive traffic without waiting to hear a membership query from the router. The typical query interval for IGMP is by default 125 s (Williamson, 2000), although this value is configurable in the multicast routers. In order to avoid flooding the LAN with IGMP messages, this value cannot be made very small. Flament et al. (1999) showed that when IGMP is used, a host will wait on average for 65 s in order to continue to receive multicast traffic after the handoff. This is because IGMP was not designed for roaming clients in a wireless environment.

The typical leave latency is about 2 minutes after the host has moved to a new subnet; that is, traffic still flows to the previous cell even after the client has moved out, thus wasting bandwidth in the previous cell.

In layer 2, when the destination cell is part of the same subnet, the multicast stream continues in both cells. Although a layer 2 triggering delay is avoided, this nevertheless contributes to a waste of bandwidth if there is no active participant in the adjacent cell. However, a multicast switch stops the multicast traffic from flowing to a neighboring cell if there is no mobile in the target cell. If a client moves between access points within a subnet, CGMP or IGMP snooping (Williamson, 2000) takes care of triggering the multicast stream in layer 2. CGMP works in conjunction with IGMP and controls the multicast traffic flow in layer 2.

Unlike unicast traffic, the layer 3 configuration time (e.g., acquisition of a new IP address) in the client does not affect the multicast stream delivery delay. On other hand, several other components, such as detection of a new cell or subnet, the triggering time, and time needed for the router to join the upstream router, affect the multimedia delivery. For example, as per Figure 9.3, the media traffic is discontinued for a period of $T_1 - T_2$, where T_1 denotes the time when media traffic is received by the mobile before handoff, and T_2 denotes the time when media traffic is received by the mobile after handoff. Router 2 in Figure 9.3 keeps on sending IGMP router query messages at regular intervals (e.g., at times t_2, t_3, and t_4). After the layer 2 and layer 3 handoff delay, denoted by h_1, the mobile waits for an IGMP router query. According to Figure 9.3, the router query is received at time t_2 after the handoff. After receiving the router query, the mobile waits for a random period of time a before it sends an IGMP query response. Once router 2 receives the client report, it joins the upstream multicast tree. There is a delay of b for joining the upstream multicast tree. Thus, the total handoff delay due to the multicast join latency is $t_2 - T_1 - h_1 + a + b$. Since the waiting time for an IGMP query message can range between 0 and 125 s, the handoff in IP multicasting will result in a large gap in streaming traffic during handoff.

9.3 Key Principles

The following are some of the key principles that should be considered when one is aiming to optimize the delivery of multicast traffic and reduce the handoff delay and packet loss:

1. Reduction in the join latency reduces the time taken for data delivery after the mobile's handoff to a new cell or subnet.
2. Tunnel overhead can be eliminated by avoiding dependence on the home agent often used in the home-subscription-based approach.

3. Reduction in the leave latency reduces the additional bandwidth consumption in the previous network. The leave latency in the previous network can be reduced by the use of proxies that send unsolicited leave messages on behalf of mobiles.
4. Parallel operations among the handoff functions reduce the overall triggering delay for media delivery.
5. Proactive join to a multicast tree reduces the join latency after the handoff to the new network.
6. Fast-handoff techniques can be applied in multiple layers based on the movement of the mobile.

9.4 Related Work

Several papers have discussed group join and leave behavior on the Internet, the effects of channel surfing, and the effects of mobility on multicast streams. The process of joining or leaving a specific multicast group while changing cell or subnet is similar to surfing on a TV or radio by flipping channels, as studied by Ferguson (1994). Almeroth and Ammar (1997) and Almeroth (2000) described multicast group behavior on the Internet, and cited results about surfing delay based on an analysis of temporal statistics for MBone (Multicast Backbone) (Eriksson, 1994). These results show that within a time interval of 2 minutes, a user leaving one session either joins another session or becomes inactive. Although this is very similar to a mobility event for multicast, where a user leaves one group and rejoins the same group in the next cell, that study did not take account of the mobility of the users and the associated handoff parameters.

Many of the architectural issues associated with mobile hosts in a multicast environment have been described by Xylomenos and Polyzos (1997), Varshney and Chatterji (1999), and Acharya et al. (1995). Wu (1999) and McAuley et al. (1999) proposed ways of taking care of fast delivery of a multicast stream when the end hosts move within a domain. Wu (1999) proposed a handover solution with preregistration in order to provide fast handoff for multicast streams while moving between subnets. This is accomplished by sending a unicast signal to a neighboring station about the multicast address that it is subscribed to in order for that neighboring station to be able to join the multicast tree even before the client moves into the neighboring cell. This solution assumes that there is a mobility support agent (MSA) in each subnet that invokes a join message on behalf of the mobile.

McAuley et al. (1999) proposed a mobile multicast proxy where the proxy's clients do not themselves directly participate in the multicast tree, but the multicast proxy participates in the formation of a multicast tree for the groups that its clients are members of. In this case, the multicast proxy performs a function similar to that of a designated router; however, the multicast proxy can be outside the member's subnet and can forward multicast messages to its receivers using unicast, multicast, or a limited-scope broadcast.

There are also proposals to extend Mobile IP to support mobility for multicast users. However, a Mobile IP-based bidirectional tunneling solution puts the multicasting burden on the home agent (HA). In this case, a user desiring to join a particular multicast group joins that group through the HA using IGMP. When the user moves to a foreign network, the HA is responsible for tunneling multicast packets to the user. When an HA has a number of users in the same multicast group visiting the same foreign network, tunneling multiple multicast packets to the foreign network is inefficient. If multiple HAs have users in the same visited network that are part of the same multicast group, multiple copies of the same multicast packets are also tunneled.

Mobile multicast (MOM) (Williamson et al., 1998) uses a Mobile IP-based approach to take care of mobility for multicast traffic. Williamson et al. (1998) made a proposal for reducing the problem that arises in bidirectional tunneling when many HAs tunnel the same multicast packet to a foreign network. In this case, one HA is elected to tunnel multicast packets to the foreign network. Range-based MOM (Lin and Wang, 2000) takes the MOM approach one step further and elects a multicast agent close to the foreign agent (FA) to tunnel multicast packets to the foreign network.

In order to avoid the duplication of multicast packets being tunneled to foreign networks, one proposed solution is remote subscription. In this case, a user desiring to join a multicast group does so in each visited network through the FA. However, this requires that after each handoff the user must rejoin a multicast group. In addition, the multicast trees used to route multicast packets are updated after every handoff to track the multicast group members. The remote-subscription mechanism has been briefly introduced in Chapter 2. In order to limit the tree updates or limit duplication of multicast packets, proxy or agent-based solutions have been proposed. For example, in the Mobicast solution (Tan and Pink, 2000a), users still rejoin the multicast group in each visited network. This architecture adopts the concept of a domain foreign agent (DFA) to shield all mobility within the foreign domain from the main multicast delivery tree. In this scenario, the DFA sends or receives the multicast traffic to or from a multicast group. When the mobile host is receiving the multicast traffic, the DFA uses a translated multicast address in its network to prevent multicast updates due to mobility.

Mysore and Bhargavan (1997) proposed a scheme to deal with the loss of transient data for the mobile hosts by assigning a location-independent unique multicast address to each mobile host. However, this proposal does not discuss the mobility of multicast sessions in a hierarchical environment where mobiles are assigned a locally scoped multicast address.

Multicasting with local scoping becomes more attractive for mobile users experiencing intradomain handoff because of its ease of deployment and its ability to provide more flexible services such as localized advertisements, news broadcasts, and location-specific information in the wireless environment. Multicasting with local scoping also deals with the global multicast address assignment problem. The proposed approach expedites stream delivery in a hierarchical multicast environment. It takes advantage of localized scope-based IP multicasting techniques in a wireless environment that could be applicable to radio and TV surfing on a mobile Internet.

9.5 Mobility in a Hierarchical Multicast Architecture

Many of the existing solutions are layer-2- and layer-3-based techniques and have not considered application layer techniques. They also do not take into account localized multicasting in a hierarchically scoped environment, where the mobile hosts could be operating in a private network with limited scope. Localized multicasting implies that the clients are assigned multicast addresses with local scoping where a time-to-live (TTL) decides a limit on a very few subnets. This approach avoids multicast address exhaustion and reduce the overlap of multicast addresses.

The proposed fast-handoff techniques are based on application layer triggers and apply many of the fast-handoff techniques, namely proactive and parallel operations, defined in Chapter 6.

We have designed and prototyped a hierarchical scope-based multicast content distribution network called MarconiNet (Dutta and Schulzrinne, 2001). This architecture uses the IETF protocols SDP (Session Description Protocol) (Handley and Jacobson, 1998), SAP (Session Announcement Protocol) (Handley et al., 2000), and SIP (Rosenberg et al., 2002) and benefits from the use of RTCP to provide many flexible features, such as localized advertisements, news broadcasts, location-specific information, quality-of-service guarantees, and optimized intradomain handoff for mobile users. There are four main functional components in this architecture, namely the radio station client (RSC), the radio antenna server (RAS) or local station, the advertisement

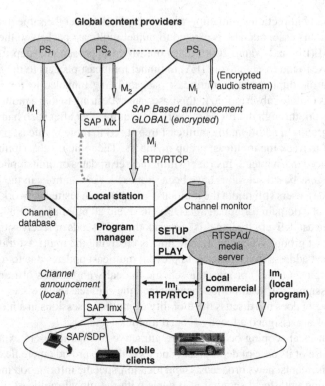

Figure 9.4 Hierarchical scope-based streaming architecture

or media server, and the Internet multimedia client (IMC). Figure 9.4 shows a functional diagram of the hierarchical streaming architecture that we have implemented, and with which we have experimented with fast-handoff techniques for multicast mobility. This architecture assumes that there is multicast connectivity throughout the network, but if there is a lack of multicast connectivity in certain parts of the network, then there are some possible application layer solutions (Finlayson, 2003) that can be deployed. We have described different ways of supporting mobility between multicast-enabled networks and nonmulticast-enabled networks (Dutta et al., 2003a) by using local proxy-based and UDP-based tunnels.

The proposed multicasting architecture consists of two tiered (hierarchically scoped) IP multicast sessions. At the higher of the two levels, a global multicast association exists between the broadcasting stations (RSCs) and the local stations (RASs). At the lower level of the hierarchy, a locally scoped multicast session is created for each broadcasting station between the server and the listening clients (IMCs) that can be privately scoped. The local server interacts with the advertisement server to provide stream control using protocols such as SIP and RTSP (Real Time Streaming Protocol) (Schulzrinne et al., 1998). We have prototyped several different functional components of the architecture, such as local and global program management, a channel monitor, application layer triggering, security, and handoff involving multiple servers. However, in this chapter, we focus the experimental analysis on fast-handoff techniques only.

We have described the details of the functional modules associated with this architecture in Dutta and Schulzrinne (2004). In the following, we briefly describe the functions of each of these modules.

9.5.1   Channel Announcement

A global streaming server (e.g., a radio/TV station or an individual broadcaster) can potentially broadcast its programs to a global audience. Thus, a global station RSC_i sends its programs live on a unique multicast address M_i, which is globally scoped and encrypted over RTP/UDP (where M_i differs for each station). These global broadcasting stations send their session announcements using a subset of SDP parameters to a global multicast address M_x, which is also encrypted. This common global multicast address M_x contains a list of programs broadcast by the main radio stations (RSCs). We have designed a Java-based interface called JSDR that provides a hierarchical searching functionality comparable to the traditional SDR tool (Handley, 1996).

9.5.2   Channel Management

The channel management module manages global and local programs in the local server. Each RAS (or local server) gets a global encryption key that it uses to listen to the global common multicast address M_x and obtain a listing of the channels. The local server broadcasts part of the list to the local domain, and hence creates a local announcement database. The subset of channel descriptions announced by each global station provides sufficient data for building a local channel database. This local announcement database contains a list of supported channels, each with their appropriate attributes, such as the name, duration, type of content, and place of origin of each program. The local station sends this program index to a locally scoped common multicast address lm_x for announcement using SAP (Handley et al., 2000). SAP helps to announce the multicast session directory. A SAP announcer periodically multicasts an announcement packet to a well-known multicast address and port, and a SAP listener on a multicast client learns of the multicast scopes it is within and listens on a well-known SAP address and port for those scopes. SAP uses SDP (Handley and Jacobson, 1998) to describe all the session parameters. SDP is intended for describing multimedia sessions for the purposes of session announcement, session invitation, and other forms of multimedia session initiation. The announcement on lm_x is not encrypted, since it gives the local stations the ability to figure out what is being relayed by them. The RAS also maintains a pair of multicast addresses for each channel. This keeps the mapping of the globally scoped multicast address M_i on which the radio station sends its programs and the locally scoped multicast address lm_i where it is relayed to. The RAS receives the audio stream on the global multicast address M_i and redirects it to the local multicast address lm_i for the IMCs. Local programs are sent on a specific locally scoped multicast address lm_l.

The client continues to send RTCP packets to the management server as long as it is receiving audio streams over RTP on a particular multicast address. Information from the RTCP packets can be used for billing, audio quality feedback, and also membership information for a particular multicast group.

Figure 9.5 shows a screenshot of the channel manager that we implemented in our test bed. This shows a listing of the global programs, local programs, and local advertisement insertions.

9.5.3   Channel Tuning

Internet multimedia clients tune to the locally scoped common multicast address lm_x to determine the currently available programs using a JSDR tuner that is based on SAP and SDP. According to the SAP specification, the antenna server updates the announcement information every few minutes or so. The client can tune to a particular channel to get details of the program that is available at that time.

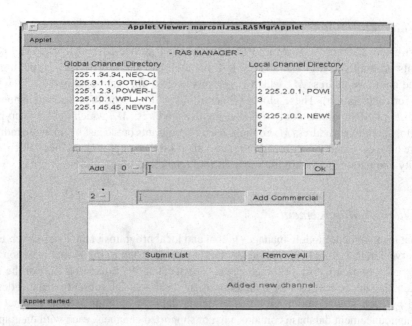

Figure 9.5   Channel manager at the local server

9.5.4   Local Advertisement Insertion

Localized information insertion can either be provisioned or be event-driven based on an external event such as an emergency notification. It is assumed that each global broadcasting station is aware of the starting times and durations of commercial breaks ahead of time or has control over the times of breaks.

Through an RTCP report, the global station notifies the local stations about a commercial break. On receiving the signal for a commercial break, the management server at the local station requests the local RTSP server to play a local advertisement on a specific locally scoped multicast address lm_l assigned to that station. A set of RTSP commands such as SETUP, PLAY, and STOP are used to control the stream delivery on the locally scoped multicast address. During this time, the local server stops forwarding the RTP stream from M_i to lm_i in the local domain. The local advertisement runs for a specific time based on information conveyed by RTCP reports. After the advertisement time is over, the local server begins relaying the global program.

9.5.5   Channel Monitor

The channel monitor provides statistics about how many clients are tuned to a specific multicast address. For each local channel being diverted, an additional RTCP signaling channel is created with a different port. Each listener periodically sends RTCP SDES (Source Description) packets to notify the local station (RAS) about who is listening to what. SDES is one of the five types of packets that RTCP offers with packet type 2. The SDES packet is a three-level structure composed of a header and zero or more chunks, each of which is composed of items describing the source identified in that chunk. SDES packets can provide participant identification and supplementary details, such as location, e-mail address, and telephone number. The RAS maps each listener to the desired channel, which in turn increases the number of listeners for that particular channel. The listener-to-channel mapping is destroyed via RTCP BYE packets or via the RTCP timeout feature. This also decreases the number of listeners for the associated channel. Figure 9.6 shows a screenshot of the channel monitor that gives statistics about how many clients are tuned to a particular channel.

9.5.6 Security

The proposed architecture offers four levels of security overall, namely global announcement encryption, global multicast stream encryption, local audit encryption, and user authentication. By using global announcement encryption, one can separate global announcements from local announcements. The local IMCs do not get access to the global announcements and can only view the local announcements. By using a global encryption key during announcements by global radio stations, the scheme does not allow the local Internet multimedia clients to find out about the global channels, and thus gives control to the local stations over announcing only a subset of these channels to the local clients. The security model for global multicast streams should effectively prevent IMCs, as well as nonpaying RASs, from receiving broadcast content. Thus, each radio station (RSC) must maintain a secret key and encrypt all outgoing content so that only a ciphertext stream is transmitted. The basic strategy is to generate a symmetric encryption key at the station and securely distribute this key to a particular RAS upon payment. Global multicast stream encryption can also be extended to a local section as a second-level hierarchy. Advertising companies can also be authenticated so that unauthorized companies cannot hijack the local advertisement insertion system.

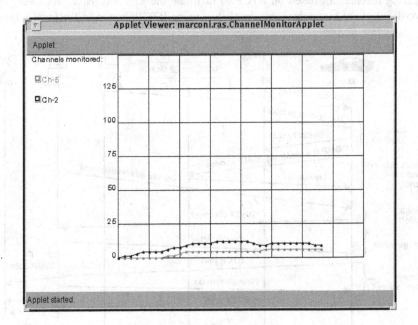

Figure 9.6   Channel monitor

9.6 Optimization Techniques for Multicast Media Delivery

In this section, we present a few proposed mobility optimization techniques for multicast streaming, and describe experiments with these techniques in the hierarchical scope-based multicast system that we discussed in Section 9.4 and implemented in a test bed. Figure 9.7 shows a protocol flow for multicast media delivery during handoff that uses a combination of RTCP- and IGMP-based triggering.

In the following sections, we describe four different optimization techniques that provide faster multicast stream delivery by reducing the join latency. The proposed techniques are based on remote-subscription-based approaches and do not use Mobile IP, and thus avoid tunnels between the home network and visited networks.

Figure 9.8 shows the experimental test bed where we implemented the functional components and demonstrated several fast-handoff techniques for multicast streaming applications. I_a, I_b, I_c, and I_d are the globally routable subnets connected to the primary interfaces of the local servers S1, S2, S3, and S4, respectively, whereas i_a, i_b, i_c, and i_d are the local subnets connected to the secondary interfaces of the respective servers. The access points are connected to the secondary interfaces of the respective servers. In this case, the mobile (represented as an automobile) performs a handoff between the cells and is thus subject to layer 2 and layer 3 handoff.

9.6.1 Reactive Triggering

The proposed reactive triggering technique uses a combination of application layer triggering using RTCP and IGMP. Normally, when a client moves to a new subnet, it sends a join request via IGMP. According to Kaur et al. (1999), IGMP can also be modified to provide an aggregate group report to reduce the join latency. As part of the proposed technique, we have implemented an application layer triggering mechanism based on RTCP to facilitate the join and leave processes. Triggering in the lower level of the hierarchy is accomplished using RTCP; however, the local server triggers

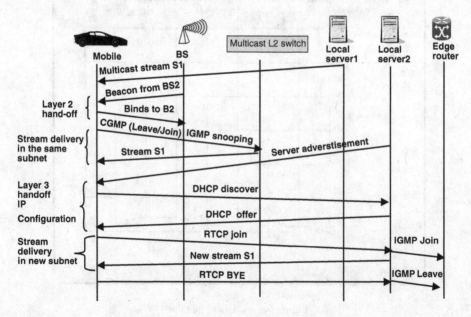

Figure 9.7   Handoff flow for multicast traffic

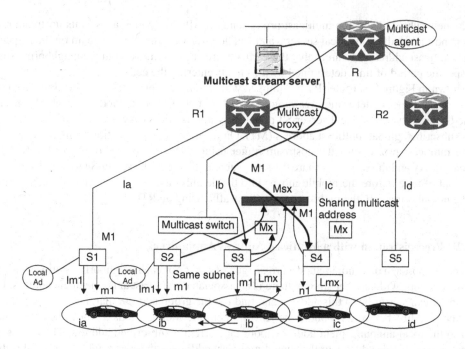

Figure 9.8 Fast handoff for multicast stream

the upstream router using IGMP. Using an RTCP-based triggering technique offers a solution in the user space, compared with IGMP, which works in the network layer. Figure 9.7 illustrates the communication flow among the mobile, the base station, the local servers, and the upstream router. It illustrates reactive triggering of a multicast stream using RTCP-based application layer triggering techniques in the lower level of the hierarchy. In this specific figure, the mobile node moves from one cell to another cell and in the process is subjected to layer 2 and layer 3 handoff. In this approach, the mobile does not need to wait for an IGMP report, nor does it wait for a random time to send an IGMP query response. The mobile also uses an RTCP BYE packet when it decides to leave a specific multicast group. A BYE packet is an RTCP packet of type 203 and is generated when a participant leaves a session or when it changes its SSRC. This application layer process helps to reduce the leave latency caused by the IGMP leave process.

9.6.2 Proactive Triggering

The proactive join method reduces the join latency for a client that is about to hand off, at the cost of additional bandwidth in the adjacent cell for a certain duration. We describe two kinds of proactive join schemes in the following: preregistration with a multicast proxy agent and preregistration with a multicast address announcer.

9.6.2.1 Preregistration with a Multicast Proxy Agent

In the first approach, we propose the use of proxy agents in each subnet. These proxy agents join the upstream multicast tree on behalf of the local downstream servers even before the clients move into the new subnet. Thus, the multicast proxy sends IGMP query messages to its upstream router beforehand on behalf of the local servers, and can help forward the global stream on the respective global multicast addresses to areas that these clients are about to move to in the neighboring subnet for a specific period of time determined by the clients entering the cell.

As shown in Figure 9.8, router R1 is equipped with a multicast proxy agent. This multicast proxy agent is on the same subnet as the streaming server S4. A multicast proxy module can also be installed in a dedicated server. The local server S3 proactively notifies the proxy agent about the impending host's subscribed global multicast address (M1). On receiving the notification from the local server S3, the multicast proxy joins the upstream router using IGMP on behalf of server S4. Since the multicast proxy and the server S4 share the same subnet, this helps the server S4 to join the multicast tree ahead of time before the mobile moves into the neighboring cell. After the mobile has moved into the neighboring cell, the mobile receives the traffic using an RTCP trigger.

9.6.2.2   Preregistration with a Multicast Address Announcer

The second approach does not include a proxy agent. Rather, for each neighboring station sharing an area that overlaps with another station, there is an associated multicast announcement address. Each local server can subscribe to this address and find out the group address that the incoming client is subscribed to. Just before a mobile node leaves a cell, the mobile sends a "movement imminent" signal to the local announcement address about the currently subscribed address. The local server in turn announces this address to the shared multicast addresses that the neighboring local stations subscribe to in the globally scoped address space. In the absence of this association, the neighboring server sends an IGMP message to the upstream router and redirects the stream to the local cells even before the client has moved to the new cell. This helps to minimize the interruption of multicast data.

This mechanism is also illustrated in Figure 9.8. In this figure, M_{sx} is the announcement address where the neighboring servers (e.g., S2, S3, and S4) learn about the multicast address of the impending host. Each of the neighboring local servers learns about the subscribed address of the mobile by tuning to the announcement address. This specific method avoids an additional proxy agent on each subnet.

9.6.3   Triggering during Configuration of a Mobile

Using this approach, the mobile's group membership information can also be passed during the client's configuration in the new network. Right after the node has handed over to the new subnet, it can send the request for the previously subscribed multicast address as part of a DHCP discover message during the layer 3 configuration process. During the process of obtaining an IP address from the DHCP server, the client can send an unsolicited *join* to the server for the desired locally scoped multicast address. The server, in turn, can join the upstream multicast router for the desired multicast group. Thus, the server can join the desired multicast group at the same time as the client is in the process of getting its new IP address configured. This process allows the client to join a multicast group during the client's configuration itself. This is an example of an optimization technique where the client performs two operations in parallel, namely configuration and join, during the process of configuring its layer 3 identifier.

There are trade-offs associated with each of these optimization techniques. The application layer triggering technique is only applicable to RTP-based traffic, as it depends heavily upon the RTCP report. The preregistration technique helps to reduce the join latency at the cost of bandwidth in the previous cell and subnet. The join operation during the mobile's configuration process provides a more efficient solution, as it does not need any additional network elements, but needs more resources owing to the parallel operations of the join operation and layer 3 configuration.

9.7 Experimental Results and Performance Analysis

In this section, we highlight the experimental results in which we optimized the join latency, and compare these results with results from the nonoptimized version. We also perform an analytical comparison of the optimized versions with the nonoptimized version.

9.7.1 Experimental Results

We conducted a series of handoff experiments for multicast traffic involving cell and subnet mobility to study the effect of handoff on multicast traffic and how the proposed optimization techniques can improve the handoff performance. We used multimedia applications such as RAT (Robust Audio Tool) (Sasse et al., 1995) and VIC (Video Conferencing) and measured the time for movement detection and IP address acquisition and the join and leave latencies by using network layer IGMP, CGMP, and application layer RTCP signaling. We focused mainly on improving the join and leave latencies in the test bed. Figure 9.9 shows the test bed where we experimented with fast handoff

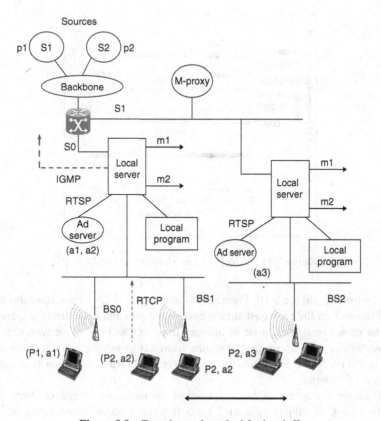

Figure 9.9 Experimental test bed for handoff

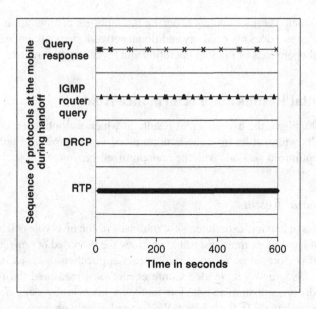

Figure 9.10 Effect of layer 2 handoff on multicast

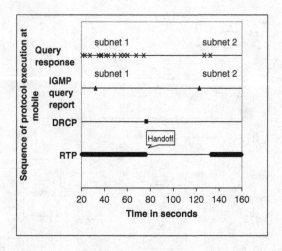

Figure 9.11 Effect of layer 3 handoff on join latency

using a multicast proxy. Figure 9.10, Figure 9.11, and Figure 9.12 demonstrate the effect of layer 2 and layer 3 handoff on the multicast stream and on the join and leave latencies, and Figure 9.13 shows how the proactive optimization techniques improve the handoff performance by reducing the packet loss. Figure 9.14 illustrates the results from a ping-pong experiment, where the mobile moved back and forth between subnets in rapid succession. We describe each of these figures in more detail in the following.

Figure 9.10 shows the effect of layer 2 handoff on multicast stream delivery. As shown in Figure 9.8, the mobile is subject to layer 2 handoff when it moves from server S2 to server S3.

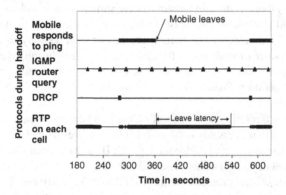

Figure 9.12 Effect of layer 3 handoff on leave latency

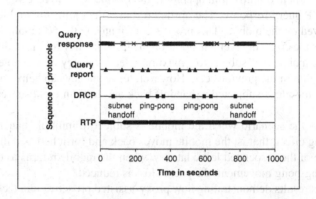

Figure 9.13 Effect of proactive join technique on multicast traffic

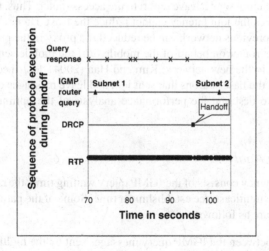

Figure 9.14 Effect of ping-pong movement on multicast traffic

During layer 2 handoff, there is no join latency, as the subnet does not change during the mobile's movement, and we did not use a multicast switch that would have contributed to the layer 2 join latency, but there is a waste of bandwidth due to the leave latency in the previous network. Figure 9.10 shows the experimental results for multicast mobility during layer 2 handoff. Since the mobile does not change subnet, it does not need to use any configuration protocol such as DRCP (Dynamic Rapid Configuration Protocol) to configure itself, and thus the DRCP packet sequence does not appear in this figure.

Figure 9.11 shows the effect of layer 3 handoff on the join latency of multicast stream delivery. As per Figure 9.8, this scenario reflects when the mobile client moves from the local server S3 to S4 and in the process changes its subnet and IP address (e.g., the IP address changes from I_a to I_b). As shown in Figure 9.11, a join latency of 60 s was observed during subnet movement. This figure shows the sequence of execution of several protocols, namely RTP, DRCP, IGMP router query, and client query response, during the handoff. As soon as the layer 2 handoff is over, the mobile uses DRCP to obtain the new IP address. The router sends periodic IGMP queries to a well-known multicast address. In the absence of an unsolicited IGMP join operation, the mobile waits to receive the IGMP query and then sends the IGMP query response. The upstream router then joins the multicast tree before the RTP traffic is received by the mobile. This process contributes to the 60 s multicast join latency.

Figure 9.12 shows the effect of layer 3 handoff on the leave latency for the multicast stream during subnet handoff. The mobile was subject to a maximum leave latency of 3 minutes. This contributes to a waste of bandwidth in the previous cell. However, leave latency can help to reduce the effect of join latency when a mobile is subjected to handoff back and forth in rapid succession, often known as "ping-pong."

Figure 9.14 shows the scenario when the mobile is subject to multiple handoffs. In addition, it shows the ping-pong effect; that is, the mobile moves back and forth between the two subnets very frequently. Because of the associated leave latency, when the mobile returns to the previous subnet as a result of the ping-pong movement, the packet loss is reduced.

Figure 9.13 shows results demonstrating how proxy-assisted proactive join techniques can reduce the join latency to almost zero. Although the mobile node moved back and forth between subnets during the experiment, we have shown only one instance of a subnet move. Since the mobile leaves the previous network, it cannot send a leave report to the access router. Thus, this proactive optimization technique that reduces the join latency cannot reduce the leave latency in the previous subnet. The leave latency in the previous network can be reduced if a proxy in the previous network sends a leave report to the access router on behalf of the mobile or if the mobile sends a leave report when it is anticipating a move to the new network. Kim and Han (2004) described the use of a multicast handoff agent (MHA) in the base stations that sent join and leave messages on behalf of the mobile.

In the next section, we describe the performance analysis of the optimized and nonoptimized versions.

9.7.2 Performance Analysis

The multicast handoff latency consists of the IGMP query waiting time, the random backoff time for the IGMP report, and the multicast tree establishment time. Some of the parameters that we used for performance evaluation are as follows:

- T_Q, the time interval between the IGMP query messages sent by the multicast router;
- T_c, the RTCP interval from the client;
- T_R, the random backoff time before the response is sent by the client;

- T_h, the layer 2 and layer 3 handoff delay;
- T_d, the transmission delay of the wired link;
- T_w, the transmission delay of the wireless link;
- P_m, the probability that there is at least one mobile in the target network subscribed to the multicast group as the mobile in the current network.

Certain assumptions were made for the performance analysis. For the sake of simplicity, the processing times for IGMP messages and for proactive operations were ignored. The packet transmission delays for both signaling (e.g., IGMP query and IGMP response) and media (e.g., RTP packets) were assumed to be the same.

9.7.2.1 Nonoptimized Version

In the case of the nonoptimized version, the total handoff latency for a multicast stream, including the media delivery delay, can be defined as

$$T_L = T_h + (2(T_d + T_w) + T_R + T_Q) \times (1 - P_m). \tag{9.1}$$

Although the layer 2 and layer 3 handoff latencies are common to all four handoff approaches, the join latency will vary between each of the proposed optimized versions. In the case of an nonoptimized handoff that uses IGMP-based handoff, the join latency is denoted as

$$T_{join} = (T_R + T_Q + T_d + T_w) \times (1 - P_m). \tag{9.2}$$

9.7.2.2 Reactive Triggering using RTCP

In the case of the RTCP-based reactive triggering approach, the client does not need to generate an IGMP query response. The client sends an RTCP join as soon as it detects the new network. The local server sends an unsolicited join to its upstream router without waiting for the periodic IGMP router query report. This process eliminates the delay contributed by the interval between periodic IGMP queries. Thus, the join latency during RTCP-based triggering approach is denoted as

$$T_{join} = (T_c + T_d + T_w) \times (1 - P_m), \tag{9.3}$$

and the handoff latency, including media delivery, is denoted as

$$T_L = T_h + (T_c + 2(T_d + T_w)) \times (1 - P_m). \tag{9.4}$$

9.7.2.3 Proxy-Based Proactive Join

In the proactive join scenario, where the local server joins on behalf of the mobile before the mobile moves into the target network, the join latency is denoted as

$$T_{join} = T_w \times (1 - P_m), \tag{9.5}$$

and the handoff latency, including media delivery, is denoted as

$$T_L = T_h + 2 \times T_w \times (1 - P_m). \tag{9.6}$$

9.7.2.4   Join during Handoff

In the case where the mobile joins the multicast tree during the configuration process, resulting in parallel operations, there is no additional join latency; the join latency is included as part of the layer 2 and layer 3 handoff delay. Thus, the total handoff latency, including media delivery, in the case of a multicast join during the configuration process is

$$T_L = T_h + (2 \times T_d + T_w) \times (1 - P_m). \tag{9.7}$$

For a sample router configuration, we took the following values to obtain results, and compared the nonoptimized version with all three optimized versions and also compared the three optimized versions with each other:

- T_R = IGMP query interval = 60 s;
- T_c = RTCP join interval = 5 s;
- T_Q = random backoff time = 10 s;
- T_h = L2 and L3 handoff delay = 3 s;
- T_d = packet transmission delay on the wired side = 1 ms;
- T_w = packet transmission delay on the wireless access side = 5 ms.

Figure 9.15 shows a comparison of the three optimized versions with the nonoptimized version. Figure 9.16 compares the handoff latencies among the three optimized versions, reactive, proactive, and parallel. As is evident from both of these figures, when there is another client subscribed to the same multicast address in the neighboring subnet (i.e., the probability of the presence of a multicast group is set to 1), the three optimized approaches and the nonoptimized approach do not suffer from any join latency. However, when there is no other client present in the neighboring subnet (i.e., the probability of the presence of a multicast group is 0), proactive and parallel operations produce the same results and work better than the reactive and nonoptimized versions.

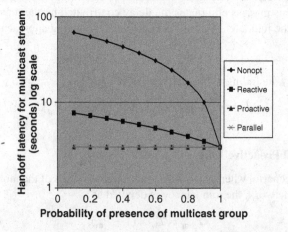

Figure 9.15 Comparison of nonoptimized vs. optimized techniques

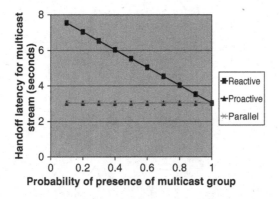

Figure 9.16   Comparison of optimized techniques

9.8   Concluding Remarks

The proposed hierarchical scope-based multicast streaming architecture provides local control while allowing content to be distributed over the Internet by using local proxies at the edges of the network. These proxies receive the multicast traffic on a global multicast address and transmit it on a locally scoped multicast address. Having the ability to use locally scoped addresses will alleviate the global multicast address allocation problem and the problem of overlapping multicast receivers in neighboring networks. The use of local proxies and a real-time localized advertisement insertion mechanism using RTCP feedback provides a good way to manage local and global programs based on the demography and interests of local communities. Application layer triggering techniques avoid the join delays caused by IGMP-based router query reports, which may be up to 2 minutes during a mobile's movement across subnets.

Although the proxy-assisted proactive technique provides better performance than the application layer reactive technique, based on the movement pattern of the mobile, one can use either the proxy-assisted proactive multicasting technique or the application layer reactive triggering technique, or a combination of both. For example, a proactive multicasting technique would work best when a mobile can discover neighboring networks ahead of time and can predict the target network that it may move to, such as in the case of a vehicle moving on a highway. In a city-like environment, where the movement of the mobile cannot be predicted easily, application layer reactive triggering techniques would work better. On the other hand, even if the multicast join during the mobile's configuration provides comparable performance to proactive triggering, one needs to extend the configuration protocol, such as DHCP, and it is also essential that the local streaming server is equipped with DHCP server functionality. Thus, based on the specific performance requirements, the network topology, and the movement pattern of the mobile, the mobile can adopt either of the proposed optimization techniques.

10

Cooperative Roaming

In a wireless network, mobile nodes (MNs) repeatedly perform tasks such as layer 2 (L2) handoff, layer 3 (L3) handoff, and authentication. These tasks are critical, particularly for real-time applications such as VoIP. We propose a novel approach, namely cooperative roaming (CR), in which MNs can collaborate with each other and share useful information about the network in which they move.

We show how we can achieve seamless L2 and L3 handoffs regardless of the authentication mechanism used and without any changes to either the infrastructure or the protocol. In particular, we provide a working implementation of CR and show how, with CR, MNs can achieve a total L2 + L3 handoff time of less than 16 ms in an open network and of about 21 ms in an IEEE 802.11i network. We consider behaviors typical of IEEE 802.11 networks, although many of the concepts and problems addressed here apply to any kind of mobile network.

10.1 Introduction

Enabling VoIP services in wireless networks presents many challenges, including quality of service (QoS), terminal mobility, and congestion control. In this chapter, we focus on IEEE 802.11 wireless networks and address issues introduced by terminal mobility.

In general, a handoff happens when an MN moves out of the range of one access point (AP) and enters the range of a new one. We have two possible scenarios:

1. If the old and new APs belong to the same subnet, the MN's IP address does not have to change at the new AP. The MN performs an L2 handoff.
2. If the old AP and the new AP belong to different subnets, the MN has to go through the normal L2 handoff procedure and also has to request a new IP address in the new subnet, that is, it has to perform an L3 handoff.

Figure 10.1 shows the steps involved in an L2 handoff process in an open network. As we have shown (Shin et al., 2004b), and Mishra et al. (2003b) have also shown, the time needed by an MN to perform an L2 handoff is usually on the order of a few hundred milliseconds, thus causing a noticeable interruption in any ongoing real-time multimedia session. In either an open 802.11 network or an 802.11 network with WEP enabled, the discovery phase constitutes more than 90% of the total

Mobility Protocols and Handover Optimization: Design, Evaluation and Application, First Edition.
Ashutosh Dutta and Henning Schulzrinne.
© 2014 John Wiley & Sons, Ltd. Published 2014 by John Wiley & Sons, Ltd.

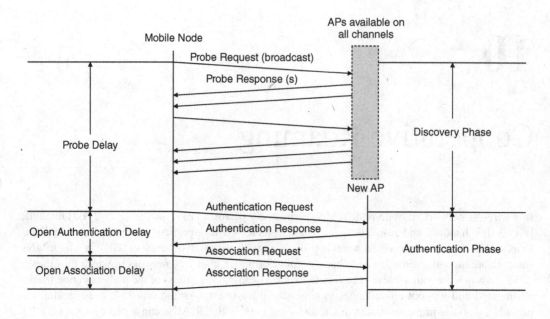

Figure 10.1 Layer 2 handoff procedure

handoff time (Mishra et al., 2003b; Shin et al., 2004b). In an 802.11 network with either WPA (Wi-Fi Protected Access) or 802.11i enabled, the handoff delay is dominated by the authentication process that is performed after associating with the new AP. In particular, no data can be exchanged amongst MNs before the authentication process has been completed successfully. In the most general case, both an authentication delay and a scanning delay are present. These two delays are additive, so, in order to achieve seamless real-time multimedia sessions, both delays have to be addressed and, if possible, removed.

When an L3 handoff occurs, the MN has to perform a normal L2 handoff and update its IP address. We can break the L3 handoff up into two logical steps: subnet change detection and new IP address acquisition via DHCP (Droms, 1997). Each of these steps introduces a significant delay.

In this chapter, we focus on the use of station cooperation to achieve seamless L2 and L3 handoffs. We refer to this specific use of cooperation as cooperative roaming. The basic idea behind CR is that MNs subscribe to the same multicast group, creating a new plane for exchanging information about the network, and help each other in various tasks. For example, an MN can discover surrounding APs and subnets by just asking other MNs for this information. Similarly, an MN can ask another MN to acquire a new IP address on its behalf so that the first MN can get an IP address for the new subnet while still in the old subnet.

For the sake of brevity and clarity, we do not consider handoffs between different administrative domains in this chapter, nor do we consider AAA-related issues, although CR could easily be extended to support them. Incentives for cooperation are also not considered, since they are a standard problem for any system using some form of cooperation (e.g., file sharing) and represent a separate research topic (Antoniadis et al., 2005; Buragohain et al., 2003; Feldman et al., 2004; Schosser et al., 2006; Wongrujira and Seneviratne, 2005).

The rest of the chapter is organized as follows. In Section 10.2, we present the state of the art for handoffs in wireless networks; in Section 10.3, we briefly describe how IPv4 and IPv6 multicast addressing is used in the present context; and Section 10.4 describes how, with cooperation, MNs can achieve seamless L2 and L3 handoffs. Section 10.5 introduces cooperation in the L2 authentication process to achieve seamless handoffs regardless of the particular authentication mechanism used. Section 10.6 considers security, and Section 10.7 shows how streaming media can be supported in CR. In Section 10.8, we analyze CR in terms of bandwidth and energy usage, Section 10.9 presents our experiments and results, and Section 10.10 shows how we can achieve seamless application layer mobility with CR. In Section 10.11, we apply CR to load balancing, and in Section 10.12 we present a more scalable way to use multicast. Section 10.13 presents an alternative to multicast and, finally, Section 10.14 concludes the chapter.

10.2   Related Work

The network community has done a lot of work on L2 and L3 handoffs in wireless networks. As of the writing of this chapter, many standards such as IEEE 802.11f (IEEE, 2003a) and IEEE 802.11e (IEEE, 2005a) have been ratified and others, such as IEEE 802.11k (IEEE, 2003b), IEEE 802.11r (IEEE, 2006a), and IEEE 802.21 (IEEE, 2005b), are emerging, with the aim of solving some of the problems a wireless environment introduces. All of these approaches, however, introduce significant changes into the infrastructure and the protocol. In particular, they have always been structured so that each MN is thought of as a stand-alone entity.

802.11f focuses on ways in which APs can share information with each other using the definition of an Inter Access Point Protocol (IAPP). This can be particularly useful for the transfer of users' credentials during handoffs, for example.

The 802.11e protocol addresses QoS problems in wireless local area networks (LANs). In particular, several different traffic classes are defined with their own medium access parameters, giving real-time traffic higher priority in accessing the wireless medium than best-effort traffic.

The 802.11k protocol utilizes MNs to collect topology information and other useful statistics about the network and conveys this information back to the APs. The APs then build a neighbor report containing all of the information about the various APs and their neighbors. These reports are then sent to the MNs so that each MN can have information about its neighboring APs. The way these reports are built is not specified and often involves each MN having to scan different channels.

No draft has been ratified by the 802.11r working group as yet. 802.11r addresses the need for fast L2 roaming in 802.11 networks, considering different authentication mechanisms as well as QoS. In 802.11r, fast basic service set (BSS) transitions can only take place between APs in the same mobility domain. A mobility domain is a set of BSSs in the same extended service eet (ESS). Within a mobility domain, APs can exchange key material and context using encapsulation over the distribution system. 802.11r does not specify how an MN discovers the best candidate AP to connect to next. Scanning, neighbor reports, and other means can be used. 802.11r supports prekeying and resource reservation between MNs and APs and defines a key hierarchy to extend pairwise master keys (PMKs) to multiple APs.

The IEEE 802.21 (Media Independent Handover) standard (IEEE, 2005b) introduces link-layer enhancements for performing intelligent handoffs between heterogeneous networks such as IEEE 802.11 and cellular networks, including both wireless and wired networks. The handoff process can be initiated by either the client or the network, and just like in IEEE 802.11k, MNs provide information about available networks and other network statistics to the infrastructure by scanning. The infrastructure then builds and stores information such as neighborhood cell lists and available services, thus helping with optimum cell selection. Furthermore, new link-layer primitives are defined in order to provide applications with consistent information regardless of the access technology used by the MN.

In all these approaches, MNs always behave as stand-alone entities and often have to scan the medium before a handoff; that is, they cause interruptions in any ongoing multimedia session. Furthermore, seamless handoffs with these approaches, when possible, require changes in the network and in the clients. CR is a client-only approach and can represent either an alternative or a complement to the current standards.

More recently, cooperative approaches have been proposed in the network community. Liu et al. (2005) showed how cooperation amongst MNs can be beneficial for all the MNs in the network in terms of bit rate, coverage, and throughput. Each MN builds a table in which possible helpers for that MN are listed. If an MN has a poor link with the AP and its bit rate is low, it sends packets to the helper, which relays them to the AP. The advantage in doing this is that the link from the MN to the helper and then from the helper to the AP is a high-bit-rate link. In this way, the MN can use two high-bit-rate links via the helper instead of the low-bit-rate one directly to the AP.

Fretzagias and Papadopouli (2004) introduced a location-sensing mechanism based on cooperative behavior among stations. Stations share location information about other stations and about landmarks so as to improve position prediction and reduce training.

Other work uses a cooperative approach, but mostly in positioning applications (Koutsonikolas et al., 2006) and sensor networks (Buttyán et al., 2005) and in the physical (Hunter and Nosratinia, 2004; Stefanov and Erkip, 2005) and application (Papadopouli and Schulzrinne, 2001) layers.

Aside from cooperation approaches and standardization efforts in the IEEE 802.11 working groups, many other approaches have been proposed in order to achieve fast handoffs in wireless networks. However, most of these approaches, such as those of Hsieh et al. (2003b) and Ote et al. (2003), require changes to either the infrastructure or the protocol or both. One good example of such a situation is Mobile IP (MIP). MIP has been standardized for many years now; however, it has never had significant deployment, in part because of the considerable changes required in the infrastructure. Fast-handoff approaches in the MIP context usually require additional network elements (Wu et al., 2002; Yokota et al., 2002) and/or changes to the protocol (Perkins and Johnson, 1998).

Ramani and Savage (2005) suggested an algorithm called syncscan, which does not require changes to either the protocol or the infrastructure. It does require, however, that all the APs in the network are synchronized, and it accelerates only unauthenticated L2 handoffs.

In this chapter, we propose a novel approach that works in an already deployed wireless environment, that is, an environment with heterogeneous networks, where new network elements cannot necessarily be introduced into the infrastructure, where all the APs are not necessarily synchronized amongst themselves, where any kind of authentication mechanism can be used, and where different subnets may be present. We use a cooperative approach amongst MNs for spreading information regarding the network topology without any infrastructure support. Our approach requires changes only to the wireless card driver, DHCP client, and authentication supplicant; no changes to the infrastructure or the protocol are required. This allows us to solve many of the problems associated with terminal mobility, regardless of the network the user moves to.

10.3   IP Multicast Addressing

CR works for both IPv4 and IPv6. In IPv4, we make extensive use of UDP-over-IP multicast packets. Different values for the time-to-live (TTL) are used according to how far we want multicast packets to reach into the IP network. This also depends on the density of MNs supporting the protocol. For example, if an MN does not receive any response after sending a request with a TTL value of 1 (same subnet), it will send the same request again but with a TTL value of 2 (next subnet), and so on. We must note, however, that the probability of an MN finding the information it needs becomes smaller as the search moves to more distant subnets. On the other hand, a small TTL can be used to limit the propagation of CR multicast frames in very congested environments.

In IPv6, we would use multicast scopes instead of IPv4 multicast. No significant changes would be required.

10.4 Cooperative Roaming

In this section, we show how MNs can cooperate with each other in order to achieve seamless L2 and L3 handoffs.

10.4.1 Overview

Shin et al. (2004b) introduced a fast MAC layer handoff mechanism for achieving seamless L2 handoffs in environments such as hospitals, schools, campuses, enterprises, and other places where MNs always encounter the same APs. Each MN saves information regarding the surrounding APs in a cache. When an MN needs to perform a handoff and it has valid entries in its cache, it uses the information in the cache directly without scanning. If it does not have any valid information in its cache, the MN uses an optimized scanning procedure called *selective scanning* to discover new APs and build the cache. In the cache, APs are ordered according to their signal strength registered when the scanning was performed, that is, just before changing AP. APs with stronger signal strength appear first. As mentioned in Section 10.1, in open networks the scanning process is responsible for more than 90% of the total handoff time. The cache reduces the L2 handoff time to only a few milliseconds (see Table 10.1), and cache misses due to errors in movement prediction introduce only a few milliseconds of additional delay (Shin et al., 2004b). Such an approach, however, works only in open networks or networks with WEP enabled. Other forms of authentication are not supported.

Earlier, we extended (Forte et al., 2006c) the mechanism introduced by Shin et al. (2004b) to support L3 handoffs. Here, MNs also cache L3 information such as their own IP address, the default router's IP address, and the subnet identifier. A subnet identifier uniquely identifies a subnet. By caching the subnet identifier, the MN detects a subnet change much faster and L3 handoffs are triggered every time the new and old APs have different subnet identifiers. Faster L3 handoffs can be achieved because the IP address and default router for the next AP and subnet are already known and can be immediately used. The approach of Forte et al. (2006c) achieves seamless handoffs in open networks only, it utilizes the default router's IP address as a subnet identifier, and it uses a suboptimal algorithm to acquire L3 information.

Table 10.1 L2 handoff time (ms)

Original handoff	457.8	236.8	434.8	317.0	566.7	321.6	241.0	364.0	216.7	273.9	343.0
Selective scanning	140.3	101.1	141.7	141.9	141.3	139.7	143.4	94.7	142.9	101.5	128.9
Cache	2.7	2.4	4.2	3.7	4.4	2.6	2.6	2.3	2.7	2.9	3.0

	Current AP	Next best AP	Second best AP
BSSID	MAC A	MAC B	MAC C
Channel	6	11	1
Subnet ID	160.39.5.0	160.39.10.0	160.39.10.0

Figure 10.2 . Example of an MN's cache structure

Here, we consider the same caching mechanism as used in Forte et al. (2006c). In order to support multihomed routers, however, we use the subnet address as a subnet identifier. By knowing the subnet mask and the default router's IP address, we can calculate the network address of a given subnet. Figure 10.2 shows the structure of the cache. Additional information such as the last IP address used by the MN, the lease expiration time, and the default router's IP address can be extracted from the DHCP client lease file, available in each MN.

In CR, an MN needs to acquire information about the network if it does not have any valid information in the cache or if it does not have L3 information available for a particular subnet. In such a case, the MN asks other MNs for the information it needs so that the MN does not have to find out about neighboring APs by scanning. In order to share information, in CR, all MNs subscribe to the same multicast group. We call an MN that needs to acquire information about its neighboring APs and subnets a requesting MN (R-MN). By using CR, an R-MN can ask other MNs if they have such information by sending an INFOREQ multicast frame. The MNs receive such a frame check if they have the information the R-MN needs and, if so, they send an INFORESP multicast frame back to the R-MN containing the information the R-MN needs.

10.4.2   L2 Cooperation Protocol

In this section, we focus on the information exchange needed in an L2 handoff. The information exchanged in the INFOREQ and INFORESP frames is a list of {BSSID, Channel, Subnet ID} entries, one for each AP in the MN's cache (see Figure 10.2). When an R-MN needs information about its neighboring APs and subnets, it sends an INFOREQ multicast frame. Such a frame contains the current content of the R-MN's cache, that is, all APs and subnets known to the R-MN. When an MN receives an INFOREQ frame, it checks if its own cache and the R-MN's cache have at least one AP in common. If the two caches have at least one AP in common and if the MN's cache has some APs that are not present in the R-MN's cache, the MN sends an INFORESP multicast frame containing the cache entries for the missing APs. MNs that have APs in common with the R-MN have been in the same location as the R-MN and so have a higher probability of having the information that the R-MN is looking for.

The MN sends the INFORESP frame after waiting for a random amount of time to be sure that no other MNs have already sent such information. In particular, the MN checks the information contained in INFORESP frames sent to the same R-MN by other MNs during this random waiting time. This prevents many MNs from sending the same information to the R-MN all at the same time.

When an MN other than the R-MN receives an INFORESP multicast frame, it performs two tasks. First, it checks whether someone is lying by providing wrong information and, if so, it tries to fix the problem (see Section 10.6.1); secondly, it records the cache information provided by the frame in its cache even though the MN did not request such information. By collecting unsolicited information, each MN can build a bigger cache in less time and in a more efficient manner, requiring fewer frame exchanges. This is very similar to what happens in software such as BitTorrent, where a client downloads different parts of a file from different peers. Here, we collect different cache chunks from different MNs.

In order to improve efficiency and minimize frame exchange further, MNs can also decide to collect information contained in INFOREQ frames.

10.4.3 L3 Cooperation Protocol

In an L3 handoff, an MN has to detect a change in subnet and also has to acquire a new IP address. When an L2 handoff occurs, the MN compares the cached subnet identifiers for the old and new APs. If the two identifiers are different, then the subnet has changed. When a change in subnet is detected, the MN needs to acquire a new IP address for the new subnet. The new IP address is usually acquired by using the DHCP infrastructure. Unfortunately, the typical DHCP procedure can take up to one second (Forte et al., 2006c).

CR can help MNs acquire a new IP address for the new subnet while still in the old subnet. When an R-MN needs to perform an L3 handoff, it needs to find out which other MNs in the new subnet can help. We call such MNs assisting MNs (A-MNs). Once the R-MN knows the A-MNs for the new subnet, it asks one of them to acquire a new IP address on its behalf. At this point, the selected A-MN acquires the new IP address via DHCP and sends it to the R-MN, which is then able to update its multimedia session before the actual L2 handoff and can start using the new IP address immediately after the L2 handoff, hence not incurring any additional delay (see Section 10.10).

We now show how A-MNs can be discovered and explain in detail how they can request an IP address on behalf of other MNs in a different subnet.

10.4.3.1 A-MN Discovery

By using IP multicast, an MN can talk directly to different MNs in different subnets. In particular, the R-MN can send an AMN_DISCOVER multicast packet containing the new subnet ID. Other MNs receiving such a packet check the subnet ID to see if they are in the subnet specified in the AMN_DISCOVER packet. If so, they reply with an AMN_RESP unicast packet. This packet contains the A-MN's default router IP address and the A-MN's MAC and IP addresses. This information is then used by the R-MN to build a list of available A-MNs for that particular subnet.

Once the MN knows which A-MNs are available in the new subnet, it can cooperate with them in order to acquire the L3 information it needs (e.g., new IP address and router information), as described in the following.

10.4.3.2 Address Acquisition

When an R-MN needs to acquire a new IP address for a particular subnet, it sends a unicast IP_REQ packet to one of the available A-MNs for that subnet. This packet contains the R-MN's MAC address. When an A-MN receives an IP_REQ packet, it extracts the R-MN's MAC address from the packet and starts the DHCP process by inserting the R-MN's MAC address into the CHaddr field of some

DHCP packets.[1] The A-MN also has to set the broadcast bit in the DHCP packets in order for it to receive DHCP packets with a MAC address different from its own address in the CHaddr field. All of this allows the A-MN to acquire a new IP address on behalf of the R-MN. This procedure is completely transparent to the DHCP server. Once the DHCP process has been completed, the A-MN sends an IP_RESP multicast packet containing the default router's IP address for the new subnet, the R-MN's MAC address, and the new IP address for the R-MN. The R-MN checks the MAC address in the IP_RESP packet to be sure that the packet is not for a different R-MN. Once it has verified that the IP_RESP packet is for itself, the R-MN saves the new IP address together with the new default router's IP address.

If the R-MN has more than one possible subnet to move to, it follows the same procedure for each subnet. In this way, the R-MN builds a list of {router, new IP address} pairs, one pair for each of the possible next subnets. After moving to the new subnet, the R-MN renews the lease for the new IP address. The R-MN can start this process at any time before the L2 handoff, keeping in mind that the whole process might take one second or more to complete and that lease times of IP addresses are usually on the order of tens of minutes or more.[2]

By reserving IP addresses before moving to the new subnet, we could waste IP addresses and exhaust the available IP pool. Usually, however, the lease time in a mobile environment is short enough to guarantee sufficient reuse of IP addresses.

Acquiring an IP address from a different subnet other than the one the IP is for could also be achieved by introducing a new DHCP option. Using this option, the MN could ask the DHCP server for an IP address for a specific subnet. This would, however, require changes to the DHCP protocol.

10.5 Cooperative Authentication

In this section, we propose a cooperative approach to authentication in wireless networks. The proposed approach is independent of the particular authentication mechanism used. It can be used for VPN, IPSec, 802.1x, or any other kind of authentication. We focus on the 802.1x framework used in Wi-Fi Protected Access (WPA) and IEEE 802.11i (IEEE, 2004).

10.5.1 Overview of IEEE 802.1x

The IEEE 802.1x standard defines a way to perform access control and authentication in IEEE 802 LANs and, in particular, in IEEE 802.11 wireless LANs using three main entities: a supplicant, an authenticator, and an authentication server.[3] The supplicant is the client that has to perform the authentication in order to gain access to the network; the authenticator, among other things, relays packets between the supplicant and the authentication server; and the authentication server, typically a RADIUS server (Rigney et al., 2000), performs the authentication process with the supplicant by exchanging and validating the supplicant's credentials. The critical point, in terms of handoff time in the 802.1x architecture, is that during the authentication process the authenticator allows only EAP over LAN (EAPOL) traffic to be exchanged with the supplicant. No other kind of traffic is allowed.

---

[1] If supported, the client-ID field must be used instead (Lemon and Sommerfield, 2006).
[2] The DHCP client lease file can provide information about current lease times.
[3] The authentication server is not required in all authentication mechanisms.

10.5.2   Cooperation in the Authentication Process

A well-known property of the wireless medium in IEEE 802.11 networks is that the medium is shared and therefore every MN can hear packets that other stations (STAs) send and receive. This is true when the MN and the STAs are connected to the same AP – that is, are on the same channel. Liu et al. (2005) made use of this particular characteristic and showed how MNs can cooperate with each other by relaying each other's packets so as to achieve the optimum bit rate. In this section, we show how a similar approach can be used for authentication purposes.

For simplicity, we suppose in the following discussion that one authenticator manages one entire subnet, so that authentication is required after each L3 handoff. In such a scenario and in this context, we also refer to a subnet as an authentication domain (AD). In general, an MN can share information about ADs in the same way as it shares information about subnets. In doing so, the MN knows whether or not the next AP belongs to the same AD as the current AP. In an L2 or L3 handoff, we have an MN which performs handoff and authentication, a correspondent node (CN) which has an established multimedia session with the MN, and a relay node (RN) which relays packets to and from the MN. Available RNs for a particular AD can be discovered following a procedure similar to the one described earlier for the discovery of A-MNs (see Section 10.4.3). The difference here is that the RN and MN have to be connected to the same AP after the handoff. In this scenario, we assume that the RNs are a subset of the available A-MNs. The basic idea is that while the MN is authenticating in the new AD, it can still communicate with the CN via the RN which relays packets to and from the MN (see Figure 10.3).

Let us look at this mechanism in more detail. Before the MN changes AD/AP, it selects an RN from the list of available RNs for the new AD/AP and sends a RELAY_REQ multicast frame to the multicast group. The RELAY_REQ frame contains the MN's MAC and IP addresses, the CN's IP address, and the selected RN's MAC and IP addresses. The RELAY_REQ will be received by all of the STAs subscribed to the multicast group and, in particular, it will be received by both the CN[4] and the RN. The RN relays packets for the MN identified by the MAC address received in the RELAY_REQ frame. After performing the handoff, the MN needs to authenticate before it can resume any communication via the AP. However, because of the shared nature of the medium, the MN will start sending packets to the RN as if it was already authenticated. The authenticator will

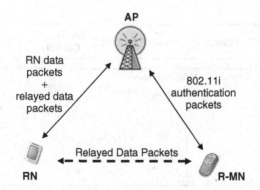

Figure 10.3   Layer 2 handoff with authentication in CR

---

[4] In congested environments, where smaller TTL values may be preferred, a separate unicast RELAY_REQ frame can be sent to the CN.

drop the packets, but the RN can hear the packets on the medium and relay them to the CN using its own encryption keys, that is, using its secure connection with the AP. The CN is aware of the relaying because of the RELAY_REQ frame, and so it will start sending packets for the MN to the RN as well. While the RN is relaying packets to and from the MN, the MN will perform its authentication via 802.1x or some other mechanism. Once the authentication process is over and the MN has access to the infrastructure, it can stop the relaying and resume normal communication via the AP. When this happens and the CN starts receiving packets from the MN via the AP, it will stop sending packets to the RN and resume normal communication with the MN. The RN will detect that it does not need to relay packets for the MN any longer and will return to normal operation.

In order for this relaying mechanism to work with WPA and 802.11i, the MN and RN have to exchange unencrypted L2 data packets for the duration of the relay process. These packets are then encrypted by the RN using its own encryption keys and are sent to the AP. By responding to an RN discovery, RNs implicitly agree to provide relaying for such frames. Such an exchange of unencrypted L2 frames does not represent a security concern, since packets can still be encrypted in higher layers and the relaying happens for a very limited amount of time (see Section 10.6.2). One last thing worth mentioning is that by using a relay, we remove the bridging delay in the L2 handoff (Mishra et al., 2003b; Shin et al., 2004b). Usually, after an MN changes AP, the switch continues sending packets for the MN to the old AP until it has updated the information regarding the new AP on its ports. The bridging delay is the amount of time needed by the switch to update this information on its ports. When we use a relay node in the new AP, this relay node is already registered to the correct port on the switch, and therefore no update is required on the switch side and the MN can immediately receive packets via the RN.

10.5.3   Relay Process

In the previous section, we have shown how an MN can perform authentication while having data packets relayed by the RN. In this section, we explain in more detail how relaying is performed.

Figure 10.4 shows the format of a general IEEE 802.11 MAC layer frame. Among the many fields, we can identify a *Frame Control* field and four *Address* fields. For the relay process, we are interested

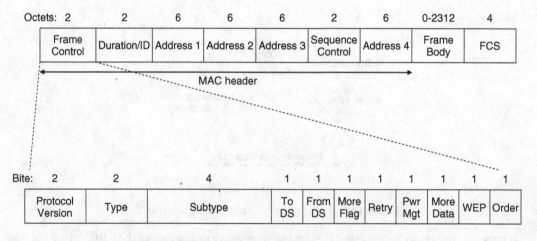

Figure 10.4   General IEEE 802.11 MAC layer frame format

Table 10.2   IEEE 802.11 MAC layer frame: content of address fields

To DS	From DS	Address 1	Address 2	Address 3	Address 4
0	0	DA	SA	BSSID	N/A
0	1	DA	BSSID	SA	N/A
1	0	BSSID	SA	DA	N/A
1	1	RA	TA	DA	SA

in the four *Address* fields and the *To DS* and *From DS* one-bit fields that are part of the *Frame Control* field. The *To DS* bit is set to one in data frames that are sent to the distribution system (DS).[5] The *From DS* bit is set to one in data frames exiting the DS. The four *Address* fields have different meanings according to the particular combination of *To DS* and *From DS* bits. Table 10.2 shows the different meanings of the *Address* fields for each combination of the *To DS* and *From DS* bits. The elements appearing in Table 10.2 are the destination address (DA), source address (SA), BSSID, receiver address (RA), and transmitter address (TA).

In infrastructure mode, when an MN sends a packet, this packet is always sent first to the AP even if both the source and the destination are associated with the same AP. For such packets, the MN sets the *To DS* bit. Other MNs on the same channel can hear the packet but discard it because, as the *To DS* field and *Address* fields suggest, these packets are meant for the AP. When the AP has to send a packet to an MN, it sets the *From DS* bit. All MNs that can hear this packet discard it, except for the MN the packet is for.

When both of the fields *To DS* and *From DS* have a value of one, the packet is sent on the wireless medium from one AP to another AP. In ad hoc mode, both of these fields have a value of zero and the frames are exchanged directly between MNs with the same independent basic service set (IBSS).

Chandra et al. (2004) presented an optimal way to continuously switch a wireless card between two or more infrastructure networks or between infrastructure and ad hoc networks so that the user has the perception of being connected to multiple networks at the same time even though they are using one single wireless card. This approach works well if no real-time traffic is present. When we consider real-time traffic and its delay constraints, continuous switching between different networks and, in particular, between infrastructure and ad hoc mode is no longer a feasible solution. Although optimal algorithms have been proposed for this (Chandra et al., 2004), the continuous switching of the channel and/or operating mode takes a nonnegligible amount of time, which becomes particularly significant if any form of L2 authentication is present in the network. In such cases, the time needed by the wireless card to continuously switch between networks can introduce significant delay and packet loss.

The approach we propose is based on the idea that ad hoc mode and infrastructure mode do not have to be mutually exclusive, but rather can complement each other. In particular, MNs can send ad hoc packets while in infrastructure mode so that other MNs on the shared medium, that is, on the same channel, can receive such packets without involving the AP. These packets use the 802.11 ad hoc MAC addresses as specified in IEEE (1999). That is, both of the fields *To DS* and *From DS* have a value of zero and the *Address* fields are set accordingly, as specified in Table 10.2. In doing so, MNs can send and receive packets directly to and from other MNs without involving the AP and without having to switch to ad hoc mode.

---

[5] A DS is a system that interconnects BSSs and LANs to create an ESS (IEEE, 1999).

This mechanism allows an RN to relay packets to and from an R-MN without significantly affecting any ongoing multimedia session that the RN might have via the AP. Such an approach can be useful in all those scenarios where an MN in infrastructure mode needs to communicate with other MNs in infrastructure or ad hoc mode (Chandra et al., 2006) and a continuous change between infrastructure mode and ad hoc mode is either not possible or convenient.

10.6   Security

Security is a major concern in wireless environments. In this section, we address some of the problems encountered in a cooperative environment, focusing on CR.

10.6.1   Security Issues in Roaming

In this particular context, a malicious user might try to propagate false information among the cooperating MNs. In particular, we have to worry about three main vulnerabilities:

1. A malicious user might want to redirect STAs to fake APs where their traffic can be sniffed and private information can be compromised.
2. A malicious user might try to perform denial-of-service (DoS) attacks by redirecting STAs to distant or nonexistent APs. This would cause the STAs to fail the association with the next AP during the handoff process. The STA would then have to rely on the legacy scanning process to reestablish network connectivity.
3. In L3, a malicious user might behave as an A-MN and try to disrupt an STA's service by providing invalid IP addresses.

In general, we have to remember that the cooperative mechanism described here works on top of any other security mechanism that has been deployed in the wireless network (e.g., 802.11i or WPA). In order for a malicious user to send and receive packets to and from a multicast group, it has to have, first of all, access to the network and thus be authenticated. In such a scenario, a malicious user is an STA with legitimate access to the network. This means that MAC spoofing attacks are not possible, as a change in MAC address would require a new authentication handshake with the network. This also means that once a malicious user has been identified, it can be isolated.

How can we attempt to isolate a malicious node? Since the INFORESP frame is multicast, each MN that has the same information as that contained in such a frame can check that the information in the frame is correct and that no one is lying. If it finds that the INFORESP frame contains wrong information, it immediately sends an INFOALERT multicast frame. This frame contains the MAC address of the suspicious STA. Such a frame is also sent by an R-MN that has received a wrong IP address, and contains the MAC address of the A-MN that provided that IP address. If *more than one* alert for the *same* suspicious node is triggered by *different* nodes, the suspicious node is considered malicious and the information it provides is ignored. Let us look at this last point in more detail.

One single INFOALERT does not trigger anything. In order for an MN to be categorized as bad, there has to be a certain number of INFOALERT multicast frames sent by *different* nodes, all regarding the *same* suspicious MN. This number can be configured according to how paranoid someone is about security but, regardless, it has to be more than one. Let us assume this number to be five.

If a node receives five INFOALERT multicast frames from five different nodes regarding the same MN, then it marks that MN as bad. This mechanism could be compromised if either a malicious user can spoof five different MAC addresses (and this is not likely, for the reasons we have explained earlier) or there are five different malicious users that are correctly authenticated in the wireless network and can coordinate their attacks. If this last situation occurs, then there are bigger problems in the network to worry about than handoff policies. Choosing the number of INFOALERT frames required to mark a node as malicious to be very large has advantages and disadvantages. It gives more protection against the exploitation of this mechanism for DoS attacks, as the number of malicious users trying to exploit INFOALERT frames would have to be high. On the other hand, it also makes the mechanism less sensitive to detecting a malicious node, as the number of INFOALERT frames required to mark the node as bad might never be reached or might take too long to reach. So, there is clearly a trade-off.

Regardless, in any of the three situations described at the beginning of this section, an MN targeted by a malicious user would be able to easily recover from an attack by using legacy mechanisms typically used in noncooperative environments, such as active scanning and DHCP address acquisition.

10.6.2   Cooperative Authentication and Security

In order to improve security in the relay process, we can introduce some countermeasures that nodes can use to prevent exploitation of the relay mechanism. The main concern in having an STA relay packets for an unauthenticated MN is that such an MN might try to repeatedly use the relay mechanism and never authenticate with the network. In order to prevent this, we have introduced the following countermeasures:

1. Each RELAY_REQ frame allows an RN to relay packets for a limited amount of time. After this time has passed, the relaying stops. The relaying of packets is required only for the time needed by the MN to perform the normal authentication process.
2. An RN relays packets only for those nodes which have sent a RELAY_REQ packet to it while still connected to their previous AP.
3. RELAY_REQ packets are multicast. All the nodes in the multicast group can help in detecting bad behavior, such as one node repeatedly sending RELAY_REQ frames.

All of these countermeasures work if we can be sure of the identity of a node; however, in general, this is not always the case, as malicious users can perform MAC spoofing attacks, for example. However, as we have explained in Section 10.6.2, MAC spoofing attacks are not possible in the present framework.

That said, we have to remember that before an RN can relay packets for an MN, it has to receive the proper RELAY_REQ packet from the MN. This packet has to be sent by the MN while still connected to the old AP. This means that the MN has to be authenticated with the previous AP in order to send such a packet. Furthermore, once the relaying timeout has expired, the RN will stop relaying packets for that MN. At this point, even if the MN can change its MAC address, it will not be able to send a new RELAY_REQ, as it has first to authenticate again with the network (e.g., using 802.11i), and therefore no relaying would take place. In the special case in which the old AP belongs to an open network,[6] a malicious node could perform MAC spoofing and exploit the relay

---

[6] Under normal conditions this is very unlikely, but it might happen for handoffs between different administrative domains, for example.

mechanism in order to have access to the secure network. In this case, securing the multicast group by performing authentication and encryption at the multicast group level could prevent this kind of attack, although it may require infrastructure support.

In conclusion, we can consider the three countermeasures introduced at the beginning of this section to be more than adequate for avoiding exploitation of the relaying mechanism.

10.7   Streaming Media Support

SIP can be used, among other things, to update new and ongoing media sessions. In particular, the IP address of one or more of the participants in a media session can be updated. In general, after an MN performs an L3 handoff, a media session update is required to inform the various parties about the MN's new IP address (Schulzrinne and Wedlund, 2000a).

If the CN does not support cooperation, the relay mechanism described in Section 10.5.2 does not work and the CN keeps sending packets to the MN's old IP address, being unaware of the relay process. This is the case, for example, when an MN establishes a streaming video session with a stream media server. In this particular case, assuming that the media server supports SIP, a SIP session update is performed to inform the media server that the MN's IP address has changed. The MN sends a re-INVITE to the media server, updating its IP address to the RN's IP address. In this way, the media server starts sending packets to the RN and relaying can take place as described earlier.

Once the relaying is over, if the MN's authentication was successful, the MN sends a second re-INVITE including its new IP address; otherwise, once the timeout for relaying expires, the relaying process stops and the RN terminates the media session with the media server.

SIP and media session updates will be discussed further in Section 10.10.

10.8   Bandwidth and Energy Usage

By sharing information, the MNs in the network do not have to perform individual tasks such as scanning, which would normally consume a considerable amount of bandwidth and energy. This means that sharing data among MNs is usually more energy and bandwidth efficient than having each MN perform the corresponding individual task. We discuss the impact of CR on energy and bandwidth in this section.

In CR, the bandwidth usage and energy expended are determined mainly by the number of multicast packets that each client has to send to acquire the information it needs. The number of multicast packets is directly proportional to the number of clients supporting the protocol that are present in the network. In general, more clients mean more requests and more responses. However, having more clients that support the protocol ensures that each client can collect more information with each request, which means that, overall, each client will need to send fewer packets. Furthermore, by having the INFORESP frames as multicast frames, many MNs will benefit from each response, not just the MN that sent the request. This will minimize the number of packets exchanged, in particular the number of INFOREQs sent.

To summarize, with an increasing number of clients, suppression of multicast takes place, so the number of packets sent remains constant.

In general, sending a few long packets is more efficient than sending many short ones. As explained in Section 10.4.2, for each AP the information included in an INFOREQ or INFORESP packet is a cache entry (see Figure 10.2), that is, a triple {BSSID, Channel, Subnet ID} with a total size of $6 + 4 + 4 = 14$ bytes. Considering that the size of an MTU (maximum transmission unit) is 1500 bytes, that each cache entry takes up about 14 bytes, and that IP and UDP headers together take up a total of 28 bytes, each INFOREQ and INFORESP packet can carry information about no more than 105 APs in a maximum of 1472 bytes.

Henderson et al. (2004) analyzed the behavior of wireless users in a campus-wide wireless network over a period of 17 weeks. They found that:

- Users spent almost all of their time at their home location. The home location was defined as the AP where they spent most of their time, and all the APs within 50 m of this one.
- The median number of APs visited by a user was 12, but the median differed between device types, with values of 17 for laptops, 9 for PDAs, and 61 for VoIP devices such as VoIP phones.

This shows that most devices will spend most of their time at their home location, which means that they will mostly deal with a small number of APs. However, even if we consider the median number of APs that clients used throughout the tracing period of 17 weeks, we can see that when laptops and PDAs were used, each MN would have had to know about the nearest 9–17 APs. For VoIP devices that were always on, the median number of APs throughout the tracing period was 61. In our implementation, each INFOREQ and INFORESP packet carries information about 105 APs at most. Regardless of the type of device used, the information about the next possible APs fits into a single INFOREQ or INFORESP frame. Also, once the cache has been filled, spending most of the time at the home location means that no new data is needed for most of the time, thus minimizing the number of INFOREQ and INFORESP packets exchanged.

The relay mechanism introduced in Section 10.5 for cooperative authentication introduces some bandwidth overhead. This is because, for each packet that has to be sent by the MN to the CN and vice versa, the packet occupies the medium twice; once when being transmitted between the MN and RN and once when being transmitted between the RN and AP. This, however, happens only for the few seconds needed by the MN to authenticate. Furthermore, both of the links MN–RN and RN–AP are maximum-bit-rate links, so the *time on air* for each data packet is small.

10.9 Experiments

In this section, we describe the implementation details and measurement results for CR.

10.9.1 Environment

All of the experiments were conducted at Columbia University on the seventh floor of the Schapiro building. We used four IBM Thinkpad laptops: three IBM T42 laptops using Intel Centrino Mobile technology with a 1.7 GHz Pentium processor and 1 GB RAM and one IBM laptop with an 800 MHz Pentium III processor and 384 MB RAM. The Linux kernel version 2.4.20 was installed on all of the laptops. All of the laptops were equipped with a Linksys PCMCIA Prism2 wireless card. Two of them were used as wireless sniffers, one of them was used as a roaming client, and one was used as a "helper" to the roaming client; that is, it replied to INFOREQ frames and behaved as an A-MN. For cooperative authentication, the A-MN was also used as an RN. Two Dell Dimension 2400 desktops were used, one as a CN and the other as a RADIUS server (Rigney et al., 2000). The APs used for the experiments were a Cisco AP1231G, which was an enterprise AP, and a Netgear WG602, which was a SOHO/home AP.

10.9.2   Implementation Details

In order to implement the cooperation protocol, we modified the wireless card driver and the DHCP client. A cooperation manager was also created in order to preserve state information and coordinate the wireless driver and DHCP client. For cooperative authentication, the WPA supplicant was also slightly modified to allow relaying of unencrypted frames. The HostAP (Malinen, 2005a) wireless driver, an open-source WPA supplicant (Malinen, 2005b), and the ISC DHCP client (Internet System Consortium, 2005) were chosen for the implementation. The modules involved and their interaction are depicted in Figure 10.5. A UDP packet generator was also used to generate small packets with a packetization interval of 20 ms in order to simulate voice traffic. For the authentication measurements, we used FreeRADIUS (FreeRADIUS Project, 2006) as the RADIUS server.

10.9.3   Experimental Setup

For the experiments, we used the Columbia University 802.11b wireless network, which was organized as one single subnet. In order to test L3 handoff, we introduced another AP connected to a different subnet (Figure 10.6). The two APs operated on two different nonoverlapping channels.

The experiments were conducted by moving the roaming client between two APs belonging to different subnets, thus having the client perform L2 and L3 handoffs in either direction. Packet exchanges and handoff events were recorded using the two wireless sniffers (Kismet) (Kershaw et al., 2005), one per channel. The trace files generated by the wireless sniffer were analyzed later using Ethereal (Combs et al., 2004).

In this experimental setup, we did not consider the possibility of a large[7] presence of other MNs under the same AP, since air-link congestion is not relevant to handoff measurements. Delays due to collisions, backoff, propagation, and AP queuing are irrelevant since they usually are on the order

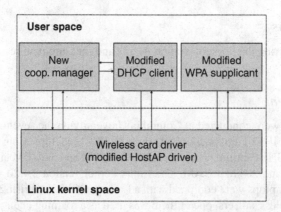

Figure 10.5   Interaction of modules in the implementation

---

[7] Other MNs were present in the Columbia wireless network during the experiments.

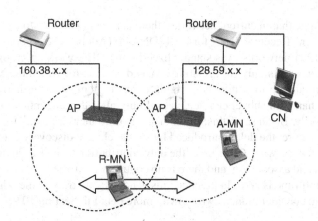

Figure 10.6 L3 handoff environment. (Source: Schulzrinne 2006. Reproduced with permission of Henning Schulzrinne.)

of microseconds under normal conditions. If these delays were very high because of a high level of congestion, we would have to worry primarily about the MN not being able to make or continue a call because the AP had reached its maximum capacity. Handoff delay would, at this point, become a second-order problem. Furthermore, in that scenario, the MN should avoid handing off to a very congested AP in the first place, as part of a good handoff policy (see Section 10.11). Updating the information at the home agent or SIP registrar is trivial and does not have the same stringent delay requirements that mid-call mobility has, and therefore it will not be considered here.

10.9.4   Results

In this section, we show the results obtained in our experiments. In Section 10.9.4, we consider an open network with no authentication in order to show the benefits of CR in an open network. In Section 10.9.4, authentication is added and, in particular, we consider a wireless network with IEEE 802.11i enabled.

We define the L2 handoff time as scanning time + open authentication and association time + IEEE 802.11i authentication time. The last contribution to the L2 handoff time is not present in open networks. Similarly, we define the L3 handoff time as subnet discovery time + IP address acquisition time.

In the following experimental results, we show the drastic improvement achieved by CR in terms of handoff time. In L2, such an improvement is possible because, as we have explained in Section 4.1, the MNs build a cache of neighboring APs so that scanning for new APs is not required and the delay introduced by the scanning procedure during the L2 handoff is removed. Furthermore, by using relays (see Section 10.5), an MN can send and receive data packets *during* the authentication process, thus eliminating the 802.11i authentication delay. In L3, the MNs cache information about which AP belongs to which subnet, hence immediately detecting a change in subnet by comparing the subnet IDs of the old and new APs. This provides a way to detect a subnet change and at the same time makes the subnet discovery delay insignificant. Furthermore, with CR, the IP address acquisition delay is completely removed, since each node can acquire a new IP address for the new subnet while still in the old subnet (see Section 10.4.3).

It is important to note that in current networks[8] there is no standard way to detect a change in subnet in a timely manner.[9] Recently, DNA for IPv4 (DNAv4) (Adoba et al., 2006a), aimed at detecting subnet changes in IPv4 networks, was standardized by the DHC working group in the IETF. This mechanism, however, works only for previously visited subnets for which the MN still has a valid IP address, and can take up to hundreds of milliseconds to complete. Furthermore, if L2 authentication is used, a change in subnet can be detected only after the authentication process has been completed successfully. Because of this, in the handoff time measurements for the standard IEEE 802.11 handoff procedure, the delay introduced by subnet change discovery was not considered.

To summarize, in theory, when CR is used, the only contribution to the L2 handoff time arises from open authentication and association, and there is no contribution to the L3 handoff time whatsoever; that is, the L3 handoff time is zero. In practice, this is not exactly true. Some other sources of delay have to be taken into consideration, as we show in more detail in Section 10.9.4.

10.9.4.1   L2 and L3 Roaming

Here, we present the handoff time when an MN performs an L2 and L3 handoff without any form of authentication; that is, the MN is moving in an open network. In such a scenario, before the L2 handoff occurs, the MN tries to build its L2 cache if it has not already done so. The MN also searches for any available A-MN that might help it to acquire an IP address for the new subnet. The scenario is the same as the one depicted in Figure 10.6.

Figure 10.7 shows the handoff time when CR was used. In particular, we show the L2, L3, and total L2 + L3 handoff times over 30 handoffs. As we can see, the total L2 + L3 handoff time has a maximum value of 21 ms in experiment 18. Also, we can see that even though the L3 handoff time is higher on average than the corresponding L2 handoff time, there are situations where these two become comparable. For example, we can see in experiment 24 that the L2 and L3 handoff times

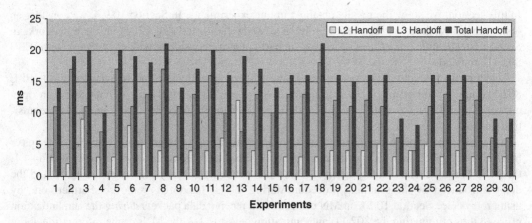

Figure 10.7   Measured L2 and L3 handoff times with CR in an open network. (Source: Schulzrinne 2006. Reproduced with permission of Henning Schulzrinne.)

---

[8] In the IETF, the DNA working group is standardizing the detection of network attachments for IPv6 networks only (Narayanan, 2006).

[9] Router advertisements are typically broadcast only every few minutes.

Table 10.3 Performance overview for CR (average)

IP_REQ to IP_RESP	867.0 ms
L2 handoff	4.2 ms
L3 handoff	11.4 ms
Total handoff	15.6 ms
Packet loss	1.3 packets

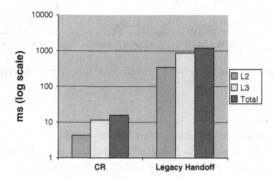

Figure 10.8 Average handoff times for CR and IEEE 802.11b in an open network. (Source: Schulzrinne 2006. Reproduced with permission of Henning Schulzrinne.)

are equal and in experiment 13 that the L2 handoff time exceeds the corresponding L3 handoff time. The main causes of this variance will be presented in Section 10.9.4.

Figure 10.7 and Table 10.3 show that, on average, with CR the total L2 + L3 handoff time is less than 16 ms, which is less than half of the 50 ms requirement for assuring a seamless handoff when real-time traffic is present.

Table 10.3 shows the average values of the IP address acquisition time, handoff time, and packet loss during the handoff process. The time between IP_REQ and IP_RESP is the time needed by the A-MN to acquire a new IP address for the R-MN. This time can give a good approximation to the L3 handoff time that we would have without cooperation. As we can see, with cooperation we reduce the L3 handoff time to about 1.5% of what we would have without cooperation. Table 10.3 also shows that the packet loss experienced during an L2 + L3 handoff is negligible when using CR.

Figure 10.8 shows the averages over 30 handoffs of the L2, L3, and L2 + L3 handoff times for CR and for the legacy 802.11 handoff mechanism. The total L2 + L3 handoff time is less than 16 ms for CR, while it is about 1210 ms for the legacy 802.11 handoff mechanism. CR has reduced the total handoff time to 1.3% of the handoff time introduced by the standard 802.11 handoff procedure.

10.9.4.2 L2 and L3 Roaming with Authentication

Here, we show the handoff times when IEEE 802.11i was used together with EAP-TLS and PEAP/MSCHAPv2.

Figure 10.9 shows the average over 30 handoffs of the delay introduced into an L2 handoff by the certificate/credentials exchange and the session key exchange. Different key lengths for the

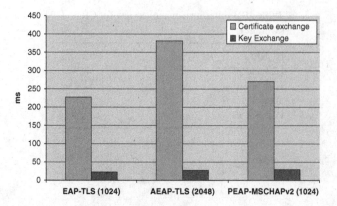

Figure 10.9 Authentication delay in IEEE 802.11i. (Source: Schulzrinne 2006. Reproduced with permission of Henning Schulzrinne.)

generation of the certificates were also considered.[10] As expected, the exchange of certificates takes most of the time. This is the reason why mechanisms such as fast reconnect (Aboba et al., 2004; 2006b) improve L2 handoff times considerably, although these times are still on the order of hundreds of milliseconds.

Generally speaking, any authentication mechanism can be used together with CR. Figure 10.10 shows the average over 35 handoffs of the total L2, L3, and L2 + L3 handoff times. In particular, we show the handoff times for EAP-TLS with 1024- and 2048-bit keys, for PEAP/MSCHAPv2 with a 1024-bit key, and CR. The average L2 + L3 handoff times are 1580, 1669, 1531, and 21 ms, respectively. By using CR, we achieve a drastic improvement in the total handoff time. As we can see, CR reduces the handoff time to 1.4% or less of the handoff time introduced by the standard

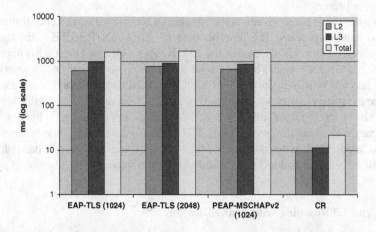

Figure 10.10 Handoff time in IEEE 802.11i networks. (Source: Schulzrinne 2006. Reproduced with permission of Henning Schulzrinne.)

---

[10] The length of the certificates affects the handoff time much more than the length of the session keys.

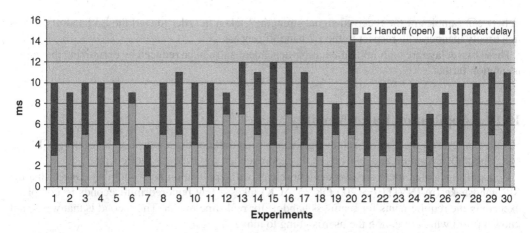

Figure 10.11 CR L2 handoff time in IEEE 802.11i networks. (Source: Schulzrinne 2006. Reproduced with permission of Henning Schulzrinne.)

802.11 mechanism. This significant improvement is possible because in L2 with CR we bypass the whole authentication handshake by relaying packets. In L3, we are able to detect a change in subnet in a timely manner and acquire a new IP address for the new subnet while still in the old subnet.

Figure 10.11 shows in more detail the two main contributions to the L2 handoff time when a relay is used. We can see that, on average, the time needed for the first data packet to be transmitted after the handoff is more than half of the total L2 handoff time. Here, by "data packet" we are referring to a packet sent by our UDP packet generator. By analyzing the wireless traces collected in our experiments, we found that the first data packet after the handoff was not transmitted immediately after the L2 handoff was completed, because the wireless driver needed to start the handshake for the authentication process. This means that the driver already had a few packets in the transmission queue that were waiting to be transmitted when our data packet entered the transmission queue. This, however, concerns only the first packet to be transmitted after the L2 handoff was completed successfully. Subsequent data packets did not encounter any additional delay due to relaying.

10.9.4.3 Measurement Variance

We encountered a high variance in the L2 handoff time. In particular, most of the delay was between the authentication request and the authentication response, before the association request. In all the measurements taken, this behavior appeared to be particularly prominent when moving from the Columbia AP to the Netgear AP. This behavior, together with the results presented by Mishra et al. (2003b), led us to the conclusion that the variance was caused by the cheap hardware used in the low-end Netgear AP.

In L3, ideally, the handoff time should be zero, as we acquire all the required L3 information while still in the old subnet. The L3 handoff time shown in Figure 10.7 can be divided roughly into two main components: *signaling delay* and *polling delay*. The signaling delay is due to the various signaling messages exchanged among the different entities involved in setting up the new L3 information in the kernel (wireless driver and DHCP client); the polling delay is introduced by the polling of variables in between samples of the received signal strength,[11] taken in order to start the L3 handoff process in a timely manner with respect to the L2 handoff process.

These two delays are both implementation-dependent and can be reduced by optimizing the implementation further.

10.10   Application Layer Mobility

Here, we suggest a method for achieving seamless handoffs in the application layer using SIP and CR. The implementation and analysis of this approach are reserved for future work.

Generally speaking, there are two main problems with application layer mobility. One is that the SIP handshake (re-INVITE $\Rightarrow$ 200 OK $\Rightarrow$ ACK) takes a few hundred milliseconds to complete, exceeding the requirements for seamless handoff for real-time media. The second is that we do not know a priori which direction the user is going to move in.

In order to solve these two problems, we have to define a mechanism that allows the MN to start the application layer handoff before the L2 handoff and to do this in such a way that the MN does not move to the wrong AP or subnet after updating the SIP session. Furthermore, the new mechanism also has to work in the event of the MN deciding not to perform an L2 handoff at all after performing the SIP session update, that is, after updating the SIP session with the new IP address.

The SIP mobility mechanism (Schulzrinne and Wedlund, 2000a) and CR can be combined. In particular, we consider an extension of the relay mechanism discussed in Section 10.5.2. Let us assume that the MN performing the handoff has already acquired all the necessary L2 and L3 information as described in Sections 10.4.2, 10.4.3, and 10.5. This means that the MN has a list of possible RNs and IP addresses to use after the L2 handoff, one for each of the various subnets it could move to next. At this point, before performing any L2 handoff, the MN needs to update its multimedia session. The uplink traffic does not cause particular problems, as the MN already has a new IP address to use and can start sending packets via the RN immediately after the L2 handoff. The downlink traffic is more problematic, since the CN will continue sending packets to the MN's old IP address as it is not aware of the change in the MN's IP address until the session has been updated.

The basic idea is to update the session so that the same media stream is sent, at the same time, to the MN and to all of the RNs in the list previously built by the MN. In this way, regardless of which subnet/AP the MN will move to, the corresponding RN will be able to relay packets to it. If the MN does not change AP at all, nothing is lost, as the MN is still receiving packets from the CN. After the MN has performed the L2 handoff and has connected to one of the RNs, it may send a second re-INVITE via the RN so that the CN sends packets to the current RN only, without involving the other RNs any longer. Once the authentication process has been successfully completed, communication via the AP can resume. At this point, one last session update is required so that the CN can send packets directly to the MN without any RN in between.

In order to send multiple copies of the same media stream to different nodes, that is, to the MN performing the handoff and its RNs, the MN can send the CN a re-INVITE with an SDP format as described in RFC 3388 (Camarillo et al., 2002) and shown in Figure 10.12. In this particular format, multiple "m" lines are present together with multiple "c" lines, grouped together by using the same flow identification (FID). A station receiving a re-INVITE with an SDP part as shown in Figure 10.12 sends an audio stream to a client with IP address 131.160.1.112 on port 30000 (if the PCM μ-law codec is used) and to a client with IP address 131.160.1.111 on port 20000. In order for the same

---

[11] The received signal strength is measured by the wireless card driver.

```
v=0
o=Laura 289083124 289083124 IN IP4 five.example.com
t=0 0
c=IN IP4 131.160.1.112
a=group:FID 1 2 3
m=audio 30000 RTP/AVP 0
a=mid:1
m=audio 30002 RTP/AVP 8
a=mid:2
m=audio 20000 RTP/AVP 0 8
c=IN IP4 131.160.1.111
a=mid:3
```

Figure 10.12   RFC 3388 SDP format

media stream to be sent to different clients at the same time, all clients have to support the same codec (Camarillo et al., 2002). In our case, we have to remember that RNs relay traffic to MNs; they do not play such traffic. Because of this, we can safely say that each RN supports any codec during the relay process, and hence a copy of the media stream can always be sent to an RN by using the SDP format described by Camarillo et al. (2002).

It is worth noting that in the session update procedure described here, no buffering is necessary. As we have explained in Section 10.9.4 and shown in Table 10.3, the L2 + L3 handoff time is on the order of 16 ms for open networks, which is less than the packetization interval for typical VoIP traffic. When authentication is used (see Figure 10.10), the total L2 + L3 handoff time is on the order of 21 ms. In both cases the packet loss is negligible, hence making buffering of packets unnecessary.

10.11   Load Balancing

CR can also play a role in AP load balancing. Today, there are many problems with the way MNs select the AP to connect to. The AP is selected according to the link signal strength and signal-to-noise ratio (SNR), and other factors such as effective throughput, number of retries, number of collisions, packet loss, bit rate, and bit error rate are not taken into account. This can cause an MN to connect to an AP with the best SNR but a low throughput, a high number of collisions, and high packet loss because that AP is highly congested. If the MN disassociates or the AP deauthenticates it, the MN looks for a new candidate AP. Unfortunately, with very high probability, the MN will pick the same AP because its link signal strength and SNR are still the "best" available. The information regarding the congestion of the AP is completely ignored, and this bad behavior keeps repeating itself. This behavior can create situations where all users end up connecting to the "best" AP, creating the scenario depicted earlier in Section 10.9.3 and at the same time leaving other APs underutilized (Forte et al., 2006b; Jardosh et al., 2005).

CR can be very helpful in such a context. In particular, we can imagine a situation where an MN wants to gather statistics about the APs that it might move to next, that is, the APs that are present in its cache. In order to do so, the MN can ask other nodes to send statistics about those APs. Each node can collect different kinds of statistics, such as available throughput, bit rate, packet loss, and retry rate. Once these statistics have been gathered, they can be sent to the MN that has requested them. The MN, at this point, has a clear picture of which APs are more congested and which others can support the required QoS, therefore making a smarter handoff decision. By using this approach, we can achieve an even distribution of traffic flows among neighboring APs.

The details of this mechanism are reserved for future study, but can easily be derived from the procedures introduced earlier for achieving fast L2 and L3 handoffs.

10.12   Multicast and Scalability

For simplicity, we have considered in this chapter one single multicast group to which all MNs subscribe. In reality, however, this introduces significant problems in terms of scalability. In order to address scalability concerns, we can think of using different multicast groups, each restricted to, for example, a subnet, an ESS, or a BSS.

The problem in this case is that two MNs belonging to two different multicast groups need to know the multicast IP addresses of each other in order to cooperate. We can follow two approaches to solve this problem. In the first approach, MNs cache their multicast address and share it in the same way as they share other information such as their subnet ID. In the second approach, each MN computes its own multicast address and other MNs' multicast addresses as a hash of one or more known parameters such as the current BSSID and subnet ID. In doing so, each MN can compute the multicast address of another MN by knowing the other MN's current BSSID and subnet ID, for example.

10.13   An Alternative to Multicast

Using IP multicast packets can become inefficient in highly congested environments with a dense distribution of MNs. In such environments, a good alternative to multicast may be represented by ad hoc networks. Switching back and forth between infrastructure mode and ad hoc mode has already been used by MNs in order to share information for fault diagnosis (Chandra et al., 2006). As we pointed out in Section 10.5.3, continuously switching between ad hoc and infrastructure mode introduces synchronization problems and channel-switching delays, making this approach unusable for real-time traffic. However, even if non-real-time traffic is present, synchronization problems could still arise when a switch to ad hoc mode occurred while there was a live TCP connection on the infrastructure network, for example. Spending a longer time in ad hoc mode might cause the TCP connection to time out; on the other hand, waiting too long in infrastructure mode might cause loss of data in the ad hoc network.

In CR, MNs can exchange L2 and L3 information contained in their cache by using the mechanism for relaying described in Section 10.5.3. Following this approach, MNs can exchange information directly with each other without involving the AP and without having to switch their operating mode to ad hoc. In particular, an MN can send broadcast and unicast packets such as INFOREQ and INFORESP packets with the *To DS* and *From DS* fields set to zero (see Section 10.5.3). Because of this, only the MNs in the radio coverage of the first MN will be able to receive such packets. The AP will drop these packets since the *To DS* field is not set.

Ad hoc multihop routing can also be used when needed. This may be helpful, for example, in the case when R-MNs acquire a new IP address for a new subnet while still in the old subnet (see

Section 10.4.3), when the current and new APs use two different channels. In such a case, a third node on the same channel as the R-MN could route packets between the R-MN and the A-MN by switching between the two channels of the two APs, thus leaving the R-MN and A-MN operations unaffected. In this case we would not have synchronization problems, since the node, when switching between the two channels, would have to switch only twice: once after receiving the IP_REQ packet from the R-MN in order to send it to the A-MN, and a second time after receiving the IP_RESP packet from the A-MN in order to send it to the R-MN.

An ad-hoc-based approach, such as the relay mechanism presented in Section 10.5.3, does not require any support in the infrastructure, and it represents an effective solution in congested and densely populated environments. On the other hand, ad hoc communication between MNs would not work very well in networks with a small population of MNs, where each MN might be able to see only a very small number of other MNs at any given time.

MNs with two wireless cards could use one card to connect to the ad hoc network and share information with other MNs, while having the other card connected to the AP. The two cards could also operate on two different access technologies such as cellular and 802.11.

If it is possible to introduce some changes into the infrastructure, we can minimize the use of multicast packets by using the SIP presence model (M. Day, 2000). In this model, we introduce a new presence service in which each subnet is a presentity. Each subnet has a contact list of all the A-MNs available in that subnet, for example, so that the presence information is represented by the available A-MNs in the subnet. When an R-MN subscribes to this service, it receives presence information about the new subnet, namely its contacts, which are the available A-MNs in that subnet.

This approach could be more efficient in scenarios with a small number of users supporting CR. On the other hand, it would require changes in the infrastructure by introducing additional network elements. The presence and ad hoc approaches are reserved for future study.

10.14   Conclusions and Future Work

In this chapter, we have defined a cooperative roaming protocol. This protocol allows MNs to perform L2 and L3 handoffs seamlessly, with an average total L2 + L3 handoff time of about 16 ms in an open network and about 21 ms in an IEEE 802.11i network without requiring any changes to either the protocol or the infrastructure. Each of these values is less than half of the 50 ms requirement for real-time applications such as VoIP to achieve seamless handoff. Furthermore, this protocol is able to provide such a fast handoff regardless of the particular authentication mechanism used, while still preserving security and privacy.

MN cooperation has many advantages and does not introduce any significant disadvantages, as in the worst-case scenario the MNs can rely on the standard IEEE 802.11 mechanisms and achieve performance similar to a scenario with no cooperation.

Node cooperation can be useful in many other applications:

- In a multi-administrative-domain environment, CR can help in discovering which APs are available for which domain. In this way, an MN might decide to go to one particular AP/domain rather than some other AP/domain, according to roaming agreements, billing, and so on.
- In Section 10.11, we have shown how CR can be used for load balancing. Following a very similar approach but using other metrics such as collision rate and available bandwidth, CR can also be used for admission control and call admission control.

- CR can help in propagating information about service availability. In particular, an MN might decide to perform a handoff to one particular AP because of the services that are available at that AP. Such services might be a particular type of encryption or authentication, a minimum guaranteed bit rate or available bandwidth, or the availability of other types of networks such as Bluetooth, UWB, and 3G cellular networks, for example.
- CR also provides advantages in terms of adaptation to changes in the network topology. In particular, when an MN finds stale entries in its cache, it can update its cache and communicate such changes to the other MNs. This applies also to virtual changes in the network topology (i.e., changes in the power levels of the APs), which might become more common with the deployment of IEEE 802.11h equipment.
- CR can also be used by MNs to negotiate and adjust their transmission power levels so as to achieve a minimum level of interference.
- Ramani and Savage (2005) described a passive scanning algorithm according to which an MN knows the exact moment when a particular AP will send its beacon frame. In this way, the MN collects statistics for the various APs using passive scanning but without having to wait for the whole beacon interval on each channel. This algorithm, however, requires all of the APs in the network to be synchronized. By using a cooperative approach, we can have the various MNs share information about the beacon intervals of their APs. In this way, we only need to have the MNs synchronized amongst themselves (e.g., via NTP), without any synchronization being required on the network side.
- Interaction between nodes in an infrastructure network and nodes in an ad hoc/mesh network can be provided in the following ways:
  1. An MN in ad hoc mode can send information about its ad hoc network. In this way, MNs in the infrastructure network can decide if it is convenient for them to switch to the ad hoc network (this would also free resources in the infrastructure network). This can, for example, happen if there is a lack of coverage or if there is high congestion in the infrastructure network. Also, an MN might switch to an ad hoc network if it has to recover some data available in the ad hoc network itself (e.g., in a sensor network).
  2. If two parties are close to each other, they can decide to switch to an ad hoc network discovered earlier and talk to each other without any infrastructure support. They might also create an ad hoc network on their own using a default channel if no other ad hoc network is available.

In future work, we will look in more detail at application layer mobility, load balancing, and call admission control. We will investigate the possibility of having some network elements such as APs support A-MN and RN functionalities; this would be useful in scenarios where only a few MNs support CR. Finally, we will look at how IEEE 802.21 (IEEE, 2005b) could integrate and extend CR.

11

System Evaluation

In this chapter, we evaluate the overall handoff system, where many of the optimization techniques that we have described in Chapter 6 function together to build a complete handoff system. We first illustrate experimental results from a few systems for both intertechnology and intratechnology handoffs, and then validate some of the optimization techniques using Petri net modeling. We demonstrate how Petri nets can model some behavioral properties of the handoff system, such as deadlocks. We also compare the performance of a few scheduling techniques that could be applied to handoff.

11.1 Summary of Key Contributions and Indicative Results

Currently, the optimization techniques for each of the handoff components in different layers are implemented independently. These optimized components have not been used in an integrated fashion to build a complete optimized handoff system. The existing experimental results from these optimization techniques have been derived for each of the individual handoff components only (e.g., discovery, authentication, and configuration). There is also no work that compares model-based results with experimental results to verify predictions of systems performance. There is no existing system evaluation technique that can verify the correctness of a handoff system or detect system anomalies of a handoff system such as deadlocks.

We have built several indicative handoff systems using the reactive, proactive, and cross-layer optimization techniques that we developed for each of the handoff components as described in Chapter 6. We have built equivalent mobility models for each of these indicative systems and compared experimental results with results from the equivalent mobility models to determine how changes in certain handoff parameters affect the overall handoff system.

Indicative results from the systems evaluation of our experimental systems demonstrate the effectiveness of the optimization techniques that we have developed. These results also demonstrate feasible scenarios where some of these optimization techniques can work together to produce a partial or complete handoff system. Comparing the results from Petri-net-based mobility models with those obtained from experiments for these optimization techniques demonstrates the correctness of the predictions. Petri-net-based behavioral analysis of these optimization techniques can determine the existence of system anomalies such as deadlocks.

Mobility Protocols and Handover Optimization: Design, Evaluation and Application, First Edition.
Ashutosh Dutta and Henning Schulzrinne.
© 2014 John Wiley & Sons, Ltd. Published 2014 by John Wiley & Sons, Ltd.

A handoff system that uses many of these optimization techniques for different handoff components implies a possible sequence of these handoff components (i.e., one can determine if some of these components can work in parallel) and allows one to verify how cross-layer techniques can be used to expedite many of the handoff operations that work in sequence. Verifying the results from the experimental systems with those from the corresponding mobility models demonstrates the effectiveness of these mobility models. These mobility models can also evaluate the systems performance based on the sequence of handoff operations and the availability of resources. They can also determine how the level of concurrency and additional resources may affect the systems performance. A behavioral analysis of the mobility model can demonstrate certain operational aspects of the system such as deadlocks that would not be evident otherwise from the experimental results.

In the rest of the chapter, we discuss experimental results from a few indicative systems, results from Petri-net-based models for several of the optimization techniques that we developed, the verification of systems performance for different handoff sequences, and deadlock detection and avoidance.

11.2   Introduction

Systems evaluation and validation of the various optimization techniques associated with a handoff event can be implemented through experimental analysis, simulation, and analytical modeling. While experimental results are limited by several constraints, such as system parameters (namely memory, CPU power, and other network parameters such as bandwidth), Petri net models can be used to validate such experimental results and perform systems evaluation with the ability to vary the system parameters. Thus, in order to validate various optimization techniques associated with the handoff system, we have applied both experimental and modeling approaches. Optimization techniques for many of the handoff components have been described in Chapter 6, each of these techniques with its own experimental results. However, in order to validate the system performance, we built a handoff system by implementing optimization techniques for these handoff components that work together in an integrated manner.

11.3   Experimental Validation

In this section, we describe experimental results from a handoff system that supported multiple types of handoff, namely *intratechnology* and *intertechnology*, where the mobile used a single interface and multiple interfaces, respectively. Both of these kinds of handoff are described in Chapter 2. These handoff systems used the set of optimization techniques that we described earlier in Chapter 6.

In particular, we describe results from three experimental systems that used several optimization techniques to reduce the handoff delay and packet loss. These three experimental systems used (1) our Media Independent Preauthentication framework, (2) cross-layer-trigger-assisted preauthentication, and (3) optimized handoff in an IMS-based network.

11.3.1   The Media Independent Preauthentication Framework

The experimental results showed that without any optimization, 4 s of handover delay was observed when the mobile moved between two homogeneous access networks (e.g., between 802.11 networks) and 200 packets were lost owing to this handover delay. The situation was worse for heterogeneous handover, where it could take up to 15 s for the mobile to authenticate and establish connectivity in a CDMA network.

We have prototyped a mobility system called Media Independent Preauthentication (MPA) (Dutta et al., 2005e) that utilizes many of the techniques that we have developed to optimize the basic handoff operations, namely network discovery, authentication, configuration, security association, and binding update. This system also reduces the link-layer handoff delay by avoiding scanning and applies cross-layer optimization techniques, such as detection of "link up" and "link down" events. In addition, it reduces packet loss by implementing a dynamic buffering and copy-and-forward mechanism (Dutta et al., 2006e) at the edges of the network.

We have applied these proactive techniques to optimize both network layer and application layer mobility protocols for both a single interface and multiple interfaces (Dutta et al., 2005c,e).

Koodli (2005) and Gwon et al. (2003) developed mobility systems that utilize proactive handoff techniques for Mobile IPv6 and Mobile IPv4, respectively. However, these systems do not address proactive discovery or preauthentication mechanisms. Also, these mechanisms require signaling exchanges between neighboring routers that work only if the routers have established trust relationships between them. We have performed a comparative analysis of the Media Independent Preauthentication mechanism and the proactive fast-handoff mechanism of FastMIPv6 (Koodli, 2005) and described the details of the results (Dutta et al., 2007b). We now give a brief comparative analysis of MPA and other proactive handover techniques. In this section, we highlight certain added features of MPA that are different from the existing make-before-break techniques. In particular, we compare MPA with FMIPv6 and highlight the functional differences. MPA provides a make-before-break mechanism and takes care of many upper-layer handover-related functions, leaving only the layer 2 handover operations to be executed during the move. There are several other proactive schemes, such as those of Gwon et al. (2003) and Koodli (2005), that utilize make-before-break techniques and provide comparable performance.

The distinct features of MPA relative to other related make-before-break schemes are as follows: (1) MPA can work over multiple types of mobility protocols; (2) MPA provides preauthentication support for both layer 3 and layer 2, thereby reducing the delay due to authentication; (3) MPA provides flexible ways of performing preconfiguration operations such as stateless autoconfiguration and stateful preconfiguration using a DHCP relay agent; (4) when assisted by the IEEE 802.21 information discovery scheme, MPA can optimize the layer 2 handoff by avoiding scanning and IEEE 802.11i authentication; (5) the MPA framework can be applied to different types of handover such as interdomain, intradomain, intertechnology, and intratechnology; and (6) MPA provides a flexible buffering mechanism in different parts of the network that can reduce the packet loss during the handover. Now, we briefly compare MPA with FMIPv6. The IETF has defined two fast-handover protocols for MIPv6, namely hierarchical MIPv6 (Soliman et al., 2006) and Fast MIPv6 (Koodli, 2005). Both of these protocols try to reduce the packet loss and handover delay experienced in the base version of MIPv6. There is a very fundamental difference between MPA and FMIPv6. While FMIPv6 is limited to the use of MIPv6 as a binding protocol for fast handover, MPA defines a mobility framework that can work independently of the mobility protocol, and can work with a number of protocols, including MIPv4, MIPv6, and SIP-based mobility. However, in the context of MIPv6, we now provide a brief functional comparison between FMIPv6 and MPA over IPv6.

FMIPv6 provides two ways of providing fast handoff: predictive mode and reactive mode. The FMIPv6 predictive mode and MPA-based optimization over MIPv6 exhibit some similarities for certain operations such as preconfiguration and proactive binding update. Dutta et al. (2011) provided

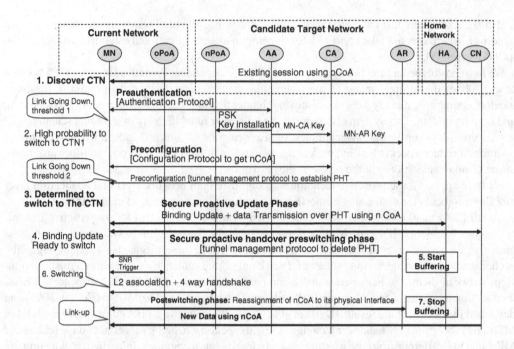

Figure 11.1   Protocol flow for Media Independent Preauthentication

a complete overview of the operation of MPA and the implementation results. Cabellos-Aparicio et al. (2005) provided some experimental results for FMIPv6 that showed that the delay due to proactive FMIPv6 was bounded by the nonoptimized layer 2 delay. Similarly, MPA over IPv6 was also bounded by the nonoptimized layer 2 delay in the absence of any assistance from IEEE 802.21's information discovery scheme. According to Cabellos-Aparicio et al. (2005), the handover latency for proactive FMIPv6 was equal to the layer 2 IEEE 802.11 handover latency and was computed to be 320 ms. On the other hand, MPA assisted by IEEE 802.21 information discovery limited the layer 2 delay to 4 ms by avoiding scanning. FMIPv6 when assisted by IEEE 802.21 information discovery can also help to reduce the layer 2 delay to a value comparable to that obtained in the case of MPA.

Figure 11.1 shows the protocol flow based on an MPA-based framework. We assume that the mobile node (MN) is already connected to a point of attachment, referred to as the old point of attachment (oPoA), and is assigned an old care-of address (oCoA). Throughout the communication flow, data packets should not be lost except for the period during the layer 2 switching procedure in step 5, but MPA procedures can help minimize the packet loss during this layer 2 handover period with the help of the information service (IS), event service (ES), and command service (CS) of IEEE 802.21. We now briefly describe the various functional phases of the preauthentication framework.

11.3.1.1   Preauthentication Phase

The mobile finds a CTN (candidate target network) through a discovery process, such as IEEE 802.21, and obtains the address and capabilities of the AA (authentication agent), CA (configuration agent), and AR (access router) in the CTN. The mobile preauthenticates with the authentication agent. If the preauthentication is successful, an MPA SA (MPA security association) is created between the mobile node and the authentication agent. Two keys are derived from the MPA SA, namely an MN–CA key and an MN–AR key, which are used to protect the subsequent signaling messages of the configuration protocol and the tunnel management protocol, respectively. The MN–CA and MN–AR keys are then securely delivered to the configuration agent and the access router, respectively. Layer 2 preauthentication can be initiated at this stage.

11.3.1.2 Preconfiguration Phase

The mobile node realizes that its point of attachment is likely to change from the old point of attachment (oPoA) to a new point of attachment (nPoA).

It then performs preconfiguration, with the configuration agent using the configuration protocol to obtain an IP address, say a new care-of address (nCoA), and other configuration parameters from the CTN. The access router uses a tunnel management protocol to establish a proactive handover tunnel with the mobile. As part of the tunnel management protocol, the mobile node registers the oCoA and nCoA as the tunnel outer address and the tunnel inner address, respectively. The signaling messages of the preconfiguration protocol are protected using the MN–CA and MN–AR keys. When the configuration agent and the access router are co-located in the same device, configuration and tunnel management may be performed by a single protocol such as IKEv2 (Kaufman, 2005). After completion of the establishment of the tunnel, the mobile is able to communicate using both the oCoA and the nCoA by the end of step 4 in Figure 11.1.

11.3.1.3 Secure Proactive Handover, Main Phase

Before the mobile switches to the new point of attachment, it starts the secure proactive handover process by executing the proactive binding update operation of a mobility management protocol such as MIPv6 or SIP-based mobility and transmitting subsequent data traffic over the tunnel. In some cases, it may cache multiple nCoAs and establish simultaneous bindings with the CH (correspondent host) and HA (home agent).

11.3.1.4 Secure Proactive Handover, Preswitching Phase

The mobile completes the binding update and becomes ready to switch to the new point of attachment. The mobile may execute the tunnel management protocol to delete or disable the proactive handover tunnel and cache the nCoA after deletion or disabling of the tunnel. A buffering module at the new access router (nAR) begins to buffer the packets (start-buffering) when it receives a signal to delete the tunnel. The mobile sends an explicit signal to stop buffering and flush the packets after the mobile has connected to the new point of attachment. Details of the buffering modules and buffering protocols are described in Chapter 6.

The decision as to when the mobile switches to the new point of attachment depends on the handover policy. In general, mobile-controlled or network-controlled policies can be used to trigger the handoff. The mobile's signal quality and location, the communication cost, and the quality of service (QoS) of the received traffic are some factors that can determine the handoff policy. The results presented in this chapter are based on the use of the signal-to-noise ratio (SNR) as the trigger for handoff.

11.3.1.5 Switching Phase

Link-layer handover occurs in this step. During this phase, any of the layer 2 security associations, including EAP-based authentication and the 802.11i-related four-way handshake, may take place. Normally, layer 2 preauthentication is taken care of by inbuilt layer 2 preauthentication support such as that provided by 802.11i. However, using the MPA scheme, layer 3 preauthentication can bootstrap layer 2 authentication, leaving only the four-way handshake to be done during this phase.

11.3.1.6 Secure Proactive Handover, Postswitching Phase

The mobile executes the switching procedure. Upon successful completion of the switching procedure and layer 2 association, the mobile immediately assigns the cached nCoA to the physical interface attached to the new point of attachment. If the proactive handover tunnel was not deleted or disabled during step 4, the tunnel can be deleted or disabled in this phase as well. After this, direct transmission of data packets using the nCoA is possible without using a proactive handover tunnel.

We have applied MPA-related optimization techniques and experimented with two mobility protocols, namely SIP-based mobility (Wedlund and Schulzrinne, 1999) and MIPv6 (Johnson et al., 2004), supporting both intratechnology and intertechnology handovers.

11.3.2 Intratechnology Handoff

In this section, we highlight experimental results for intratechnology handoff.

An intratechnology handover is defined as occurring when a mobile moves between access technologies of the same type, such as between 802.11[a,b,n] and 802.11 [a,b,n], between CDMA 1xRTT and CDMA 1xEV-DO, or between two LTE networks. In this scenario, a mobile may be equipped with a single interface (with multiple PHY types of the same technology) or with multiple interfaces. An intratechnology handover may involve intrasubnet or intersubnet movement, and thus the mobile may need to change its L3 identifier, depending upon the type of movement.

Figure 11.2 shows the topology of the experimental test bed that we used to experiment with intratechnology handoff using 802.11 access networks. The test bed emulates two different visited domains and a home domain. Each visited domain has several subnetworks. The mobile moves from one visited domain to another domain and, in the process, changes its subnet. Network A is the oPoA, where the mobile node initially resides prior to handover. Network B is the nPoA, network C is where the correspondent node (CN) resides, and network D is the home network, where the home agent and AAA server reside.

The configuration protocol was DHCP, and the authentication agent was a PANA (Protocol for carrying Authentication for Network Access) (Forsberg et al., 2008) server with a backend Diameter server to carry out EAP-TLS (Extensible Authentication Protocol TLS) (Aboba et al., 2004). The configuration agent was a DHCP relay agent and the next access router was an edge router that ran over a Linux operating system. In an IPv6 network, the nAR behaves as the configuration agent in network B.

The MPA mechanisms work independently of the underlying mobility management protocol. We have demonstrated the benefits of MPA using both SIP mobility and MIPv6 as mobility management protocols. In the case of MPA for MIPv6, the CN started an RTP session with the MN while the MN was in network 1 via the HA using an MIPv6 tunnel. MPA created a proactive handover tunnel between the MN and nAR in network B. This was an IPSec tunnel in Encapsulating Security Payload

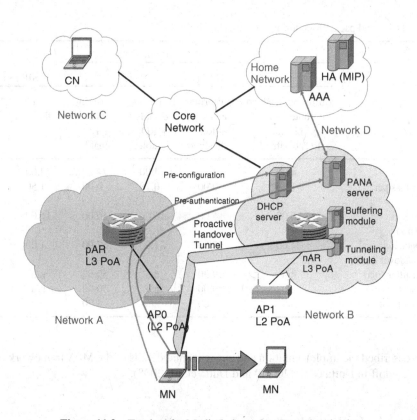

Figure 11.2   Test bed for Media Independent Preauthentication

(ESP) mode, and we used PANA to dynamically establish and terminate the IPSec tunnel. Before the handoff, the tunneled MIPv6 traffic between the MN and HA went through the IPSec tunnel created by MPA with IPSec policy settings. If the configuration agent and router are co-located, a single protocol, such as IKEv2, can take care of both of the functions of configuration and tunnel management.

In the case of SIP-based mobility, an IP–IP-based tunnel was used as a proactive handover tunnel between the mobile node and the next access router. After successful setup of a connection using SIP, voice traffic flowed between the MN and the CN. This voice traffic was carried over RTP/UDP. We used RAT (Robust Audio Tool) as the media agent, and the streaming traffic was generated using a codec with a spacing of 20 ms between packets. A SIP re-INVITE was used as the binding update over the proactive handover tunnel before the mobile moved to the target network. Preauthentication and buffering procedures were carried out similarly to MIPv6.

Table 11.1 shows experimental results from the mobility system referring to intratechnology handover involving IEEE 802.11 access networks. Both an application layer mobility protocol, such as SIP-based mobility, and a network layer mobility protocol, such as MIPv6, were used for experimental validation of these techniques. These results demonstrate how several of the proactive optimization techniques can work together to minimize handover delays, packet loss, and jitter. The results also demonstrate the effect of buffering at the edge routers in reducing packet loss at the cost of added delay. These results shown in Table 11.1 are average values taken over five runs.

Table 11.1   Delay and packet loss during proactive handoff

Handoff optimization parameters	Mobility type					
	MIPv6				SIP mobility	
	Buffering disabled + route optimization (RO) disabled	Buffering enabled + RO disabled	Buffering disabled + RO enabled	Buffering enabled + RO enabled	Buffering disabled	Buffering enabled
L2 handoff (ms)	4.00	4.30	4.00	4.00	4.00	5.00
Avg. packet loss per handoff	1.30	0.00	0.70	0.00	1.50	0.00
Avg. interpacket interval (ms)	16.00	16.00	16.00	16.00	16.00	16.00
Avg. interpacket arrival during handover (ms)	21.00	45.00	21.00	67.00	21.00	29.00
Avg. packet jitter (ms)	n/a	29.30	n/a	50.60	n/a	13.00
Buffering period (ms)	n/a	50.00	n/a	50.00	n/a	20.00
Avg. buffered packets	n/a	2.00	n/a	3.00	n/a	3.00

We have described the implementation and experimental details of the MPA framework for intrat-echnology handoff in Dutta et al. (2005e) and Dutta et al. (2008).

11.3.3   Intertechnology Handoff

In Chapter 2, we described several different types of heterogeneous handover mechanisms. However, in this section we analyze the results from intertechnology handover involving two specific access technologies, namely IEEE 802.11 and CDMA2000, using the optimization techniques described in Chapter 6.

Several types of handoff scenarios are possible involving handoff with multiple interfaces. We experimented with two different handoff scenarios: the *break-before-make* and the *make-before-break* scenarios. We describe these scenarios in the following.

11.3.3.1   Break-Before-Make Scenario

In the normal handoff scenario involving multiple interfaces, the new interface comes up only after the link to the old interface is taken down. This scenario is termed "break-before-make" and usually gives rise to undesirable packet loss and handoff delay. In the experimental test bed, without any optimization, the handoff delay and associated packet loss resulted from the PPP (Point-to-Point Protocol) configuration delay (16 s) and the binding update delay, for example SIP re-INVITE (1.5 s) for SIP-based mobility and the MIP registration delay (500 ms) for MIPv6. Lower-layer triggers such as "link down" can help expedite the handoff process in the second interface and will help to reduce the packet loss. In order to optimize the "break-before-make" scenario, we applied a fast "link down" detection technique along with preauthentication, proactive configuration techniques, buffering, and a copy-and-forwarding technique to reduce the packet loss and delay. Using this technique, the mobile was able to resume the communication in the new network within 50 ms and the packet loss was reduced to 0.

We now describe two types of make-before-break scenarios.

11.3.3.2 Make-Before-Break Scenario A

In the first type of make-before-break scenario, the second interface is prepared proactively while the mobile is still communicating using the old interface, and at some point the mobile decides to use the second interface as the active interface and completes the authentication and binding update procedures. This results in less packet loss, as it uses "make-before-break" techniques to set up the layer 2 configuration on the new network while the mobile is still connected to the old network. In a typical break-before-make handoff without any optimization, the mobile prepares the CDMA interface only after the current interface (802.11) goes down.

As part of our efforts to deal with this scenario, we have developed a make-before-break algorithm to support handoff between CDMA and 802.11 networks. Using this technique, the mobile sets up the layer 3 configuration in the CDMA network using the PPP interface while it continues to communicate using the current 802.11 interface. This technique reduces the layer 2 and layer 3 configuration-related delays and packet loss. However, it uses up more power resources, since both of the interfaces are active at the same time. We have experimented with this mechanism for both a network layer protocol, namely MIPv4, and application layer mobility, namely SIP-based mobility, over 802.11 and CDMA 1xRTT access networks. When we used the make-before-break technique, there was no packet loss in either MIP or SIP-based mobility. However, initial jitter was observed for the in-flight packets after a handover from the 802.11 to the CDMA network and out-of-order packets were received after a handover from the CDMA to the 802.11 network. We have published a complete experimental analysis of handoff involving multiple interfaces in Dutta et al. (2005c).

11.3.3.3 Make-Before-Break Scenario B

In the second type of make-before-break scenario, some of the required functions, such as network selection, context transfer of CDMA network parameters such as the PPP state, and security associations, are established ahead of time using the current interface. This scenario helps to conserve energy. By activating the second interface only after an appropriate network has been selected and the mobile has authenticated itself using the old interface, the mobile can utilize its battery efficiently for authentication.

We have confirmed from experiments (Dutta et al., 2005c) that a make-before-break technique aided by a combination of proactive optimization methodologies and cross-layer triggers can reduce the binding update and configuration delay during handover. Figure 11.3 shows a comparison of audio outputs at the mobile between optimized and nonoptimized handovers.

Figure 11.3(A) shows this comparison for homogeneous handover, whereas Figure 11.3(B) does so for handover between 802.11 and CDMA networks. Similarly to the results shown in Table 11.1, the optimized homogeneous handover does not have any media interruption, compared with 4 s media interruption for nonoptimized handover. The make-before-break optimization technique for heterogeneous handover results in no media interruption, compared with 16 s media interruption for SIP-based mobility and 18 s media interruption for nonoptimized handover.

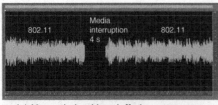

(a) Nonoptimized handoff - homogeneous

(a) MIP-based nonoptimized handoff - heterogeneous

(b) SIP-based nonoptimized handoff - heterogeneous

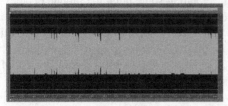

(b) MPA-assisted optimized handoff - homogeneous

(c) MPA- and 802.21-assisted optimized
handoff - heterogeneous

(A) Comparison of optimized and nonoptimized
homogeneous handoff

(B) Comparison of optimized and nonoptimized
heterogeneous handoff

Figure 11.3   Comparison of optimized vs. nonoptimized handoff

11.3.4   Cross-Layer-Trigger-Assisted Preauthentication

In this section, we describe experimental results from a handover system that used preauthentication techniques assisted by the media-independent handover functions defined by IEEE 802.21. IEEE 802.21-based cross-layer triggers help in handover preparation. Figure 11.4 shows the interaction between the MPA-related functions and the 802.21-based cross-layer triggers known as MIHF (Media Independent Handover Function). Figure 11.5 depicts the experimental integrated test bed, which used the MIHF and MPA components shown in Figure 11.4.

The experimental test bed included two types of access networks, namely EV-DO and Wi-Fi, which were connected via a core network infrastructure. The complete test bed consisted of the following entities.

A mobile node was equipped with IEEE 802.11 (Wi-Fi) and CDMA 1xEVDO interfaces. The IEEE 802.11 (Wi-Fi) and EV-DO interfaces had IP addresses IP0 and IP1, respectively. The MN ran an MPA client that supported IPSec, IKE, MOBIKE, and MIH-related services that provided cross-layer triggers.

The MPA server was equipped with several modules, including an authentication agent, a tunneling agent, a configuration agent, and a buffering module. The AA preauthenticated the MN. The tunneling agent managed an IPSec tunnel from the MN to serve as a proactive handover tunnel (PHT) and performed layer 3 handover using MOBIKE (S. Eronen, 2006). This test bed differed slightly from the MPA framework described in Section 11.2.2, as the tunneling agent was implemented on a node outside the target network (i.e., the EV-DO network), not on the AR in the target network, because we did not have control of the equipment in the operator's network. As a result, the MPA server acted as a proxy AR to the EV-DO network.

An MIH information server in the test bed was populated with information about neighboring network elements such as Wi-Fi access points and cellular network elements.

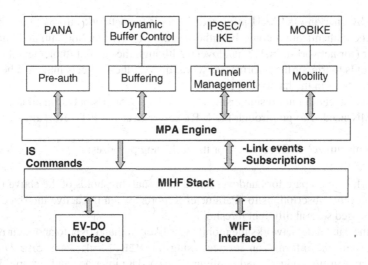

Figure 11.4 Interaction between MIHF and MPA components

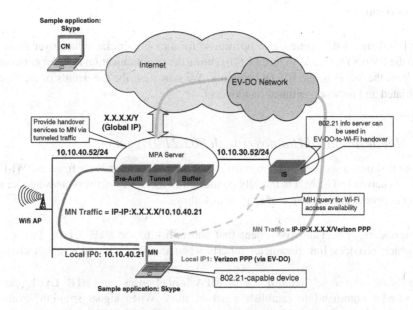

Figure 11.5 Experimental test bed for MIHF-assisted MPA

A correspondent node was connected to the Internet and communicated with the MN via a Skype (Baset and Schulzrinne, 2006) VoIP (Voice-over-IP) session. In this mobility scenario, the mobile node engaged in a VoIP session with the CN over the Wi-Fi network (path A) and then performed a handover to the EV-DO network (path B). While the MN was still connected to the Wi-Fi network, the MPA stack utilized the MIH services to trigger an authentication and configuration process with the EV-DO network in anticipation of the mobile node's move. The MPA engine learned of the target network by querying the MIH information server for network information. The MPA stack that triggered the authentication could be either on the MN (for mobile-initiated handover) or on the MPA server (for network-initiated handover). Although the current implementation used signal strength thresholds to trigger the information server query, other policies may also be implemented to trigger several steps in the handover process.

Since the tunnel agent did not reside inside the cellular operator's network, all communication to and from the MN needed to go through the MPA server over the PHT even after L2 handover, as shown in path B.

The MPA agents utilized MIH services for the following purposes:

- Identifying when to prepare for handover based on signal thresholds of the active interface. This was done by event subscriptions to "Parameter Reports" when the active interface's signal level in the MN crossed several different thresholds.
- Identifying the candidate networks the mobile was likely to hand over to and their related parameters by querying the information service, using the MIH_Link_Actions *Power Up* MIH command to power up, connect, and configure the EV-DO interface and set up a PHT once the preauthentication procedure was over.
- Using the MIH command MIH_Link_Actions *Power Down* to turn off the old link once handover was complete.

Figure 11.6(a) shows the sequence of operations for a mobile-initiated handover from the Wi-Fi network to the EV-DO network. Figure 11.6(b) shows the communication flow for network-initiated handover from the Wi-Fi to the EV-DO network. We now describe the details of the flow for both mobile-initiated and network-initiated handovers.

11.3.5 Mobile-Initiated Handover with 802.21 Triggers

Figure 11.6(a) shows a sequence diagram for a mobile-initiated handover from the Wi-Fi network to the EV-DO network. The MN is initially connected to the Wi-Fi network and then hands over to the EV-DO network. The sequence of steps is as follows.

- *(1) Subscribe request.* The MPA client first subscribes to the MIH_Link_Param_Report event, which provides link parameter reports when the Wi-Fi signal strength crosses certain values.
- *(2) Configure threshold request.* The MPA client uses an MIH_Link_Configure _Threshold command to establish a set of three Wi-Fi signal strength levels that will trigger notifications. Once a threshold level is crossed, the MIHF propagates the appropriate notification to the MPA client.
- *(3) Link parameter report (threshold 1).* When the MPA client receives the first event notification reporting that the Wi-Fi signal strength has crossed the first threshold, the MPA client prepares for a potential handover, and queries the MIH information server (steps 4 and 5) for available neighboring networks via the MN's current serving network. The information server then sends a response with the information that the cellular network is available

(Steps 6 and 7 are not described, for brevity.)

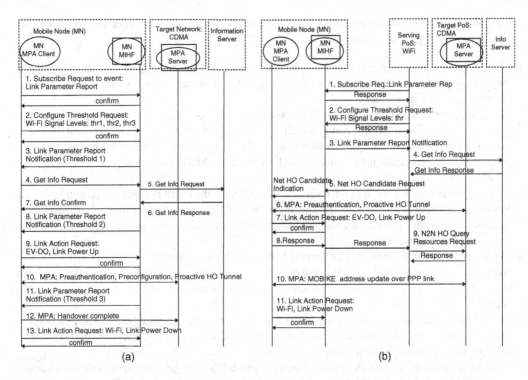

Figure 11.6   (a) Mobile-initiated and (b) network-initiated handover

- *(8) Link parameter report (threshold 2).* When the signal strength weakens further and the mobile crosses the second threshold, the MPA client receives an event notification and starts setting up the cellular connection.
- *(9) Link up request.* The MPA client brings the EV-DO interface up and establishes an EV-DO connection using an `MIH_Link_Actions` command. It is important to note that this step can be performed after Step 10 if the IP address to be assigned to the EV-DO interface can be obtained in Step 10; however, this optimization requires the EV-DO network to support MPA.
- *(10) MPA proactive handover.* The MPA client starts preauthentication and preconfiguration through the serving Wi-Fi interface.
- *(11) Link parameter report.* When the MPA client receives the third link parameter report event notification, indicating crossing of the third threshold value, the MPA client completes the handover operation via a MOBIKE address update.
- *(13) Link power down request.* The MPA client then uses an `MIH_Link_Actions` command to bring down the Wi-Fi interface.

11.3.6   Network-Initiated Handover with 802.21 Triggers

Figure 11.6(b) shows a sequence diagram for a network-initiated handover from the Wi-Fi network to the EV-DO network. In addition to the entities depicted in Figure 11.6(a), a new entity called the serving PoS (point of service) in the Wi-Fi network is used to realize a network-initiated handover. The sequence of steps is as follows.

- *(1) Subscribe request.* The serving PoS subscribes with the MN to get an `MIH_Link_Param_Report` event notification, which will provide link parameter reports when the Wi-Fi signal strength crosses a given value.
- *(2) Configure threshold request.* The serving PoS uses an `MIH_Link_Configure_Threshold` command to configure the Wi-Fi signal strength level that will trigger a layer 2 event notification. Once this threshold level is crossed, the MIHF in the mobile node will propagate the appropriate notification to the PoS using the MIH protocol to provide remote event services.
- *(3) Link parameter report.* When the serving PoS receives the event notification reporting that the Wi-Fi signal strength has crossed the specified threshold, the serving PoS queries the MIH information server (step 4) for available neighboring networks. The information server then reports that the cellular network is available.
- *(5) Net HO candidate request.* The serving PoS sends an `MIH_Net_HO_Candidate_Query` request message to the mobile indicating the candidate networks available for handover. The candidate networks are selected based on the information obtained from the information server in Step 4.
- *(6) MPA preauthentication.* Once the target PoS has been selected and the authentication server is known, the mobile node contacts the MPA server and starts preauthentication, and sets up the proactive tunnel through the serving Wi-Fi PoS.
- *(7) Link up request.* The MPA client verifies the availability of the cellular network as indicated in the `MIH_Net_HO_Candidate` request message by bringing the EV-DO interface up and establishes an EV-DO connection using an `MIH_Link_Actions` command.
- *(8) Net HO candidate response.* Once the EV-DO connection is established, the MPA client responds with an `MIH_Net_HO_Candidate_Query` response message, indicating the EV-DO network as the candidate network.
- *(9) Network-to-network HO query resource request/response.* The serving PoS (Wi-Fi) sends the target PoS (CDMA) an `N2N(Network to Network)_HO_Query_Resource` request message, to verify that the target PoS has resources before committing to the handover. Once the serving PoS gets a positive response, it can commit to the handover. Although MIH provides a command to indicate handover commitment (i.e., `MIH_Net_HO_Commit`), we used the MPA proactive handover (Step 9) as the indication of handover commitment.
- *(10) MPA proactive handover.* The MPA client completes the handover operation by a MOBIKE address update.
- *(11) Link power down request.* The MPA client then uses an `MIH_Link_Actions` command to bring down the Wi-Fi interface.

It is important that MIH handover preparation and MPA preauthentication procedures are completed before the mobile makes a handover to the target network. In the next section, we describe the timing for different handover operations.

11.3.7   Handover Preparation Time

The handover preparation time does not directly affect the handover performance and user experience. However, the amount of time the mobile needs to prepare for handover depends upon the speed of movement of the mobile (e.g., pedestrian or vehicular), the cell size (e.g., picocell or macrocell), and the type of handover (e.g., single-interface or multiple-interface). Generally, it is important to reduce the handover preparation time to make the system more resilient to sudden changes in the network characteristics.

This handover preparation time in the experimental test bed included the following components:

(i) Propagation of the link events from the link layer to the MIH user (i.e., a local MIH user in the case of an MN-initiated handover and a remote MIH user in network-initiated handover.)
(ii) Querying the information server database.
(iii) MIHF internal operations.
(iv) MPA layer 3 handover.

During the experiment, we measured the time delays for execution of the operations (ii), (iii), and (iv). While the delays we measured were for different MIH-related operations in the network-initiated handover scenario described in Figure 11.6, some of these measurements can be applied to the mobile-initiated handoff scenario as well.

Table 11.2, Table 11.3, Table 11.4, and Table 11.5 show the values measured for each of the above operations. We describe the delays associated with each of these operations in the following.

11.3.7.1 Information Service Transaction Delay

We measured various operations in the information server that constituted the transactions associated with a request. This sequence started with receiving a *Get Information* request message containing an information server query and finished with sending the corresponding response. Table 11.2 shows five values measured for each operation and their averages. The average information server transaction execution time was 26.6 ms, with a lower bound of 13 ms and an upper bound of 53 ms.

11.3.7.2 MIH Message Composition and Parsing Delay

Depending on the MIH message type, the time for message composition and parsing might vary. This depends on the number of TLVs included in each message and the TLV type, which dictates the complexity of its composition and parsing. Table 11.3 and Table 11.4 show the minimum, maximum, and average values of the times taken for different suboperations associated with message composition

Table 11.2 Processing time in the information server

	Measurement #					
	1	2	3	4	5	Average
Get info request parsing (ms)	3	3	4	4	5	3.8
Pass indication from MIHF to MIH user (ms)	2	10	2	3	2	3.8
Query processing (ms)	5	29	5	25	6	14
Get info response composition (ms)	3	2	4	3	2	2.8
Get info response sending (ms)	2	1	1	5	2	2.2
Total time (ms) for processing in the info server						26.6

Table 11.3 MIH message composition time

Measurement point	Message type	Execution time (ms) (average, min., max.)
MN	Link parameter report indication	1.6, 0, 2
Serving PoS	Register response	4.4, 3, 8
Serving PoS	Subscribe request	4.8, 3, 11
Serving PoS	Get info request	6.2, 5, 2
Serving PoS	Net HO candidate request	25.4, 10, 51
Info server	Get info response	2.8, 2, 3

Table 11.4 MIH message-parsing time

Measurement point	Message type	Execution time (ms) (average, min., max.)
MN	NET HO candidate query request	12.6, 6, 19
Serving PoS	Subscribe response	12, 7, 17
Serving PoS	Configure threshold response	40.2, 10, 54
Serving PoS	Link parameter report indication	21.2, 14, 50
Serving PoS	Get info response	11.4, 8, 17
Info server	Get info request	3.8, 3, 5

Table 11.5 Delays for MIHF-related components

			Execution time	
Measurement point	Operation description	Min. (ms)	Average (ms)	Max. (ms)
MN	Compose/transmit link parameter report ind.	10	10.4	11
Serving PoS	Recv/parse/process link parameter report ind.	20	28.8	53
Serving PoS	Compose and transmit get info request	11	14.4	22
Info server	Receive/parse/process get info request	10	21.6	44
Info server	Compose/send get info response	3	5	9
Serving PoS	Receive/parse/process get info response	10	20	28
Serving PoS	Compose/send net HO candidate request	11	31.2	56
MN	Receive/process net HO candidate request	8	15.2	22
	Total		146.6	

and parsing delay, respectively. These values were taken into account in calculating the handover preparation time.

11.3.7.3 MIH Performance for MPA Triggering

We measured the time it took to perform all the MIH-related operations in the network-initiated handover scenario that occurred, starting with the initial handover trigger (i.e., crossing the signal strength threshold in the MN and creation of the link parameter report indication), until the MPA handover operation was triggered. Table 11.4 shows the averages of five measurements of the execution time for each of the specified operations, with their corresponding lower and upper bounds. Table 11.5 shows the timing associated with each of the MIH-related operations.

In order to calculate the total time for the MIH MPA triggering operations, the following network propagation delays need to be added:

1. The round trip propagation delay between the MN and the serving PoS (MN–PoS RTT).
2. The round trip propagation delay between the serving PoS and the information server (PoS–IS RTT).

In the current experimental test bed, we estimated these delays using ping messages, which gave round trip values for ICMP messages of 1.5 ms for the MN–PoS RTT and 0.3 ms for the PoS–IS RTT, bringing the time for MIH to trigger MPA to 148.4 ms in the test bed environment.

These round trip propagation delays can be adjusted for a real network environment to estimate the performance of a realistic network. Since an MN and its serving PoS are relatively close to each other, we estimate their round trip propagation delay, MN–PoS RTT, as 5 ms. We estimate the round trip propagation delay betwen the serving PoS and the information server, PoS–IS RTT, as 30 ms as the PoS and the information server, usually located in the network core, would be separated by few network segment hops. Thus, in a realistic network, the time it would take for MIH to trigger the MPA preauthentication and handover would be approximately 146.6 ms + 5 ms (MN–PoS RTT) + 30 ms (PoS–IS RRT) = 181.6 ms. This time does not include the propagation time of the link event from the link layer to the MIHF, which we have not measured.

11.3.7.4 Delays Due to MPA Operation

The MPA-related delays can be attributed to several factors, such as delays due to preauthentication, setting up proactive handover tunnels, and sending the binding update for data redirection. In the current test bed, we measured the delays for these components. As shown in Figure 11.6, preauthentication and proactive tunnel setup took place before the PPP link was set up. Alternatively, these two operations could take place in parallel with PPP configuration operations, which may take up to 5 s. Measurement showed that the complete preauthentication operation took about 2175 ms. This time delay consists of several factors, such as four round trip signals associated with EAP-GPSK (Extensible Authentication Protocol–Generalized Pre Shared Key), generation of keys at the authentication server, and message-processing delays at the end hosts. The setup time for the proactive handover tunnel was measured to be 4730 ms, which includes the time for the IKE handshake to set up an IPSec tunnel in ESP mode, and the initial MOBIKE signaling exchange. These two operations took place over the Wi-Fi interface in the previous network. The final step in the MPA operation was the binding update and was performed using the MOBIKE address update. It took around 400 ms to complete the round trip MOBIKE signaling over a PPP link.

Our estimate of the MIH handover preparation time before triggering an MPA operation in a realistic network is less than 200 ms, which is less than 10% of the time that MPA preauthentication procedures would take. This seems to be a satisfactory time to allow proper timing of the MPA operations and handover procedure.

The information server transaction delay and MIHF performance can be improved by improving the query execution time, the message composition delay, and the message-parsing time.

11.4   Handoff Optimization in IP Multimedia Subsystem

In this section, we describe how some of the proactive optimization techniques that we have designed can improve the handoff performance in IMS (IP Multimedia Subsystem) by reducing the packet loss and handover delay. We have built a complete experimental prototype IMS system, which we briefly introduced in Chapter 6 to illustrate the route optimization technique. Here, we demonstrate how the handoff delay can be reduced when security association is performed proactively by transferring the security context between points of attachment in two neighboring networks. This specific system does not optimize layer-2- and layer-3-related operations but focuses only on optimization of the security association. Soon after this work was published (Dutta et al., 2007d), 3GPP created a new working group called the Multimedia Session Continuity (MMSC) group and produced a technical specification (3GPP., 2008) to define several different scenarios in which fast handoff would be supported for multimedia traffic in an IMS environment. Figure 6.43 shows the experimental test bed where we experimented with these optimization techniques.

11.4.1   Nonoptimized Handoff Mode

In the nonoptimized mode of operation, a new call context is created at the proxy server every time the mobile moves to the new network. A call context consists of the call data records and the media parameters associated with a specific call. During the handoff, the MN completes all the handoff functions in layers 2 and 3 as described earlier in Chapter 3. Specifically, after the MN establishes PPP access to the new network, it performs the MIP binding and then obtains the server configuration information via DHCP. Then the SIP-related handoff functions are performed, starting with SIP reregistration and reestablishment of the security association using the AKA (authentication and key agreement) procedure. If the MN moves during an active session, session maintenance is carried out, with the transmission of an encrypted SIP re-INVITE message that carries the SDP description of the ongoing session. Upon receipt of this message, the P-CSCF (proxy call session control function) creates a new call context for the same mobile and interacts with the PDSN (packet data serving node) on the visited network to allow the traffic. It is important to point out that mobility binding is taken care of by Mobile IP here, and SIP re-INVITE is used for creating the call context in the P-CSCF. The call context consists of the call data record (CDR) and other bandwidth- and port-related information for the media. Successful creation of a call context at the P-CSCF results in resumption of the media in the new access network. The message flow for the nonoptimized operational mode is shown in Figure 11.7.

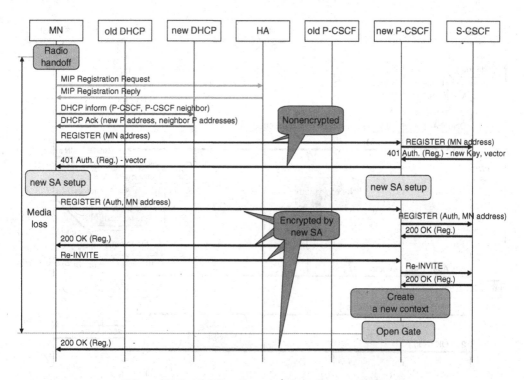

Figure 11.7   Call flow for nonoptimized handoff in IMS

11.4.2   Optimization with Reactive Context Transfer

In the reactive mode of operation, all of the layer-2- and layer-3-related operations take place as in the nonoptimized mode. The detailed message flow is provided in Figure 11.8. The difference between the two operational modes is evident from a comparison with Figure 11.7 (nonoptimized mode).

In particular, the session maintenance information message, such as a re-INVITE that carries the SDP description of the active session, does not play any role in context creation and thus does not affect the media handoff delay. The context created in the new visited network's P-CSCF is transferred from the old visited network's P-CSCF. The objective of this approach is to reduce the handoff delay by eliminating the dependence on the session maintenance messages (re-INVITE and 200 OK) after the handoff. Session maintenance is carried out to create context at the P-CSCF i.e., the outbound SIP proxy.

After the radio handoff is over and the PPP connection is complete in the new network, the MN performs the MIP binding and obtains the required configuration information using DHCP. The MN then generates a SIP REGISTER message via the new P-CSCF. When this message reaches the S-CSCF (serving call session control function), the S-CSCF informs the old P-CSCF that it should transfer the context of the active session to the new P-CSCF. At this point the old P-CSCF transfers the context to the new P-CSCF, and the context is created at the new P-CSCF in the new network. After the security association is set up via AKA between the MN and the new P-CSCF, media traffic is allowed at the PDSN, and the session resumes.

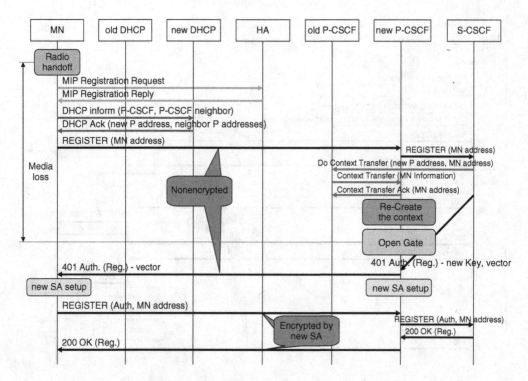

Figure 11.8   Optimized handoff with reactive context transfer

11.4.3   Optimization with Proactive Security Context Transfer

The proactive mode of handoff minimizes the delay due to security association and context creation more than the reactive mode, and thus reduces the media interruption further compared with the reactive mode. In this mode, the context creation in the new IMS network and security association with the new P-CSCF are completed while the MN is still in the old network. This technique works in conjunction with the proactive discovery of neighboring P-CSCFs while the mobile is in the previous network. As discussed in Chapter 6, carriers could use information service discovery methods such as IEEE 802.21 or IEEE 802.11u to obtain detailed information about the neighboring networks and servers.

Figure 11.9 presents the detailed message flows for proactive handoff. Prior to the MN's layer 2 handoff, some of the handoff functions (e.g., server discovery and establishment of the security association) are done proactively while the mobile is still in the old network. Specifically, the MN, utilizing a DHCP INFORM message, acquires the addresses of P-CSCFs from the neighboring IMS networks.

In this case, the DHCP server is populated with information about the SIP servers in the neighboring networks. After the MN has identified the new neighboring network it is likely to move to, it informs its current P-CSCF about the address of its new P-CSCF. The current P-CSCF transfers the context of the active session (e.g., SDP parameters and CDR parameters) to the new P-CSCF. Similarly, a new security association is established between the new P-CSCF and the mobile after the mobile sends a Move_Notify message to the S-CSCF when the movement is imminent. This

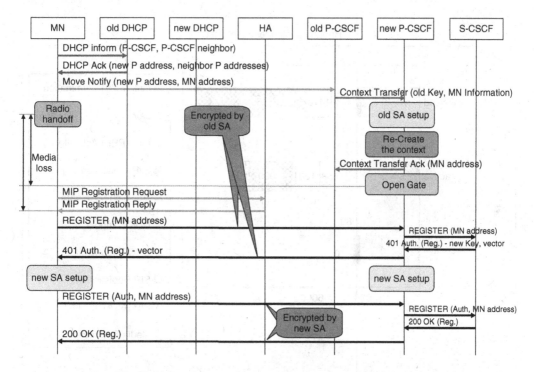

Figure 11.9 Optimized handoff with proactive context transfer

mechanism performs a proactive AKA operation by transferring the security context from the current P-CSCF to the new P-CSCF ahead of time by creating a transient AKA during the handoff. The mobile performs the normal AKA procedure after it moves to the new network. Thus, the new PDSN allows the specific mobile's media even before the mobile has moved. After the mobile reestablishes its connection in the new network and completes the MIP operations, the media continue, eliminating the delay due to the AKA procedure and context creation. The mobile node's SIP-related signaling, such as re-REGISTER or re-INVITE, does not affect the media handoff delay here. Reregistration in the new network helps to renew the transient AKA that was established in the previous network.

Based on the message flows in Figure 11.8 and Figure 11.9, it is obvious that these proactive handoff techniques reduce the handoff delay that is typically caused by reestablishment of the security association. We compare the results for each handoff operation in the following section.

11.4.4 Performance Results

The focus of this performance analysis is on highlighting the relative effectiveness of proactive security association compared with the other two handoff techniques, namely the nonoptimized and reactive modes. In Figure 11.10, we plot the delays associated with the different handoff-related functions that contribute to the handoff delay in these three handoff scenarios.

On average, the mobile was subject to 3666 ms delay for the proactive mechanism, 9685 ms delay for the reactive security association mechanism, and 12,526 ms delay for the nonoptimized handoff.

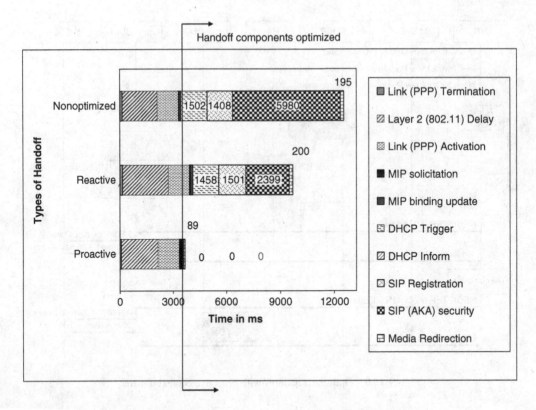

Figure 11.10   Comparison of optimized handoff components

The number of packets lost is proportional to the handoff delay and depends on the packet generation rate.

The overall handoff delay consists of delays due to different operations such as layer 2 configuration, layer 3 configuration, binding update, registration, security association, and media redirection. As is evident, proactive handoff does not contribute any delay due to DHCP-based server discovery, context transfer, or SIP-based security association, in contrast to the reactive and nonoptimized cases. On the other hand, the nonoptimized case suffers from the maximum delay due to additional signaling messages during the SIP-based security association and context creation phase. The layer 2 handoff delay, PPPoE (PPP over Ethernet) access delay, and MIP binding delay remain the same in all three handoff scenarios. In the case of reactive handoff, besides layer 2 delay, the major component of the delays comes from the SIP registration, security association, and context transfer process.

Since we used Mobile IPv4 for mobility, these results are inclusive of inherent triangular-routing delays and can be reduced further if the routing mitigation techniques described in Chapter 6 are applied. Mobility protocols such as MIPv6 and SIP-based mobility may result in a smaller binding update delay because they avoid triangular routing. The SIP-related signaling required for context transfer and security association contributes to an additional handoff time in the nonoptimized case compared with the reactive case, since it needs to create the context with additional re-INVITE signaling.

Table 11.6   Effect of emulated distance on handoff components

One-way delay	Proactive handoff			Reactive handoff			Nonoptimized handoff		
	AKA, context transfer delay (ms)	MIP update delay (ms)	L2 PPP delay (ms)	AKA, context transfer delay (ms)	MIP update delay (ms)	L2 PPP delay (ms)	AKA, context transfer delay (ms)	MIP update delay (ms)	L2 PPP delay (ms)
0	0	51	2736	1010	62	1523	3999	41	2239
50	0	152	2693	1375	161	1744	4584	145	2217
100	0	252	2650	1741	261	1964	5170	248	2194
150	0	352	2607	2107	360	2184	5756	352	2172
200	0	453	2563	2472	459	2405	6342	455	2150
250	0	553	2520	2838	558	2625	6927	559	2128
300	0	654	2477	3203	658	2845	7513	663	2106
350	0	755	2434	3569	757	3066	8099	766	2084
400	0	855	2391	3935	856	3286	8685	870	2061
450	0	956	2347	4300	955	3506	9270	973	2039
500	0	1057	2304	4666	1055	3726	9856	1077	2017

In order to gain insight into the effect of these optimization techniques in a real deployment scenario, we used a NIST delay simulator (Carson and Santay, 2003) and varied the emulated distance between the home network and the visited network by varying the delays from 0 to 500 ms, with an increment of 50 ms. We present the results in Table 11.6.

These results show only the components of the handoff delay that are affected when an additional delay is introduced between the home network and the visited network. From the analysis, it is apparent that delays related to layer 2 and layer 3 configuration are not affected by an additional emulated delay because these operations do not involve the home network. The delays due to the Mobile IP binding update increase in all three cases as the emulated distance between the home network and visited network is increased, but there is an appreciable increase in the delay for SIP-based security association in the cases of nonoptimized and reactive handoff. In the proactive-handoff case, the additional network transport delay does not have any effect on the handoff delay component related to SIP, AKA, and context transfer. The additional handoff delay in the proactive case is contributed by the increased MIP update delay only.

11.5   Systems Validation Using Petri-Net-Based Models

In this section, we introduce Petri-net-based models for some of the optimization techniques that we described in Chapter 6. We describe the application of a MATLAB-based Petri net tool (Matcovschi et al., 2003) to model some of the handoff functions described in Chapter 3, validate the optimization techniques by way of Petri net-based behavioral analysis methods, and evaluate the systems performance by using the cycle time and the Floyd algorithm. We then evaluate three different scheduling techniques for a few of the handoff operations and apply the cycle time and Floyd algorithm approaches to validate the systems performance. We also illustrate how certain sequences of transitions may give rise to deadlocks by doing a reachability and matrix analysis.

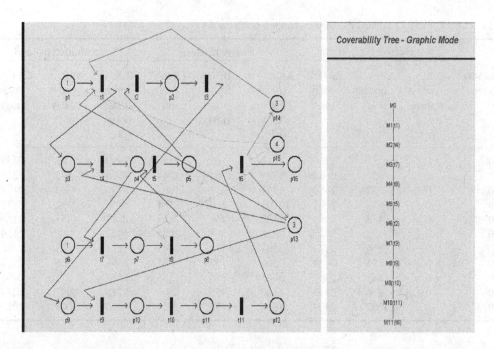

Figure 11.11   MATLAB-based model of four handoff functions

11.5.1   MATLAB®-Based Modeling of Handoff Functions

In this section, we describe the results from the MATLAB-based modeling of many of the hand-off functions. The MATLAB-based Petri net tool was used to study behavioral properties such as reachability, markings, liveness, and the systems performance of the handoff models.

Figure 11.11 shows a MATLAB-based model that illustrates a sequence of a few handoff operations, namely discovery, attachment, configuration, and authentication. This figure is the MATLAB® equivalent of the model shown in Figure 4.14. Places p13, p14, and p15 in Figure 11.11 represent resource places representing energy, bandwidth, and CPU cycles, respectively. This figure also shows the markings associated with this set of models as generated from the MATLAB® model. These markings represent a stepwise execution of the handoff events.

Figure 11.12, Figure 11.13, and Figure 11.14 illustrate MATLAB-based models for three different handoff sequences as shown in Figure 4.16, Figure 4.19, and Figure 4.20, respectively. These models illustrate coverability trees and markings as obtained from the MATLAB-based Petri net tool. Figure 11.12 shows the MATLAB-based model, reachability tree, and corresponding marking of the sequential handoff operations. Figure 11.13 shows a handoff model and parameters illustrating concurrent security and scanning operations. Figure 11.14 shows the MATLAB-based model for a handoff event where the security, L2 discovery, and L3 discovery operations are performed in parallel. The matrices A_i, A_o, and A represent the input, output, and incidence matrices, respectively, for the Petri net handoff model shown in Figure 11.12. These matrices were obtained from the MATLAB-based Petri net model. Figure 11.15 shows a MATLAB-based screenshot of these matrices. The matrices are as follows:

$$A_{\mathrm{i}} = \begin{bmatrix}
1 & 0 & 0 & 0 & 0 & 0 & 0 & 0 & 0 & 0 & 0 & 0 & 0 & 0 & 0 & 0 & 0 & 0 & 0 & 0 & 0 & 0 & 0 & 0 & 0 & 0 \\
0 & 1 & 0 & 0 & 0 & 0 & 0 & 0 & 0 & 0 & 0 & 0 & 0 & 0 & 0 & 0 & 0 & 0 & 0 & 0 & 0 & 0 & 0 & 0 & 0 & 0 \\
0 & 0 & 1 & 0 & 0 & 0 & 0 & 0 & 0 & 0 & 0 & 0 & 0 & 0 & 0 & 0 & 0 & 0 & 0 & 0 & 0 & 0 & 0 & 0 & 0 & 0 \\
0 & 0 & 0 & 1 & 0 & 0 & 0 & 0 & 0 & 0 & 0 & 0 & 0 & 0 & 0 & 0 & 0 & 0 & 0 & 0 & 0 & 0 & 0 & 0 & 0 & 0 \\
0 & 0 & 0 & 0 & 1 & 0 & 0 & 0 & 0 & 0 & 0 & 0 & 0 & 0 & 0 & 0 & 0 & 0 & 0 & 0 & 0 & 0 & 0 & 0 & 0 & 0 \\
0 & 0 & 0 & 0 & 0 & 0 & 0 & 0 & 0 & 1 & 1 & 0 & 0 & 0 & 0 & 0 & 0 & 0 & 0 & 0 & 0 & 0 & 0 & 0 & 0 & 0 \\
0 & 0 & 0 & 0 & 0 & 0 & 0 & 0 & 1 & 0 & 0 & 0 & 0 & 0 & 0 & 0 & 0 & 0 & 0 & 0 & 0 & 0 & 0 & 0 & 0 & 0 \\
0 & 0 & 0 & 0 & 0 & 0 & 1 & 0 & 0 & 0 & 0 & 0 & 0 & 0 & 0 & 0 & 0 & 0 & 0 & 0 & 0 & 0 & 0 & 0 & 0 & 0 \\
0 & 0 & 0 & 0 & 0 & 0 & 0 & 0 & 0 & 0 & 0 & 1 & 0 & 0 & 0 & 0 & 0 & 0 & 0 & 0 & 0 & 0 & 0 & 0 & 0 & 0 \\
0 & 0 & 0 & 0 & 0 & 0 & 0 & 0 & 0 & 0 & 0 & 0 & 1 & 0 & 0 & 0 & 0 & 0 & 0 & 0 & 0 & 0 & 0 & 0 & 0 & 0 \\
0 & 0 & 0 & 0 & 0 & 0 & 0 & 1 & 0 & 0 & 0 & 0 & 0 & 0 & 1 & 1 & 0 & 0 & 0 & 0 & 0 & 1 & 1 & 0 & 1 & 0 \\
0 & 0 & 0 & 0 & 0 & 0 & 0 & 0 & 0 & 0 & 0 & 0 & 0 & 0 & 0 & 0 & 1 & 0 & 0 & 0 & 0 & 0 & 0 & 0 & 0 & 0 \\
0 & 0 & 0 & 0 & 0 & 0 & 0 & 0 & 0 & 0 & 0 & 0 & 0 & 0 & 0 & 0 & 0 & 1 & 0 & 0 & 0 & 0 & 0 & 0 & 0 & 0 \\
0 & 0 & 0 & 0 & 0 & 0 & 0 & 0 & 0 & 0 & 0 & 0 & 0 & 0 & 0 & 0 & 0 & 0 & 1 & 0 & 0 & 0 & 0 & 0 & 0 & 0 \\
0 & 0 & 0 & 0 & 0 & 0 & 0 & 0 & 0 & 0 & 0 & 0 & 0 & 0 & 0 & 0 & 0 & 0 & 0 & 1 & 0 & 0 & 0 & 0 & 0 & 0 \\
0 & 0 & 0 & 0 & 0 & 0 & 0 & 0 & 0 & 0 & 0 & 0 & 0 & 0 & 0 & 0 & 0 & 0 & 0 & 1 & 0 & 0 & 0 & 0 & 0 & 0 \\
0 & 0 & 0 & 0 & 0 & 0 & 0 & 0 & 0 & 0 & 0 & 0 & 0 & 0 & 0 & 0 & 0 & 0 & 1 & 0 & 0 & 0 & 0 & 0 & 0 & 0 \\
0 & 0 & 0 & 0 & 0 & 0 & 0 & 0 & 0 & 0 & 0 & 0 & 0 & 0 & 0 & 0 & 0 & 1 & 0 & 0 & 0 & 0 & 0 & 0 & 0 & 0 \\
0 & 0 & 0 & 0 & 0 & 0 & 0 & 0 & 0 & 0 & 0 & 0 & 0 & 0 & 0 & 0 & 0 & 1 & 0 & 0 & 0 & 0 & 0 & 1 & 0 & 0 \\
\end{bmatrix},$$

$$A_{\mathrm{o}} = \begin{bmatrix}
0 & 1 & 1 & 1 & 0 & 0 & 0 & 0 & 0 & 0 & 0 & 0 & 0 & 0 & 0 & 0 & 0 & 0 & 0 & 0 & 0 & 0 & 0 & 0 & 0 & 0 \\
0 & 0 & 0 & 0 & 1 & 0 & 0 & 0 & 0 & 0 & 0 & 0 & 0 & 0 & 0 & 0 & 0 & 0 & 0 & 0 & 0 & 0 & 0 & 0 & 0 & 0 \\
0 & 0 & 0 & 0 & 0 & 0 & 0 & 1 & 0 & 0 & 0 & 0 & 0 & 0 & 0 & 0 & 0 & 0 & 0 & 0 & 0 & 0 & 0 & 0 & 0 & 0 \\
0 & 0 & 0 & 0 & 0 & 0 & 0 & 0 & 0 & 1 & 0 & 0 & 0 & 0 & 0 & 0 & 0 & 0 & 0 & 0 & 0 & 0 & 0 & 0 & 0 & 0 \\
0 & 0 & 0 & 0 & 0 & 1 & 0 & 0 & 0 & 0 & 0 & 0 & 0 & 0 & 0 & 0 & 0 & 0 & 0 & 0 & 0 & 0 & 0 & 0 & 0 & 0 \\
0 & 0 & 0 & 0 & 0 & 0 & 1 & 0 & 0 & 0 & 0 & 0 & 0 & 0 & 0 & 0 & 0 & 0 & 0 & 0 & 0 & 0 & 0 & 0 & 0 & 0 \\
0 & 0 & 0 & 0 & 0 & 0 & 0 & 0 & 0 & 0 & 0 & 0 & 0 & 0 & 0 & 1 & 0 & 0 & 0 & 0 & 0 & 0 & 0 & 0 & 0 & 0 \\
0 & 0 & 0 & 0 & 0 & 0 & 0 & 0 & 1 & 0 & 0 & 0 & 0 & 0 & 0 & 0 & 0 & 0 & 0 & 0 & 0 & 0 & 0 & 0 & 0 & 0 \\
0 & 0 & 0 & 0 & 0 & 0 & 0 & 0 & 0 & 0 & 1 & 1 & 0 & 0 & 0 & 0 & 1 & 0 & 0 & 0 & 0 & 0 & 0 & 0 & 0 & 0 \\
0 & 0 & 0 & 0 & 0 & 0 & 0 & 0 & 0 & 0 & 0 & 1 & 0 & 0 & 0 & 0 & 0 & 0 & 0 & 0 & 0 & 0 & 0 & 0 & 0 & 0 \\
0 & 0 & 0 & 0 & 0 & 0 & 0 & 0 & 0 & 0 & 0 & 0 & 1 & 0 & 0 & 0 & 0 & 0 & 0 & 0 & 0 & 0 & 0 & 0 & 0 & 0 \\
0 & 0 & 0 & 0 & 0 & 0 & 0 & 0 & 0 & 0 & 0 & 0 & 0 & 1 & 0 & 0 & 0 & 0 & 0 & 0 & 0 & 0 & 0 & 0 & 0 & 0 \\
1 & 0 & 0 & 0 & 0 & 0 & 0 & 0 & 0 & 0 & 0 & 0 & 0 & 0 & 0 & 1 & 0 & 0 & 0 & 0 & 0 & 0 & 0 & 0 & 0 & 0 \\
0 & 0 & 0 & 0 & 0 & 0 & 0 & 0 & 0 & 0 & 0 & 0 & 0 & 0 & 0 & 0 & 1 & 0 & 0 & 0 & 0 & 0 & 0 & 0 & 0 & 0 \\
0 & 0 & 0 & 0 & 0 & 0 & 0 & 0 & 0 & 0 & 0 & 0 & 0 & 0 & 0 & 0 & 0 & 0 & 0 & 0 & 0 & 0 & 0 & 0 & 0 & 0 \\
0 & 0 & 0 & 0 & 0 & 0 & 0 & 0 & 0 & 0 & 0 & 0 & 0 & 0 & 0 & 0 & 0 & 0 & 1 & 0 & 0 & 0 & 0 & 0 & 0 & 0 \\
0 & 0 & 0 & 0 & 0 & 0 & 0 & 0 & 0 & 0 & 0 & 0 & 0 & 0 & 0 & 0 & 0 & 0 & 1 & 0 & 0 & 0 & 0 & 0 & 0 & 0 \\
0 & 0 & 0 & 0 & 0 & 0 & 0 & 0 & 0 & 0 & 0 & 0 & 0 & 0 & 0 & 0 & 0 & 0 & 1 & 0 & 0 & 0 & 0 & 0 & 1 & 0 \\
0 & 0 & 0 & 0 & 0 & 0 & 0 & 0 & 0 & 0 & 0 & 0 & 0 & 0 & 0 & 0 & 0 & 0 & 1 & 0 & 0 & 0 & 0 & 0 & 1 & 0 \\
\end{bmatrix},$$

$$A = \begin{bmatrix}
-1 & 1 & 1 & 1 & 0 & 0 & 0 & 0 & 0 & 0 & 0 & 0 & 0 & 0 & 0 & 0 & 0 & 0 & 0 & 0 & 0 & 0 & 0 & 0 & 0 & 0 \\
0 & -1 & 0 & 0 & 1 & 0 & 0 & 0 & 0 & 0 & 0 & 0 & 0 & 0 & 0 & 0 & 0 & 0 & 0 & 0 & 0 & 0 & 0 & 0 & 0 & 0 \\
0 & 0 & -1 & 0 & 0 & 0 & 0 & 1 & 0 & 0 & 0 & 0 & 0 & 0 & 0 & 0 & 0 & 0 & 0 & 0 & 0 & 0 & 0 & 0 & 0 & 0 \\
0 & 0 & 0 & -1 & 0 & 0 & 0 & 0 & 1 & 0 & 0 & 0 & 0 & 0 & 0 & 0 & 0 & 0 & 0 & 0 & 0 & 0 & 0 & 0 & 0 & 0 \\
0 & 0 & 0 & 0 & -1 & 1 & 0 & 0 & 0 & 0 & 0 & 0 & 0 & 0 & 0 & 0 & 0 & 0 & 0 & 0 & 0 & 0 & 0 & 0 & 0 & 0 \\
0 & 0 & 0 & 0 & 0 & -1 & 1 & 0 & 0 & 0 & 0 & 0 & 0 & 0 & 0 & 0 & 0 & 0 & 0 & 0 & 0 & 0 & 0 & 0 & 0 & 0 \\
0 & 0 & 0 & 0 & 0 & 0 & 0 & 0 & -1 & 1 & 0 & 0 & 0 & 0 & 0 & 1 & 0 & 0 & 0 & 0 & 0 & 0 & 0 & 0 & 0 & 0 \\
0 & 0 & 0 & 0 & 0 & 0 & 0 & -1 & 1 & 0 & 0 & 0 & 0 & 0 & 0 & 0 & 0 & 0 & 0 & 0 & 0 & 0 & 0 & 0 & 0 & 0 \\
0 & 0 & 0 & 0 & 0 & 0 & -1 & 0 & 0 & 1 & 1 & 0 & 0 & 0 & 0 & 1 & 0 & 0 & 0 & 0 & 0 & 0 & 0 & 0 & 0 & 0 \\
0 & 0 & 0 & 0 & 0 & 0 & 0 & 0 & 0 & 0 & -1 & 1 & 0 & 0 & 0 & 0 & 0 & 0 & 0 & 0 & 0 & 0 & 0 & 0 & 0 & 0 \\
0 & 0 & 0 & 0 & 0 & 0 & 0 & 0 & 0 & 0 & 0 & 1 & 0 & 0 & 0 & 0 & 0 & 0 & 0 & 1 & 1 & 0 & 0 & 0 & 0 & 0 \\
1 & 0 & 0 & 0 & 0 & 0 & 0 & -1 & 0 & 0 & 0 & 0 & 0 & 0 & -1 & -1 & 0 & 0 & 0 & 0 & 0 & -1 & -1 & 0 & -1 & 0 \\
0 & 0 & 0 & 0 & 0 & 0 & 0 & 0 & 0 & 0 & 0 & 0 & 0 & 0 & 0 & 0 & -1 & -1 & 0 & 1 & 0 & 0 & 0 & 0 & 1 & 0 \\
0 & 0 & 0 & 0 & 0 & 0 & 0 & 0 & 0 & 0 & 0 & 0 & 0 & 0 & 0 & 0 & 0 & -1 & 1 & 1 & 0 & 0 & 0 & 0 & 0 & 0 \\
0 & 0 & 0 & 0 & 0 & 0 & 0 & 0 & 0 & 0 & 0 & 0 & 0 & 0 & 0 & 0 & 0 & 0 & -1 & 0 & 1 & 1 & 0 & 0 & 0 & 0 \\
0 & 0 & 0 & 0 & 0 & 0 & 0 & 0 & 0 & 0 & 0 & 0 & 0 & 0 & 0 & 0 & 0 & 0 & 0 & 0 & -1 & 1 & 0 & 0 & 0 & 0 \\
0 & 0 & 0 & 0 & 0 & 0 & 0 & 0 & 0 & 0 & 0 & 0 & 0 & 0 & 0 & 0 & 0 & 0 & 0 & 0 & -1 & 0 & 0 & 1 & 0 & 0 \\
0 & 0 & 0 & 0 & 0 & 0 & 0 & 0 & 0 & 0 & 0 & 0 & 0 & 0 & 0 & 0 & 0 & 0 & 0 & 0 & -1 & 0 & 0 & 0 & 1 & 0 \\
0 & 0 & 0 & 0 & 0 & 0 & 0 & 0 & 0 & 0 & 0 & 0 & 0 & 0 & 0 & 0 & 0 & 0 & 1 & 0 & 0 & 0 & 0 & -1 & -1 & 0 \\
\end{bmatrix}.$$

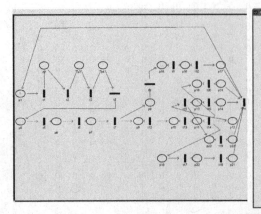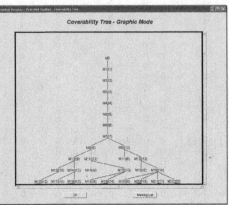

M130 = [0,0,0,0,0,0,0,0,0,0,0,0,0,0,1,0,1,0,0,0,1,0,0,1,1]
M131 = [0,0,0,0,0,0,0,0,0,0,0,0,0,1,0,0,1,0,0,0,0,1,0,1,1]
M132 = [0,0,0,0,0,0,0,0,0,0,0,0,0,1,1,0,0,0,0,0,1,0,1,1]
M133 = [0,0,0,0,0,0,0,0,0,0,0,0,0,1,0,0,1,1,0,0,1,0,1,0]
M134 = [0,0,0,0,0,0,0,0,0,0,0,0,0,1,0,0,1,0,0,0,1,1,0,1]
M135 = [0,0,0,0,0,0,0,0,0,0,0,0,0,1,0,0,1,0,0,1,0,0,1,1]
M136 = [0,0,0,0,0,0,0,0,0,0,0,0,0,1,0,0,0,1,0,0,0,1,0,1,1]
M137 = [0,0,0,0,0,0,0,0,0,0,0,0,0,1,0,1,0,0,0,0,1,0,1,1]
M138 = [0,0,0,0,0,0,0,0,0,0,0,0,0,1,0,0,1,0,0,0,1,0,1,1]

Figure 11.12   Sequential handoff operations

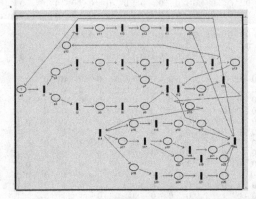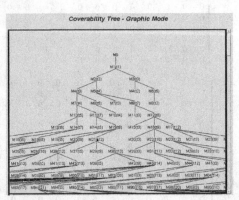

M529 = [0,0,0,0,0,0,0,0,0,0,1,0,0,0,0,0,0,0,1,0,1,0,1,0,1,0,1]
M530 = [0,0,0,0,0,0,1,1,0,0,0,0,0,0,0,0,0,1,0,1,0,1,1,0,1,0]
M531 = [0,0,0,0,0,0,1,1,0,0,0,0,0,0,0,0,0,1,0,1,1,0,0,1,1,0]
M532 = [0,0,0,0,0,0,1,1,0,0,0,0,0,0,0,0,1,1,0,0,1,0,1,1,0]
M533 = [0,0,0,0,0,0,1,1,0,0,0,0,0,0,1,0,0,0,0,1,0,1,1,1,0]
M534 = [0,0,0,0,0,0,1,1,0,0,0,1,0,0,0,0,0,1,0,1,0,1,0,0,0]
M535 = [0,0,0,0,1,0,1,0,0,0,0,0,0,0,0,0,0,1,0,1,0,1,0,1,1,0]
M536 = [0,0,0,0,0,0,0,0,0,0,0,0,0,0,0,0,0,1,0,1,0,1,1,0,1,1]
M537 = [0,0,0,0,0,0,0,0,0,0,0,0,0,0,0,0,0,1,0,1,1,0,0,1,1,1]
M538 = [0,0,0,0,0,0,0,0,0,0,0,0,0,0,0,0,0,1,1,0,0,1,0,1,1,1]
M539 = [0,0,0,0,0,0,0,0,0,0,0,1,0,0,0,0,0,1,0,1,0,1,0,1,1,1]
M540 = [0,0,0,0,0,0,0,0,0,0,1,0,0,0,0,0,0,1,0,1,0,1,0,1,0,1]
M541 = [0,0,0,0,0,0,1,1,0,0,0,0,0,0,0,0,0,1,0,1,0,1,0,1,1,0]
M542 = [0,0,0,0,0,0,0,0,0,0,0,0,0,0,0,0,0,1,0,1,0,1,0,1,1,1]

Figure 11.13   Concurrent security and scanning operations

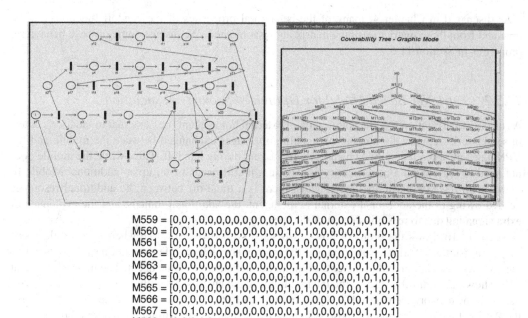

M559 = [0,0,1,0,0,0,0,0,0,0,0,0,0,0,1,1,0,0,0,0,0,1,0,1,0,1]
M560 = [0,0,1,0,0,0,0,0,0,0,0,0,0,1,0,1,0,0,0,0,0,0,1,1,0,1]
M561 = [0,0,1,0,0,0,0,0,0,1,1,0,0,0,1,0,0,0,0,0,0,0,1,1,0,1]
M562 = [0,0,0,0,0,0,0,0,1,0,0,0,0,0,1,1,0,0,0,0,0,0,1,1,1,0]
M563 = [0,0,0,0,0,0,0,0,1,0,0,0,0,0,1,1,0,0,0,0,1,0,1,0,0,1]
M564 = [0,0,0,0,0,0,0,0,1,0,0,0,0,0,1,1,0,0,0,0,0,1,0,1,0,1]
M565 = [0,0,0,0,0,0,0,0,1,0,0,0,0,1,0,1,0,0,0,0,0,0,1,1,0,1]
M566 = [0,0,0,0,0,0,0,0,1,0,1,1,0,0,0,1,0,0,0,0,0,0,1,1,0,1]
M567 = [0,0,1,0,0,0,0,0,0,0,0,0,0,0,1,1,0,0,0,0,0,0,1,1,0,1]
M568 = [0,0,0,0,0,0,0,0,1,0,0,0,0,0,1,1,0,0,0,0,0,0,1,1,0,1]

Figure 11.14 Concurrent security, L2 discovery, and L3 discovery operations

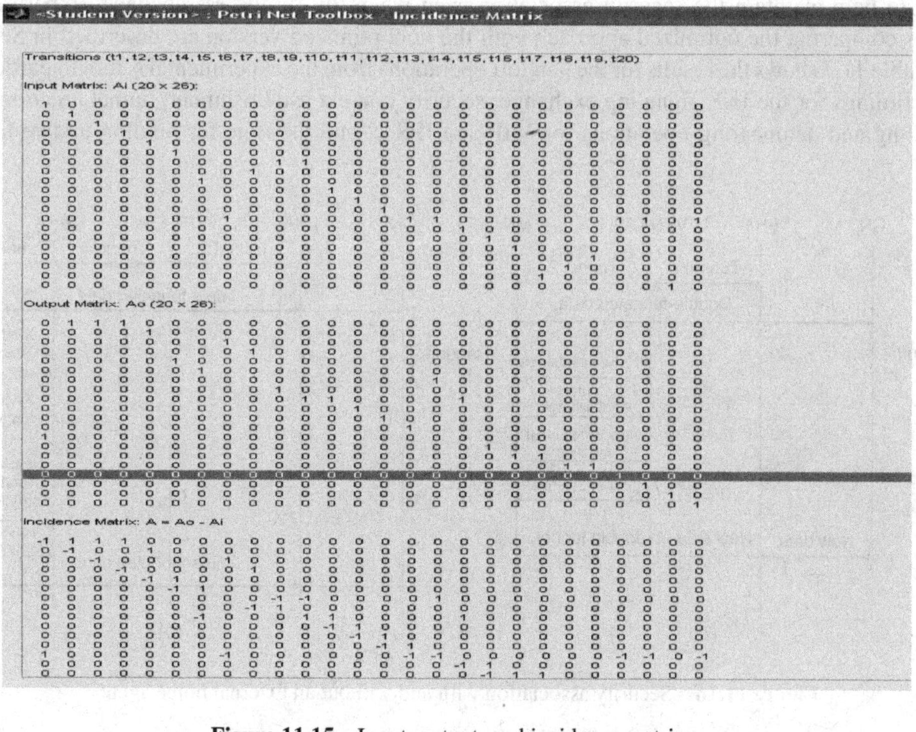

Figure 11.15 Input, output, and incidence matrices

As discussed in Chapter 4, many of the behavioral properties of the handoff functions can be obtained from these matrices using a matrix analysis method. We discuss some of these behavioral properties in Section 11.6.

11.5.2   Petri-Net-Based Model for Optimized Security Association

As discussed in Section 6.4, the addition of an external home agent (i.e., external to the user's enterprise network) reduces the number of signaling messages exchanged between the mobile and the VPN gateway when the mobile changes its IP address during handoff. Thus, while the mobile does not need to reestablish a new security association, it still needs to set up an additional Mobile IP tunnel with the external home agent. Thus, there is a trade-off between the additional resources needed owing to the additional external home agent, the additional tunnels, and the avoidance of extra signaling due to reestablishment of the security association.

Figure 11.16 shows the protocol flows associated with two cases, one where the mobile needs to reestablish the security association and the other where the mobile does not need to reestablish the security association, but at the cost of an additional external home agent. The network element i-HA shown in Figure 11.16(a) is the internal home agent and resides within the mobile's enterprise network; the network element x-HA shown in Figure 11.16(b) is the external home agent. We derive the Petri net models for both of these systems, one with and one without the external home agent, and evaluate the systems performance of each of the systems based on the experimental results.

Figure 11.17 shows the Petri net models corresponding to the call flows for the two scenarios shown in Figure 11.16. In Section 6.4, we presented the related experimental results and highlighted how the delay and packet loss are reduced when an external home agent is introduced as an anchor agent to help maintain the security association even when the IP address changes. Experimental results comparing the optimized approach with the nonoptimized version are described in Section 6.4. Table 11.7 shows the results for the handoff operations from the experimental setup. In particular, these timings for the IKE signaling exchange, security context establishment, tunnel creation, and tunneling and detunneling operations for MIP and IPSec, and the time for binding update by the

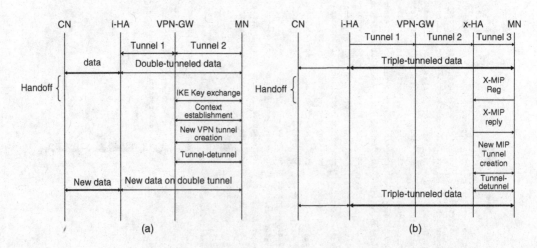

Figure 11.16   Security association with and without an external home agent

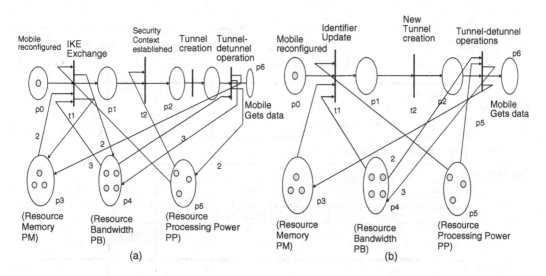

Figure 11.17   Petri net model for security association

Table 11.7   Timings for security association during handoff with xHA

Transition	Handoff operation	Time taken for operation (ms)
t1	IKE exchange	30
t2	Security context establishment	400
t3	VPN tunnel creation	6000
t4	MIP tunnel creation	10
t5	Tunneling/detunneling Mobile IP data	5
t6	Tunneling/detunneling VPN data	60
t7	External MIP update	300

external home agent were applied to the Petri net model to determine the systems performance by way of the cycle time and the Floyd algorithm.

In order to evaluate an optimization of a Petri net system, any of the three Petri-net-based methods described in Chapter 4 can be applied. We verified the optimized system described in this section by using a matrix-based solution as described in Section 4.10. The results demonstrate that in order to achieve the desired performance, the system needs to utilize more resources, such as an additional home agent that uses up more resources owing to triple encapsulation.

11.5.3   Petri-Net-Based Model for Hierarchical Binding Update

In Section 6.7.3, we showed experimentally how hierarchical binding update reduces the delay contributed by the global binding update using network layer and application layer mobility protocols. In order to achieve that, we introduced an additional anchor point closer to the mobile that took care of

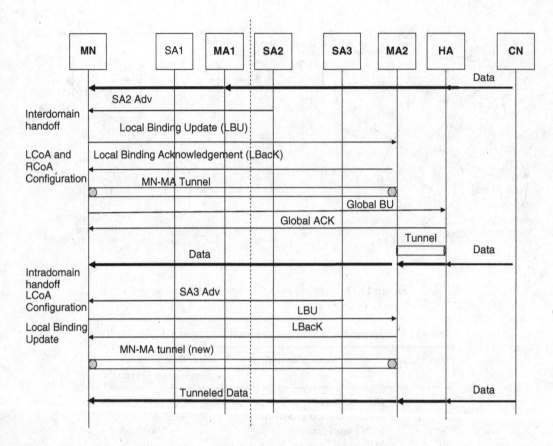

Figure 11.18 Flow for hierarchical mobility management

the hierarchical binding update. Here, we introduce an equivalent Petri net model that demonstrates the hierarchical binding update mechanism. Matrix-based analysis methods can be developed to evaluate the systems performance by using the cycle-time-based approach and the Floyd algorithm. Figure 11.18 shows the communication flow for interdomain and intradomain binding update and Figure 11.19 shows the corresponding Petri net model.

11.5.4 Petri-Net-Based Model for Media Redirection of In-Flight Data

In Section 6.8, we demonstrated experimentally how, by using different optimization techniques, packet loss due to media redirection delay can be reduced. Figure 11.20 shows an equivalent Petri net model for one of these optimization techniques, namely the mobility-proxy-based approach (Hsieh et al., 2003a). This model also allows us to verify the correctness of this optimization technique in terms of things such as such as liveness and deadlock properties and evaluate the systems performance.

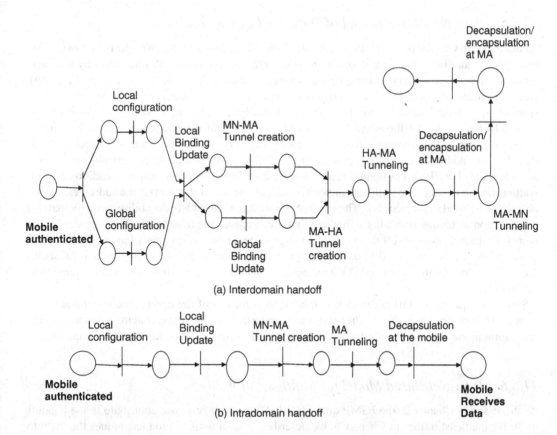

Figure 11.19   Petri net model for interdomain and intradomain binding updatee

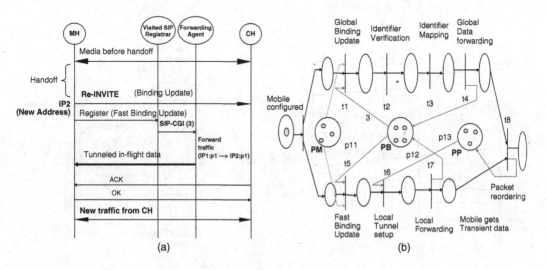

Figure 11.20   Petri net model for media forwarding

11.5.5   Petri-Net-Based Model of Optimized Configuration

As described in Chapter 6, the duplicate address detection (DAD) process takes the most time during layer 3 configuration. Chapter 6 also described several ways to reduce the time taken by the DAD process. We have used Petri nets to verify one specific optimization mechanism (Dutta et al., 2006b), where duplicate address detection is performed during the layer 3 identifier acquisition phase. This specific mechanism eliminates the time taken by the neighbor discovery process that is often performed by the client after the address is obtained. However, this mechanism adds extra load to the server or router, as this intermediate network entity (i.e., server or router) needs to collect information about the addresses that are being used (some of those addresses could be statically configured, and some may have been configured using DHCP), and the network also requires additional bandwidth owing to the periodic multicast announcement by the server that carries the addresses that are already being used in the network. The specific model that we used has the ability to verify that this optimization technique is deadlock free and that it can reduce the time for the DAD process at the cost of additional resources. If there are not enough system resources, namely bandwidth, memory, and CPU cycles, then there will be deadlock during some of these operations. Figure 11.21(a) shows the protocol flow for the optimized DAD mechanism and Figure 11.21(b) shows the associated Petri net model.

Similarly, equivalent Petri net models can be derived for any of the optimization techniques discussed in Chapter 6. Then, these Petri net models can be used to derive the systems performance and behavioral properties of the optimized system using any of the methods described in Chapter 4.

11.5.6   Petri-Net-Based Model for Multicast Mobility

As discussed in Chapter 2, the IGMP query interval and query response contribute to the handoff delay for multicast traffic. In Chapter 9, we described several optimization techniques that help to reduce the handoff delay and packet loss for multicast streaming traffic. Here, we illustrate both

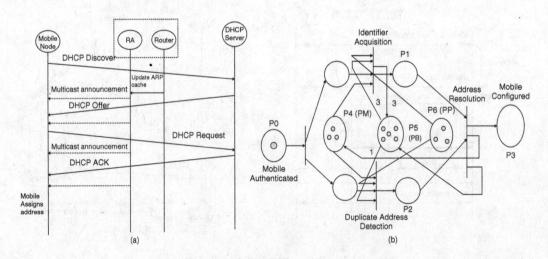

Figure 11.21   Petri net model for optimized DAD

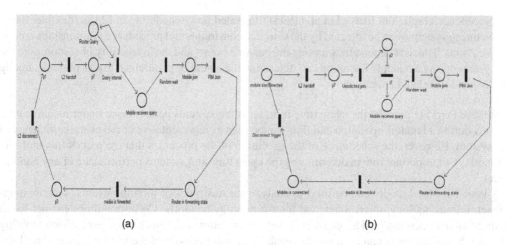

Figure 11.22   Petri net model for multicast mobility

nonoptimized and optimized versions of multicast mobility using the Petri net model. Figure 11.22(a) shows how the mobile's connectivity is delayed owing to the triggering delay resulting from the IGMP router query interval. On the other hand, Figure 11.22(b) shows how an unsolicited IGMP join helps to reduce the join latency during handoff.

11.6   Scheduling Handoff Operations

In this section, we illustrate some of the different ways in which some of the primitive handoff operations described in Chapter 3 can be scheduled, and evaluate the overall systems performance for three different schedules, namely sequential, concurrent, and proactive. A timed Petri net can be used to illustrate these ways of scheduling handoff operations. One basic approach is through the use of a heuristic search for an optimal or near-optimal schedule in the reachability tree of the Petri net model. Scheduling techniques can help derive specific schedules to optimize performance indices such as the handoff delay under certain systems resource constraints. Resource constraints can be defined as limitations on systems resources, beyond which a handoff operation will fail owing to lack of resources.

The scheduling of a handoff operation needs to take the following general guidelines into account:

1. Required systems performance indices such as handoff delay and packet loss are usually achieved under resource constraints, namely limitations on battery power, CPU cycles, and available channel resources (effective user bandwidth). Thus, the minimum cycle time achieved with a specific schedule is considered to be equivalent to the maximum systems performance.
2. A handoff schedule should not suffer from a deadlock condition where a specific operation cannot proceed because of nonavailability of data from the previous operation or because of lack of resources.
3. Both the resources and the precedence relationships among events need to be modeled to allow maximum flexibility. Thus, the Petri net model depends on both the resources required and the

precedence graph. Van Brussel et al. (1993) illustrated how scheduling in FMSs (flexible manufacturing systems) can be affected by the resource constraints and precedence relationships among the events. The data dependency among the handoff events and the availability of resources in the system during a handoff operation can determine the extent of parallelism that is possible among the handoff-related operations.

In these Petri net models, the cycle time represents the systems performance under resource constraints during a handoff operation and thus can be seen as representative of the overall efficiency of the system. However, the scheduling of the execution of the processes that are part of this mobility event plays an important role in determining the cycle time and systems performance of any handoff event.

A Petri-net-based model can be used to analyze various types of mobility events, such as intrasubnet, intratechnology, intersubnet, and intertechnology handoffs. Corresponding Petri-net-based optimization models can be derived by applying optimization techniques to the generalized mobility model. These techniques can be applied to each of the subprocesses of the handoff process in a hierarchical manner. In this section, we categorize handoff optimization techniques based primarily on the sequential, concurrent, and proactive modes of scheduling events and model these in Petri nets. Depending upon the type of scheduling technique, the systems resources expended during a specific operation will vary over time. While sequential handoff operations takes more time than proactive or concurrent operations, the optimized models using concurrent or proactive operations will need to spend more instantaneous resources during the handoff period. Instantaneous resources can be defined as the peak bandwidth, battery power, or number of CPU cycles that are used when these operations take place.

In order to conduct a performance analysis for different sequences of operations, we initially considered two handoff-related operations, namely discovery and authentication. We applied three different scheduling mechanisms to schedule these two handoff-related operations and studied the overall performance. Figure 11.23, Figure 11.24, and Figure 11.25 illustrate how these two specific handoff operations in the IEEE 802.11 environment can be represented in a Petri net model using sequential, concurrent, and proactive optimization techniques, respectively. Optimal system performance is obtained by comparing the handoff performance (cycle time) and resource utilization (number of tokens) for these handoff methodologies. We have described Petri-net-based handoff analysis and evaluated the effects of different scheduling schemes on the handoff process in Dutta et al. (2007c) and Dutta et al. (2009).

11.6.1 Sequential Scheduling

Figure 11.23 shows a Petri net model that represents the state transitions when discovery- and authentication-related operations are performed in sequence. As described in Chapter 4, the resources used during the handoff operations are represented as tokens. The number of tokens needed for each type of operation varies depending upon the amount of resources needed during each of these operations.

In general, scanning is part of the layer 2 discovery process and is followed by layer 2 authentication, a four-way handshake, and finally association with the layer 2 access point. In Figure 11.23, P_0, P_1, P_2, P_3, and P_4 are places that represent different states of discovery and authentication. Shared resources are represented by places P_B, P_M, and P_P, which represent effective user bandwidth (e.g., number of bits transferred), memory, and processing power, respectively. We have described these

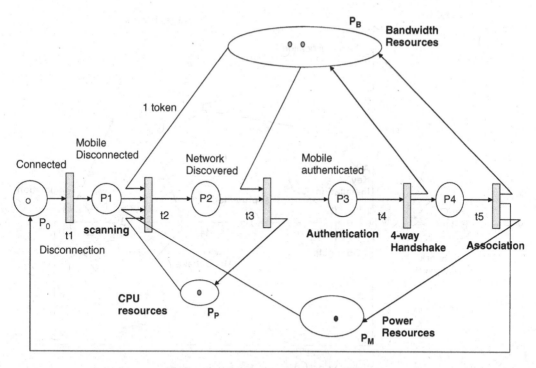

Figure 11.23   Sequential handoff operations

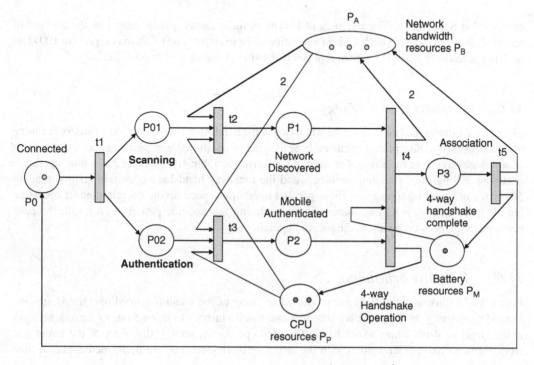

Figure 11.24   Concurrent handoff operations

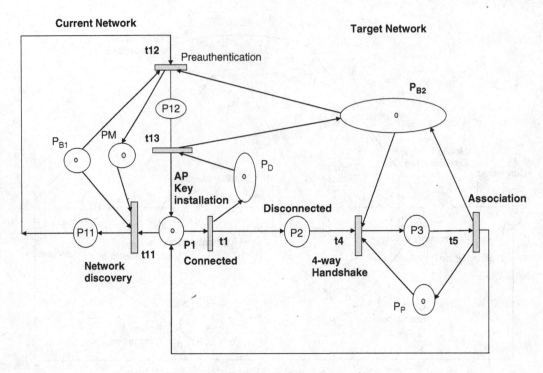

Figure 11.25 Proactive handoff operations

resources in Section 4.6. The numbers of tokens in these shared places represent the amount of resources expended during each of these operations. For example, one token may represent 100 kbits of data for resource place P_B, which represents the shared resource of bandwidth.

11.6.2 Concurrent Scheduling

Figure 11.24 shows the Petri net model when two of the handoff-related operations, namely scanning and authentication, take place concurrently at the cost of additional bandwidth resources as a result of additional signaling messages. For example, referring to Table 4.4, the layer 2 scanning operation results in 298 bytes of signaling messages and the four-way handshake operation results in about 504 bytes of signaling messages. These parallel operations speed up the overall handoff operation but the mobile consumes more shared resources during that specific period. As a result, the peak resource usage goes up owing to the parallel operations.

11.6.3 Proactive Scheduling

Figure 11.25 illustrates the Petri net model when some of the handoff-related operations are performed proactively. In the proactive scenario, the mobile intends to move from its current network to the target network. Many of the handoff-related operations, such as discovery of the target network elements and authentication with the target network elements, are performed ahead of time while the mobile is in the current network. Thus, the shared resources (e.g., channel resources in the access network) and bandwidth in the core network are utilized while the mobile is in the current network, and some additional resources such as power and CPU cycles are also used to support operations such as tunneling and proactive IP address caching. For example, according to Table 4.4, local caching of the IP address will require an additional six CPU cycles, and the tunneling operation with the target router will require an additional 60 bytes of control messages and three CPU cycles and will consume 384 nJ (nanojoules) of battery. P_{B1}, P_M, and P_D are shared resources that are used in the current network and P_{B2} and P_P are shared resources expended in the target network.

11.7 Verification of Systems Performance

The systems performance of a Petri net model of a mobility event can be verified in several ways. We illustrate two scenarios here.

In one scenario, the minimum cycle time can be obtained from the Petri net model by investigating the number of circuits,[1] the number of transitions, and the delay associated with each transition. Thus, the performance requirement expressed in terms of the cycle time C can be satisfied if and only if $CN_k - T_k \geq 0$ [2] for all circuits in the net. In the second scenario, the token loading matrix and transition matrix are obtained based on the markings of the Petri net model and the associated values of the transitions. Then, the Floyd algorithm is applied to validate the systems performance by computing the shortest distance between every pair of places.

We have calculated the cycle time and verified the systems performance based on the three scheduling techniques when applied to two basic handoff operations – discovery and authentication. Experimental results for these two operations were used to provide the transition timings for the Petri net modeling. Similar methods can be applied to compute the cycle time and overall systems performance for handoff events demonstrating different types of optimization techniques, namely hierarchical binding update, proactive discovery and configuration, and anchor-based security association, modeled using Petri nets.

11.7.1 Cycle-Time-Based Approach

We have introduced the details of the cycle-time-based approach in Chapter 4. We applied this approach to the three optimization techniques. Table 11.8 shows the times for the transitions $t1$, $t2$, $t3$, $t4$, and $t5$ for several different primitive handoff operations associated with the discovery and scanning processes obtained from our experiments (Lopez et al., 2007).

This analysis worked under the assumption that there was a handoff delay requirement of 100 ms to support a real-time application with a specific set of resources. We evaluated the overall cycle time when different schedules were applied to these handoff operations and verified if the system was conformant with the delay requirement. Table 11.9 shows the cycle times for three different handoff sequences involving discovery and security association. It is apparent from the results that although the time for concurrent scheduling of operations is smaller than the time for sequential scheduling, proactive scheduling is the only type of scheduling that satisfies the delay bound of 100 ms under this resource constraint.

---

[1] "Circuit" is defined in Chapter 4.

[2] Terms associated with the cycle time and the Floyd algorithm are defined in Section 4.11.

Table 11.8   Experimental results – layer 2 operations

Transition	Handoff operation	Time taken for operation (ms)
t1	Disconnection trigger	5
t2	Scanning	400
t3	Authentication	50
t4	Four-way handshake	10
t5	Association	5

Table 11.9   Cycle times from Petri net model

Optimization schedule	Relevant loop in Petri net	D_i	N_i	Max D/N_i
Sequential	P0T1P1T2P2T3P3T4P4T5P0	470	1	470
Concurrent	P0T1P1T3P3T4P0	420	1	420
Proactive	P1T1P2T4P3T5P1	17	1	17

11.7.2   Using the Floyd Algorithm

We have described the details of how the Floyd algorithm can be used to study systems performance in Chapter 4. In this section, we apply the Floyd algorithm to study the performance of the three handoff sequences.

Equations 11.1 and 11.2 represent the matrices obtained when the Floyd algorithm is applied to verify the systems performance of the mobility event using sequential scheduling, and Equation 11.3 represents the matrices illustrating proactive scheduling as shown in Figure 11.25. Values from the mobility event involving discovery and authentication were used to build the token loading matrix P and the transition time matrix Q. The first element of the matrix P is P_{00}, and the last element is P_{77}. The distance matrix and the matrix S were then derived from these two matrices. By inspecting the values of the matrix S in Equation 11.2, which reflects sequential scheduling, it is found that at least one of the diagonal elements is negative. Thus, this specific type of sequential scheduling cannot provide the desired systems performance of a cycle time of 100 ms. In order to achieve the desired performance level, faster facilities could be used to reduce the transition time or more tokens (resources) could be used in the shared places, thereby increasing the level of concurrency.

$$P = \begin{bmatrix} 0 & 1 & 0 & 0 & 0 & 0 & 0 & 0 \\ 0 & 0 & 0 & 0 & 0 & 0 & 0 & 0 \\ 0 & 0 & 0 & 0 & 0 & 0 & 0 & 0 \\ 0 & 0 & 0 & 0 & 0 & 0 & 0 & 0 \\ 0 & 0 & 0 & 0 & 0 & 0 & 0 & 0 \\ 0 & 0 & 2 & 2 & 2 & 0 & 2 & 2 \\ 0 & 0 & 1 & 0 & 1 & 1 & 0 & 0 \\ 0 & 0 & 1 & 0 & 0 & 0 & 0 & 0 \end{bmatrix}, \quad Q = \begin{bmatrix} w & 5 & w & w & w & w & w & w \\ w & w & 400 & w & w & w & w & w \\ w & w & w & 50 & w & 50 & 50 & 50 \\ w & w & w & w & 10 & 10 & w & w \\ 5 & w & w & w & w & 5 & 5 & w \\ w & w & 400 & 50 & 10 & w & 50 & 50 \\ w & w & 400 & w & w & w & w & w \end{bmatrix}, \quad (11.1)$$

$$
CP - Q = \begin{bmatrix}
\infty & 95 & \infty & \infty & \infty & \infty & \infty & \infty \\
\infty & \infty & -300 & \infty & \infty & \infty & \infty & \infty \\
\infty & \infty & \infty & \infty & \infty & \infty & \infty & \infty \\
\infty & \infty & \infty & -50 & \infty & -50 & -50 & -50 \\
\infty & \infty & \infty & \infty & -10 & -10 & \infty & \infty \\
-5 & \infty & \infty & \infty & -10 & -10 & \infty & \infty \\
\infty & \infty & -200 & 150 & 190 & \infty & 150 & 150 \\
\infty & \infty & -300 & \infty & 90 & 90 & \infty & \infty
\end{bmatrix}, \quad
S = \begin{bmatrix}
-270 & 195 & -105 & -155 \\
\cdot & \cdot & \cdot & \cdot \\
\cdot & \cdot & \cdot & \cdot \\
\cdot & \cdot & \cdot & \cdot \\
\cdot & \cdot & \cdot & \cdot
\end{bmatrix}, \quad (11.2)
$$

$$
P = \begin{bmatrix}
0 & 1 & 0 & 0 & 0 \\
0 & 0 & 0 & 0 & 0 \\
0 & 0 & 0 & 0 & 0 \\
0 & 0 & 1 & 0 & 0
\end{bmatrix}, \quad
Q = \begin{bmatrix}
w & 5 & w & w & w \\
w & w & 10 & w & w \\
5 & w & w & 5 & 5 \\
w & w & 10 & w & w \\
w & w & 10 & w & w
\end{bmatrix}, \quad (11.3)
$$

$$
CP - Q = \begin{bmatrix}
\infty & 95 & \infty & \infty & \infty \\
\infty & \infty & -10 & \infty & \infty \\
-5\infty & \infty & -5 & -5 \\
\infty & \infty & 90 & \infty & \infty \\
\infty & \infty & 90 & \infty & \infty
\end{bmatrix}, \quad
S = \begin{bmatrix}
90 & 95 & \cdot & \cdot & \cdot \\
105 & 100 & \cdot & \cdot & \cdot \\
\cdot & \cdot & 85 & \cdot & \cdot \\
\cdot & \cdot & \cdot & 85 & \cdot \\
\cdot & \cdot & \cdot & \cdot & 85
\end{bmatrix}. \quad (11.4)
$$

Equation 11.3 shows the matrices for proactive scheduling based on the transition times obtained from the experiments. By inspecting the matrix S in equation 11.4, it is apparent that all the diagonal elements of the matrix are positive. Thus, by applying the Floyd algorithm, it can be verified that proactive scheduling, when applied to the discovery and authentication processes, satisfies the systems performance requirement of a minimum cycle time of 100 ms. We have also used several automated tools such as TimeNet (Zimmermann et al., 1999a), STPNplay (Ryuther, 2004), and a Petri net tool (Matcovschi et al., 2003) to model the behavior of the handoff system, capture the systems performance, and evaluate the performance characteristics of the mobility protocols and associated optimizations. By analyzing these Petri-net-based models, one can easily predict the performance of a handoff system given a set of resources when various scheduling mechanisms are applied.

11.8 Petri-Net-Based Modeling for Multi-Interface Mobility

In this section, we illustrate Petri net modeling involving handover between two different types of access networks (e.g., 802.11 and CDMA), covering three scenarios, namely *parallel operations*, *break-before-make*, and *make-before-break*, using the two interfaces of a multi-interface mobile. As described in Table 3.1, each access network has different ways of discovering resources and network parameters; the authentication mechanisms and encryption algorithms are also different for each access network. Thus, the resource requirements (e.g., battery, bandwidth, and CPU) for each of the handoff operations in the two access networks are different.

11.8.1 Multihoming Scenario

When a multi-interface mobile has coverage for both CDMA and 802.11, both of the interfaces start to be configured at the same time and consume battery and CPU resources on the mobile, while bandwidth resources are consumed in each access network independently. Ideally, based on some policy such as the type of application or tariff, the mobile should make a decision regarding which interface should be used for communication.

11.8.2   Break-Before-Make Scenario

In a break-before-make handoff scenario, one interface does not start to be configured until the second interface is disconnected. In such a scenario, where one type of interface (e.g., CDMA) comes up after the other interface (e.g., 802.11) has gone down, the resource consumption during handoff will depend on whether the mobile is handing over from 802.11 to CDMA or vice versa.

11.8.3   Make-Before-Break Scenario

During a make-before-break operation, when the mobile is still communicating using the 802.11 interface, the CDMA interface starts to be activated as the signal-to-noise ratio on the 802.11 interface begins to deteriorate. While this helps to reduce the handoff delay compared with the break-before-make case, it requires more battery power in the mobile since both interfaces stay powered on for some time. In certain types of make-before-break scenario, many of the handoff-related operations, such as resource discovery, authentication, and security context transfer for the CDMA interface, are taken care of by the 802.11 interface and thus could delay the activation of the CDMA interface. Delaying the activation of the CDMA interface helps to reduce the battery usage of the mobile. Thus, it is important to consider when to start the activation of the CDMA interface based on the availability of resources in the network.

11.8.4   MATLAB®-based Petri Net Modeling for Multi-Interface Mobility

Resource modeling of multiple interfaces as part of the performance modeling of multi-interface mobility allows a more principle-based decision about which interface to use for a handoff operation. In the case of multi-interface mobility, it is not currently easy to figure out which interface needs to be turned on and when for each handoff operation (e.g., discovery or configuration). Interactions between multiple interfaces, and therefore the question of when to switch up or switch down an interface, can be investigated by Petri net modeling. Since each interface has different resource requirements for the same kind of handoff operation, a Petri net model allows one to find out which operation can be carried out by which interface given resource constraints such as battery, CPU, and bandwidth.

We modeled the above three scenarios using the MATLAB-based Petri net tool. Table 11.10 shows the amounts of resources and time needed to take care of specific handoff-related operations for two different types of interface, namely CDMA1X-EvDO and IEEE 802.11. These values are based on the experimental results from the test bed described in Chapter 6. Since these two interfaces have different access characteristics, the amounts of resources and the times required to complete each of the handoff operations also vary. In Chapter 4, we discussed the energy required to transmit and receive a bit for CDMA and 802.11. As discussed by SALAWU and Elizabeth (2009), it takes more energy to transmit a bit using a CDMA interface than it does using an 802.11 interface. The layer 2 authentication processes for 802.11 and CDMA are different and need different signaling messages and numbers of bytes exchanged. Also, PPPoE gives rise to an additional 8 bytes of header for the same handoff-related operations.

Table 11.10 Resources and timing for 802.11 and CDMA

Operations	Resources in 802.11			Resources in CDMA			Timing (ms)	
	Battery (nJ)	Bytes	CPU (cycles)	Battery (nJ)	Bytes	CPU (cycles)	802.11	CDMA
Discovery	414000	345	12	1968000	328	9	745	422
Layer 2 authentication	4126800	3439	29	1392000	232	14	106	200
Configuration	2257200	1881	22	5454000	909	12	510	850
Security association	940800	784	10	4752000	792	10	640	4500
Binding update	422400	352	18	2160000	360	18	168	599

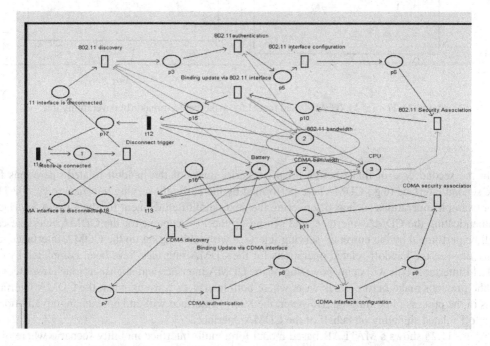

Figure 11.26 MATLAB-based model for parallel CDMA and 802.11 operations

The MATLAB-based model in Figure 11.26 illustrates a scenario where the mobile is under the coverage of both 802.11 and CDMA access. Thus, in this scenario, both of the interfaces will begin their configuration independently. However, since each interface will have its dedicated access network, a single bandwidth (channel resource) place cannot be shared by the operations related to both interfaces, whereas battery power and CPU cycles can be considered as shared resources for both. Thus, two different bandwidth places are included in the model, one for each type of access network. Each type of bandwidth resource will be shared among the handoff events within its specific access network. For example, the channel resources for the CDMA network will be shared by the identifier configuration and discovery operations in the CDMA network.

Figure 11.27 shows a MATLAB-based model for a multi-interface mobility scenario where the mobile is communicating using its 802.11 interface. Here, as the signal-to-noise ratio of the 802.11 interface decreases, the CDMA interface is in the process of being connected.

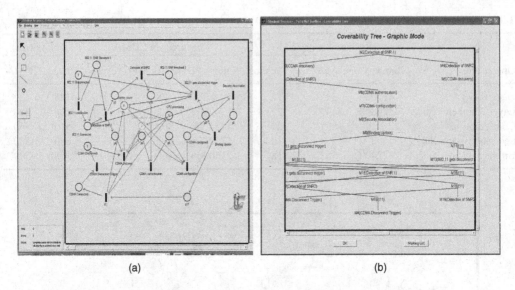

(a)   (b)

Figure 11.27   (a) MATLAB-based model for make-before-break; (b) coverability tree

In the second type of make-before-break scenario, many of the handoff-related operations for the second interface (e.g., CDMA) are performed by the currently serving interface (e.g., 802.11). For example, operations such as discovering the next point of attachment for the CDMA interface, authenticating the CDMA interface, and setting up the PPP context for the CDMA interface can still be performed by the currently serving interface without bringing up the CDMA interface until after most of the handoff-related operations for the CDMA interface have been completed by the 802.11 interface. This will delay powering on the CDMA interface and not use as much resources as in the previous make-before-break case, where both interfaces were up when the CDMA interface was in the process of coming up. However, the 802.11 interface will end up exchanging additional handoff-related signaling on behalf of the CDMA interface.

Figure 11.28 shows a MATLAB-based model for a multi-interface mobility scenario where only one interface (e.g., 802.11 or CDMA) is active at any point of time and the CDMA interface prepares itself only after the 802.11 interface is disconnected. Since the two interfaces are not active at the same time, they do not consume as much resources as in previous two scenarios, but this scenario takes the most amount of time for handoff to be completed. Unlike the make-before-break case shown in Figure 11.27, here the CDMA interface goes through the process of getting connected only after the 802.11 interface has been disconnected at a specific signal-to-noise threshold.

11.9   Deadlocks in Handoff Scheduling

Optimization intrinsically requires cross-layer designs and speculative execution in preparation for a possible handover. Thus, the sequence of handoff operations (e.g., parallel, proactive, or sequential) plays a role in determining the extent of optimization that can be achieved. The use of Petri nets allows resource dependencies and data dependencies to be explicitly represented and thus allows

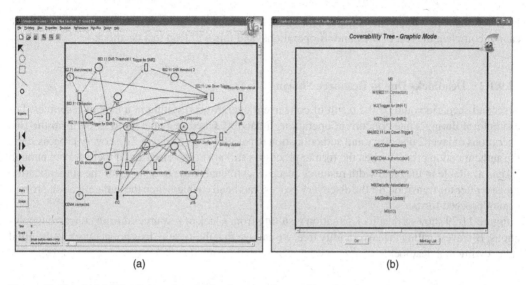

Figure 11.28 (a) MATLAB-based model for the second type of break-before-make scenario; (b) coverability tree

checking for blocking and deadlocks. The scheduling of handoff operations needs to take data dependency and resource availability into account. In this section, we describe a few scenarios that might give rise to a deadlock in the system, and ways to avoid such deadlocks by changing the schedule or adding resources.

Reachability analysis has been described in Chapter 4. Reachability analysis is one way of detecting deadlocks in a handoff system. If a specific schedule generates a coverability tree such that no subsequent transitions are allowed from a specific marking (e.g., M_i), then that specific sequence exhibits a deadlock. Thus, by inspecting the coverability tree of a handoff schedule, it is possible to determine whether a deadlock exists in that schedule.

We now illustrate a few scenarios that show how deadlocks can take place during a handoff process. Then, we propose solutions to deal with this situation. We have used the MATLAB-based Petri net tool to construct coverability trees and verify the deadlock properties, and then demonstrate how deadlocks can be eliminated.

11.9.1 Handoff Schedules with Deadlocks

In this section, we illustrate a few specific handoff schedules that exhibit a deadlock situation, and propose solutions that avoid these deadlocks. We describe a few specific scenarios in which we compare two schedules, one with a deadlock and one without.

11.9.1.1 Deadlock Due to Lack of Data

A specific handoff schedule could also lead to a deadlock if the sequence of operations does not follow the data dependency graph. In this case, some specific handoff operation cannot proceed owing to a lack of data that is expected from another handoff operation. For example, in a normal situation, if the mobile configures its layer 3 identifier and assigns it to the interface as part of the layer 3 handover without finishing an L2 handoff operation (e.g., a channel change), the mobile cannot complete the rest of the handoff operations, and this will lead to a deadlock.

11.9.1.2   Deadlocks Due to Resource Sharing

A second scenario could be the result of concurrent operations resulting in a lack of resources. For example, if during some concurrent operations a handoff schedule is designed to perform the two operations of layer 2 discovery and authentication in parallel (e.g., the layer 2 discovery process starts the authentication process when the former process is still not completed) and if there are not enough tokens available in the bandwidth resource place (P_B) to enable the transition for the authentication process after the transition for the discovery process has been enabled, then the authentication process cannot proceed further.

Figure 11.29 shows a deadlock situation resulting from a lack of resources during concurrent operations. By investigating the coverability tree, we can see that the system does not get back to its initial state, owing to a deadlock.

11.9.1.3   Deadlocks in Simultaneous Mobility

Another deadlock scenario can arise from a failure of binding update during simultaneous mobility, where the binding update from one of the clients is not completed owing to overlapping handoffs of the mobiles. The rest of the handoff operations cannot proceed, owing to noncompletion of the binding updates of the two clients. We have described the details of the simultaneous mobility problem in Chapter 8. Here, we illustrate the Petri net modeling of simultaneous-mobility scenarios.

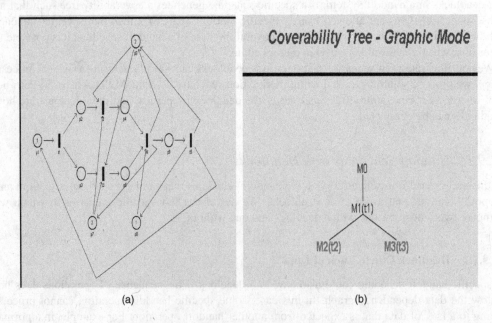

(a)   (b)

Figure 11.29   Deadlock due to resource constraints

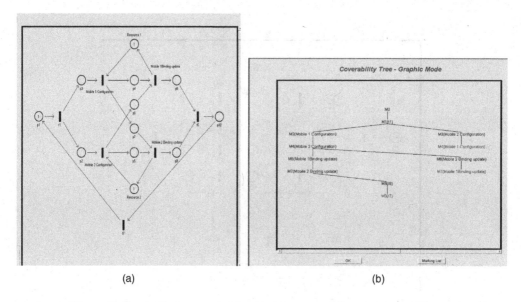

(a)   (b)

Figure 11.30   (a) Petri net model for simultaneous mobility; (b) coverability tree

Figure 11.30(a) shows a Petri net model equivalent to Figure 8.1, where there is no problem due to simultaneous mobility and both of the mobiles can communicate with each other. Figure 11.30(b) shows the corresponding coverability tree. It is apparent that successful communication between the two mobiles depends on successful completion of the configuration and binding updates of both mobiles.

Figure 11.31 illustrates an example of a Petri net for a deadlock situation in simultaneous mobility because one of the mobile is repeatedly reconfigured. Figure 11.32 shows the equivalent coverability tree, which shows the markings.

11.9.2   Deadlock Prevention and Avoidance in Handoff Schedules

There are several ways in which deadlock situations can be prevented or avoided. *Deadlock prevention* consists of falsifying one or more of the necessary conditions mentioned in Chapter 4 by using static resource allocation policies so that deadlocks are completely eliminated. Since deadlock prevention is accomplished by static policies that are known to result in poor resource utilization, and the use of reachability analysis to arrive at deadlock prevention policies can become infeasible if the state space is very large, *deadlock avoidance* techniques are sometimes preferred. Deadlock avoidance techniques attempt to falsify one or more of the necessary conditions in a dynamic way by keeping track of the current state and possible future conditions. The idea is to let the necessary conditions prevail as long as they do not cause a deadlock, but to falsify them as soon as a deadlock becomes a possibility in the immediate future. Deadlock avoidance leads to better resource utilization.

Viswanadham et al. (1990) discussed deadlock prevention and avoidance in flexible manufacturing systems using Petri net models. Similar techniques can be applied for the prevention or avoidance

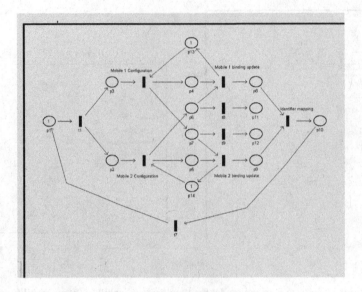

Figure 11.31   Deadlock in simultaneous mobility

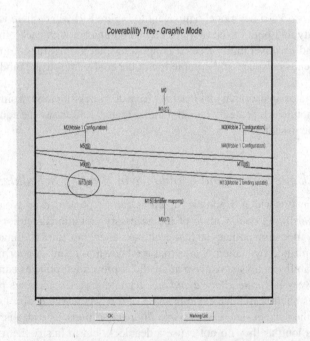

Figure 11.32   Coverability tree for deadlock in simultaneous mobility

of deadlocks in a handoff system. Each of the above deadlock scenarios can be avoided by either adding resources, changing the schedule, or introducing additional components into the network. The reachability graph of a Petri net model representing a handoff system can be used to arrive at resource allocation policies that enforce deadlock prevention.

Deadlocks in the first scenario described in Section 11.9.1 can be prevented in two ways. In the first of these ways, deadlocks can be prevented by scheduling the handoff operations so as not to enable a transition unless there is data available from the previous operation. Thus, the mobile should not be scheduled to perform a layer 3 operation unless the layer 2 handoff operation is over. The second way involves additional operations to avoid the deadlock. For example, unless a transient tunnel between the mobile and the next-hop router has been established, the rest of the operations, such as binding update and media forwarding to the mobile, cannot be completed, and this will lead to incomplete handoff of the mobile. Thus, setting up this additional tunnel is an additional operation that is needed to avoid a possible deadlock in such situation.

One way to prevent a deadlock, illustrated in Figure 11.33, is to increase the number of tokens in the resource place P_B (e.g., bandwidth resources) to take care of the concurrent operations. Figure 11.33 illustrates how the existing deadlock situation due to lack of resources during concurrent operations, as shown in Figure 11.29, is taken care of by adding additional resources. This shows how the allocation of additional resources has resulted in elimination of the deadlock. We can verify that a schedule is deadlock free by doing a reachability analysis. As shown in the coverability tree, in the absence of a deadlock, the mobile returns to its initial state.

Deadlocks in the simultaneous-mobility scenario can be avoided by installing additional components such as proxies in the network, as explained in Chapter 8. Figure 11.34 illustrates a

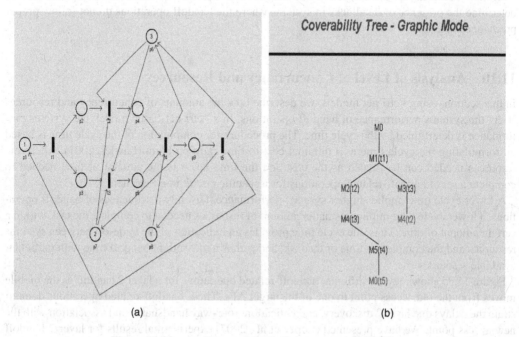

(a)   (b)

Figure 11.33 Avoidance of deadlock in concurrent operations

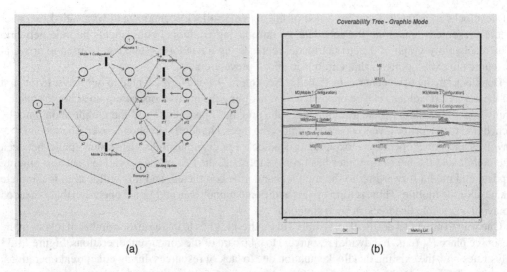

(a)   (b)

Figure 11.34   (a) Deadlock avoidance in simultaneous mobility; (b) coverability tree

MATLAB-based model and corresponding coverability tree that show how a deadlock in simultaneous mobility can be avoided by introducing a retransmission technique or a forwarding agent.

Similarly, MATLAB-based models can be used to construct an equivalent coverability tree that can determine the presence of deadlocks in systems where the handoff operations do not follow proper precedence rules.

11.10   Analysis of Level of Concurrency and Resources

In this section, using Petri net models, we describe how the amounts of concurrency and resources affect the systems performance of handoff operations. In a Petri-net-based model, the systems performance is determined by the cycle time. The procedure for computation of the cycle time is based on formulating the cycle time as a minimal cost-to-time ratio problem (Lawler, 2001), where the resources needed can be treated as the cost and the time taken is the amount of time needed to complete a set of handoff-related operations for a specific circuit in a Petri net.

A lower cycle time implies higher systems performance. Given two sequences of handoff operations, a lower cycle time implies a smaller amount of resources needed to complete the task within a certain amount of time. Thus, the cycle time provides an indication of the trade-off between systems resources and the completion time of the task, and is thus a measure of handoff time with particular available resources.

Figure 11.35 shows several different handoff-related operations for a layer 2 handoff as the mobile moves from the old access point to one of the target APs. These handoff-related operations demonstrate the delays due layer 2 discovery, authentication, four-way handshake, and association with the new access point. We have presented (Lopez et al., 2007) experimental results for layer 2 handoff operations in both roaming and nonroaming scenarios.

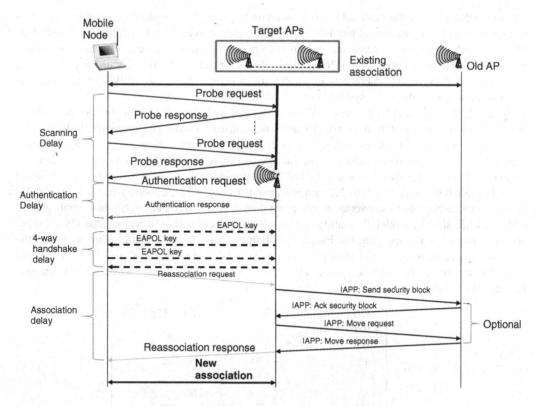

Figure 11.35 Illustration of layer 2 handoff

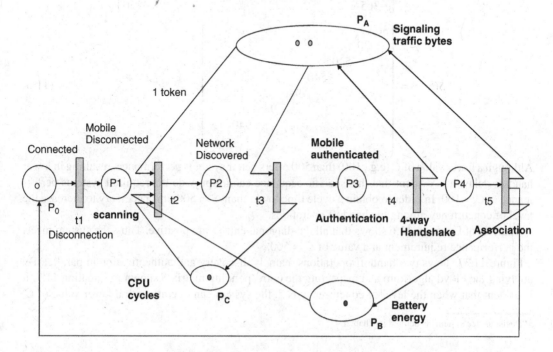

Figure 11.36 Sequential layer 2 operations

Next we describe several Petri net models based on the handoff operation shown in Figure 11.35 using four possible sequences of handoff operations and resources. We then apply the Floyd algorithm to verify if a specific sequence of handoff operations meets a given systems performance requirement in terms of cycle time. This analysis shows how changing the concurrency level (e.g., no parallel operations, two operations in parallel, or three operations in parallel) or changing the resource parameters affects the systems performance.

Figure 11.36 shows a Petri net model illustrating sequential layer 2 handoff operations, namely layer 2 discovery, authentication, four-way handshake, and association. The timings of the transitions $t1$, $t2$, $t3$, $t4$, and $t5$ were values obtained from experimental results (Lopez et al., 2007). Given a required systems performance value C, one can use the Floyd algorithm to determine whether this value of systems performance can be achieved for a certain level of concurrency of operations and resource availability. Since the Floyd algorithm can only be applied to decision-free Petri nets, [3] these Petri net models need to be converted to corresponding decision-free Petri nets where applicable.

Figure 11.37 shows a model illustrating two layer 2 handoff operations in parallel, namely scanning and authentication. By applying the Floyd algorithm, we can derive a matrix S to determine the shortest distance between every pair of places for a given value of C. For example, for the sequential model shown in Figure 11.36, Equation 11.5 shows the matrix S for a C value of 100, whereas Equation 11.6 shows this matrix for a C value of 500:

$$S(100) = \begin{bmatrix} -370 & 35 & -305 & -355 & -365 & -370 & -315 & -370 \\ -465 & -370 & . & . & . & . & . & . \\ -65 & . & -265 & . & . & . & . & . \\ -15 & . & . & -370 & . & . & . & . \\ -5 & . & . & . & -370 & . & . & . \\ 135 & . & . & . & . & -260 & . & . \\ -365 & . & . & . & . & . & -365 & . \\ -365 & . & . & . & . & . & . & -350 \end{bmatrix}, \tag{11.5}$$

$$S(500) = \begin{bmatrix} 30 & . & . & . & . & . & . & . \\ . & 30 & . & . & . & . & . & . \\ . & . & 540 & . & . & . & . & . \\ . & . & . & 540 & . & . & . & . \\ . & . & . & . & 940 & . & . & . \\ . & . & . & . & . & 935 & . & . \\ . & . & . & . & . & . & 940 & . \\ . & . & . & . & . & . & . & 135 \end{bmatrix}. \tag{11.6}$$

Although a lower value of C (e.g., 100) than 500 offers better systems performance, resulting in lower handoff delay, it is evident that this specific sequence cannot provide the required systems performance at $C = 100$. In order to obtain a cycle time lower than $C = 500$, it is necessary to increase the level of concurrency among the handoff operations.

Analysis of Equation 11.6 shows that all the diagonal entries are positive. Thus, the system meets the performance requirement at a value of $C = 500$.

Figure 11.37 shows two handoff operations, namely discovery and authentication in parallel. By applying the Floyd algorithm and inspecting the corresponding matrix S shown in Equation 11.7, it is evident that when the level of concurrency is 2, the system can operate with a lower value of C,

---

[3] "Decision-free system" is defined in Chapter 4.

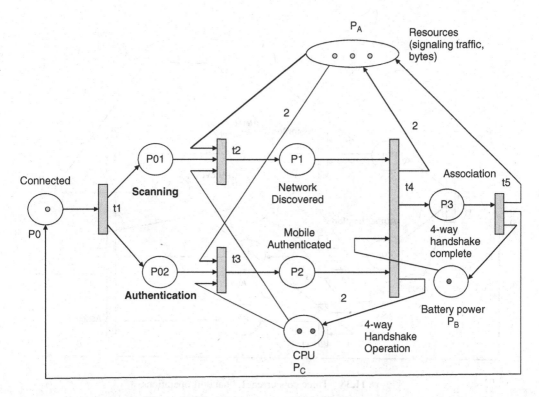

Figure 11.37 Two concurrent L2 handoff operations

namely 450 (better systems performance), compared with the C value of 500 (lower performance) for sequential operations as shown in Equation 11.6. This indicates that the handoff delay will be reduced when the system meets the performance requirements with a lower value of C.

$$S(450) = \begin{bmatrix} 30 & . & . & . & . & . & . & . & . \\ . & 30 & . & . & . & . & . & . & . \\ . & . & 380 & . & . & . & . & . & . \\ . & . & . & 940 & . & . & . & . & . \\ . & . & . & . & 840 & . & . & . & . \\ . & . & . & . & . & 435 & . & . & . \\ . & . & . & . & . & . & 940 & . & . \\ . & . & . & . & . & . & . & 435 & . \\ . & . & . & . & . & . & . & . & 840 \end{bmatrix} . \qquad (11.7)$$

We can add further concurrency as shown in Figure 11.38, where the discovery, authentication, and four-way handshake operations take place in parallel. Compared with Figure 11.37, the additional level of concurrency requires more resources in place P_A in order to avoid deadlocks.

By applying the Floyd algorithm and inspecting the corresponding matrix S shown in Equation 11.8, we can see that with this additional concurrency the system meets the systems

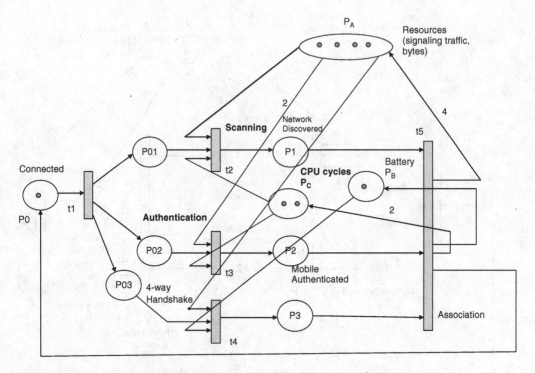

Figure 11.38   Three concurrent L3 handoff operations

performance requirement at a cycle time $C = 410$:

$$
S(410) = \begin{bmatrix}
0 & . & . & . & . & . & . & . & . & . \\
. & 0 & . & . & . & . & . & . & . & . \\
. & . & 395 & . & . & . & . & . & . & . \\
. & . & . & 390 & . & . & . & . & . & . \\
. & . & . & . & 1235 & . & . & . & . & . \\
. & . & . & . & . & 765 & . & . & . & . \\
. & . & . & . & . & . & 395 & . & . & . \\
. & . & . & . & . & . & . & 1235 & . & . \\
. & . & . & . & . & . & . & . & 395 & . \\
. & . & . & . & . & . & . & . & . & 765
\end{bmatrix}.
\tag{11.8}
$$

The Petri net models in the previous examples show how adding concurrency results in a reduction in cycle time (increased systems performance). While keeping the level of concurrency the same, the cycle time can also be reduced if additional resources are dedicated to reducing the transition timings. For example, the Petri net model shown in Figure 11.37 can be converted to the Petri net model shown in Figure 11.39 with additional resources in P_A. These additional resources will help expedite the scanning and authentication operations. Thus, the timings of $t2$ and $t3$ will be reduced.

By applying the Floyd algorithm and inspecting the matrix S in Equation 11.9, it can be verified that the model shown in Figure 11.39 meets the systems performance requirement at a value of

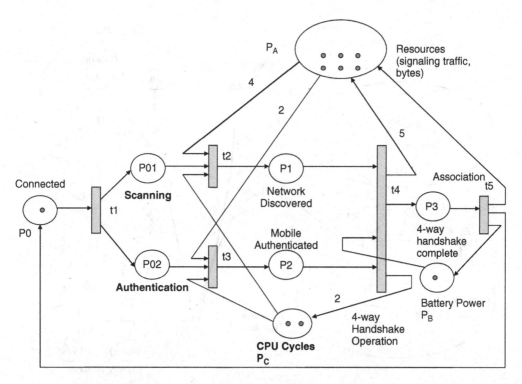

Figure 11.39   Two concurrent handoff operations with additional resources

$C = 200$:

$$S(200) = \begin{bmatrix} 80 & 195 & 195 & \cdot & \cdot & \cdot & \cdot & \cdot & \cdot \\ 195 & 180 & \cdot & \cdot & \cdot & \cdot & \cdot & \cdot & \cdot \\ \cdot & \cdot & 130 & \cdot & \cdot & \cdot & \cdot & \cdot & \cdot \\ \cdot & \cdot & \cdot & 1090 & \cdot & \cdot & \cdot & \cdot & \cdot \\ \cdot & \cdot & \cdot & \cdot & 340 & \cdot & \cdot & \cdot & \cdot \\ \cdot & \cdot & \cdot & \cdot & \cdot & 1855 & \cdot & \cdot & \cdot \\ \cdot & \cdot & \cdot & \cdot & \cdot & \cdot & 1090 & \cdot & \cdot \\ \cdot & \cdot & \cdot & \cdot & \cdot & \cdot & \cdot & 185 & \cdot \\ \cdot & \cdot & \cdot & \cdot & \cdot & \cdot & \cdot & \cdot & 340 \end{bmatrix} \qquad (11.9)$$

From this analysis of Petri net models, we have verified that a handoff system can provide a desired lower value of the cycle time C if we either increase the number of concurrent operations or increase the amount of resources for the same level of concurrency.

11.11   Trade-off Analysis for Proactive Handoff

As discussed in Chapter 4, several types of systems resources are utilized during a handoff operation. Proactive and concurrent operations reduce the handoff delay and packet loss at the cost of additional resources. For example, although proactive handoff operations offer better performance,

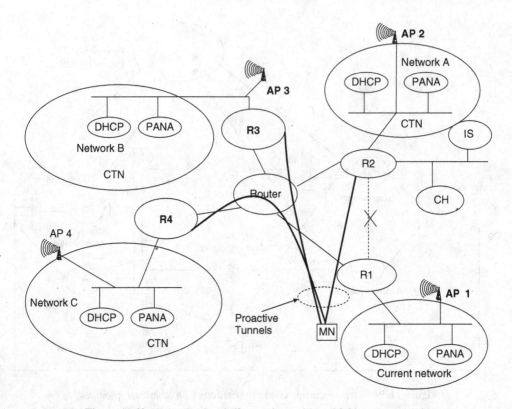

Figure 11.40   Proactive handoff operations with multiple target networks

they utilize systems resources while the mobile is in the current network and is engaged in these proactive operations. Here. we discuss the different levels of proactive operations that are needed when the mobile is about to move to a target network, and investigate the trade-off between handoff delay and resource utilization during this operation.

When the mobile is about to move to a new network, it performs a set of proactive handoff operations with the target network in order to reduce the handoff delay and packet loss. When there are multiple neighboring networks, doing proactive operations with additional networks increases the probability of successful handover. However, this also needs more systems resources because of the many handoff-related operations involved in the proactive operations, such as tunneling, preconfiguration, and preauthentication.

Figure 11.40 shows an example of how a mobile can perform proactive handoff operations with multiple target networks. This shows the current network and three neighboring networks, namely networks A, B, and C. Multiple tunnels can be set up between the mobile and the target routers, and proactive operations can take place with multiple target networks. Establishing proactive operations with multiple networks increases the probability of successful handover but ends up using more resources.

Some of the resources that are utilized during these proactive operations can be defined as follows:

1. The additional signaling needed to complete preauthentication with the neighboring networks results in additional bandwidth usage in the current network.

2. Caching the IP addresses of the neighboring networks in the mobile for a certain amount of time needs additional processing in the mobile to store these IP addresses. In addition, it also uses up temporary IP addresses from the neighboring networks by keeping these addresses in the cache.
3. There is a cost (e.g., CPU and power) associated with setting up additional transient tunnels (proactive tunnels) between the mobile and the target routers in the neighboring networks and tearing them down once the handover is over. There is also a cost (e.g., bandwidth) associated with maintaining these tunnels during the handover.
4. The binding update with multiple IP addresses obtained from the neighboring networks results in multiple transient data streams between the CN and mobile over these transient tunnels. This results in additional bandwidth usage because of duplicate data streams.

We now describe several levels of such proactive operations. The resource requirements will also vary based on the level of proactive operations used.

The very basic level of proactive operations involves authenticating the mobile with multiple authentication agents in the neighboring networks, but preconfiguration and binding update take place only after the layer 2 movement to a specific network is complete.

In the second level of proactive operations, the mobile can also complete its preconfiguration while in the previous network, but it can perform the binding update after it has moved to the new network. As in the previous case, here the mobile also does not need to set up proactive tunnels, since the binding update is done after the mobile has moved to the new network.

The third level of proactive operations involves completion of all three processes of authentication, configuration, and binding update while the mobile is in the previous network. However, this specific type of preauthentication utilizes the greatest amount of resources.

When only preauthentication and preconfiguration are done ahead of time with multiple networks, the mobile sends only one binding update to the CN or the home agent after it has moved to the new network.

In the case of a proactive binding update, a binding update with multiple contact addresses is sent. These contact addresses are the care-of addresses obtained ahead of time from the neighboring networks during the preconfiguration phase. The following is an illustration of this specific case when the mobile sends multiple binding updates while in the previous network but ends up moving to a specific target network. The MN sends a binding update to the CH with multiple potential care-of addresses such as $c1$, $c2$, and $c3$, which were obtained from three neighboring networks. This allows the CN to send transient multiple streams of data to the mobile over the preestablished tunnels. After the mobile has moved to the target network, it sends another binding update to the CN with the care-of address of the mobile in the network that the mobile has moved into. This way, the CN stops sending media to other neighboring networks that it does not end up moving to. One of the issues with multiple streams is the consumption of extra bandwidth for a small period of time.

Alternatively, one can apply a buffering technique at the neighboring target access routers or at the home agent. Transient data can be forwarded to the mobile after it has moved in. The forwarding of data can be triggered by the mobile either as part of Mobile IP registration or by a separate buffering protocol.

We have experimentally verified multiple preauthentication techniques using location-assisted handoff (Dutta et al., 2007a). In this specific experiment, we created multiple possible target networks. Based on the relative location of the mobile with respect to several access points in the target networks, the mobile completed proactive authentication with one or more target networks. We experimented with two different levels of preauthentication processes as described above. In one case, the mobile performed multiple authentication, but sent a binding update to one access network after setting up a proactive tunnel with the target network. In the second case, the mobile completed preauthentication with all three neighboring networks and sent binding updates over proactive tunnels.

We now discuss some guidelines for roaming clients that use preauthentication mechanisms to reduce handoff delay. These guidelines can help to determine the extent of the preauthentication operation that is needed based on the specific type of movement of the client. As discussed in Chapter 5, IEEE 802.11i and 802.11r take advantage of a preauthentication mechanism in layer 2. Thus, many of the guidelines observed for 802.11i-based preauthentication and 802.11r-based fast roaming could also be applicable to clients that use MPA-based preauthentication techniques. However, since MPA operations are not limited to movement within a specific subnet and involve intersubnet and interdomain handover, these guidelines need to take into account other factors such as the movement pattern of the mobile and the cell size.

The time needed to complete the preauthentication mechanism is an important parameter, since the mobile node needs to determine how far ahead of time the mobile needs to start the preauthentication process so that it can finish the desired operations before the handover to the target network starts. The time needed to complete the preauthentication operations will vary depending upon the speed of the mobile (e.g., whether the mobile is moving at a vehicular or pedestrian speed) and on the cell size (e.g., Wi-Fi or cellular). The cell residence time is defined as the average time the mobile stays in a cell before the next handoff takes place. This time is dependent upon the coverage area and the velocity of the mobile. Thus, the cell residence time is an important factor in determining the desirable preauthentication time that should be considered for a mobile.

Since the preauthentication operation involves six suboperations as described earlier in the chapter, and each suboperation takes some discrete amount of time, it is possible that only part of these suboperations may be completed before the handoff, depending upon the available delay budget. For example, the mobile could complete only the network discovery and network layer authentication processes before the handoff and postpone the rest of the operations until after the handover is complete. On the other hand, if the mobile is in a slow-moving vehicle and the adjacent cells are widely spaced, the mobile could complete all the desired MPA-related operations. Finishing all the MPA-related operations ahead of time reduces the handoff delay but adds other constraints such as cell residence time.

We give a numerical example here of a preauthentication process. The variables are as follows:

- D = coverage diameter;
- v = velocity of the mobile;
- RTT = round trip time from AP to AAA server, including the processing time for authentication, T_{auth};
- T_{psk} = time spent in installing keys proactively on the target APs.

Given the values $D = 100$ ft, $T_{psk} = 10$ ms, and $RTT = 100$ ms, if the mobile needs to do only the preauthentication procedure associated with MPA, then the following can be calculated for a successful MPA procedure before the handoff is complete:

$$2RTT + T_{psk} < D/v, v = 100 \text{ ft}/(200 \text{ ms} + 10 \text{ ms}) \approx 500 \text{ ft/s}.$$

Similarly, for a similar cell size, if the mobile is involved in both preauthentication and preconfiguration operations as part of the MPA procedure, and it takes an amount of time $T_{config} = 190$ ms to complete the layer 3 configuration, including IP address configuration, then for a successful MPA operation,

$$2RTT + T_{psk} + T_{config} < D/v, v = 100 \text{ ft}/(200 \text{ ms} + 10 \text{ ms} + 190 \text{ ms}) \approx 250 \text{ ft}/s.$$

Thus, compared with only the preauthentication part of the MPA operation, in order to be able to complete both the preauthentication and the preconfiguration operations successfully, either the mobile needs to move at a slower pace or it needs to expedite these operations for this given cell size. Thus, the extent of the MPA operations will be constrained by the velocity of the mobile.

Alternatively, if the mobile does complete all of the preauthentication procedure much ahead of time, it uses up resources accordingly by way of reserving IP addresses from neighboring networks, using resources for tunnel setup, and using additional bandwidth to carry the preauthentication-related signaling. Thus, during the preauthentication procedure, there is always a trade-off between the performance benefit (e.g., low delay or less packet loss) and systems resources. This is also largely governed by the network characteristics, the cell size, and the speed of movement of the mobile.

11.12   Concluding Remarks

The systems evaluation of some indicative handoff systems presented in this chapter demonstrates how many of the handoff components can work together using the proposed proactive, reactive, and cross-layer optimization techniques. For example, we have built the Media Independent Preauthentication handoff system, which uses several of the proposed proactive optimization techniques, namely prehandoff triggers, proactive network discovery, network-layer-assisted layer 2 preauthentication, proactive layer 3 configuration, and proactive binding update, and uses a dynamic buffering mechanism to reduce the handoff delay and minimize packet loss. Based on the available systems resources and the required handoff performance, a network architect for an enterprise or wireless service provider can pick and choose from a set of optimization techniques (e.g., proactive, reactive, or cross-layer) for different handoff components and build a customized mobility system. In some cases, the network architect may have the freedom to change the system parameters, but in other cases may not be able to change the system parameters or may not be able to install new network elements. Thus, based on the performance requirements and specific systems limitations, a service provider can choose the specific optimization technique that is optimal for their desired operation. The proposed mobility models can be used to design optimized handoff systems where the systems performance can be predicted ahead of time before service providers deploy them on a wide scale.

12

Conclusions

In this chapter, we discuss some of the general principles of mobility optimization, summary of contributions, and possible future work in the area of handoff optimization.

12.1 General Principles of Mobility Optimization

Currently, there are many mobility protocols available, each with its own strengths and weaknesses. Each of these mobility protocols has, historically, evolved its own optimization techniques without regard for any generic framework. Thus, it is desirable to have a set of guidelines for protocol developers, mobile-network designers, and architects who plan to use these mobility protocols and associated optimization techniques based on their usage requirements and applicability. In this chapter, we summarize the fundamental components and factors that drive this systems optimization and describe the fundamental principles of systems optimization for mobility management. Some of these are protocol design methodologies, and some are guidelines for any service provider or enterprise that might wish to deploy these mobility protocols and relevant optimization techniques.

- Since the current mobility protocols and associated optimization techniques are ad hoc in nature, it is useful to have a systematic analysis of the mobility event when designing appropriate optimization techniques.
- Since mobility involves various layers of the protocol stack, it is important to discover the type of mobility that a mobile will be subject to, such as layer 2, layer 3, or application layer mobility. The type of mobility is determined by the mobile node's mobility pattern, such as cell handoff, subnet handoff, or domain handoff, the type of application supported on the mobile node, and the type of access network.
- Since layer 2 handoff optimization techniques are access-dependent, it is important to consider the access characteristics of each network, such as the channel access algorithm (e.g., CSMA/CA, OFDM, or TDMA). For example, a CDMA network will have different access characteristics from an 802.11 network. The amount of resources used (e.g., channel bandwidth) will vary with the type of access network.

Mobility Protocols and Handover Optimization: Design, Evaluation and Application, First Edition.
Ashutosh Dutta and Henning Schulzrinne.
© 2014 John Wiley & Sons, Ltd. Published 2014 by John Wiley & Sons, Ltd.

- Each mobility event (e.g., handoff) can be considered to consist of a set of primitive functions, such as discovery, configuration, authentication, security association, registration, binding update, and media delivery. Optimizations of these primitive functions can take place independently of each other but often benefit from cross-layer triggers.
- A mobility event can be considered as a discrete-event dynamic system, where each of the abstract functions can be considered as a specific discrete event. Optimizing each of the discrete events can contribute to the overall optimization of the system.
- The scheduling of the primitive functions that are part of these handoff events plays an important role in the overall systems behavior, including systems performance and resource usage.
- The scheduling of the handoff primitives needs to take account of the data dependency among the abstract operations. The data dependency will determine the extent of parallelism that is possible during the handoff operations.
- Deadlocks need to be avoided during any mobility operation. Deadlocks are typically caused by a lack of data from previous primitive operations or a lack of resources needed for an operation. Thus, the scheduling of the primitive events should ensure that there are enough resources available for parallel or speculative operations of any kind and that data is available.
- It is important to consider the type of transport (e.g., RTP or TCP) supported by an application running on the mobile when it is subjected to handoff, as each of these applications has different performance requirements in terms of packet loss, delay, and jitter.
- Since there are several mobility protocols available and each of these protocols is suitable for a specific type of application (e.g, RTP- or TCP-based transport) and a specific type of handoff (e.g., layer 2, layer 3, or interdomain handoff), a policy-based mobility management scheme can be appropriate in many cases.
- Since the primitive handoff operations in each layer take place independently of the operations in other layers, cross-layer triggers from lower layers can help to expedite the handoff operations in the upper layers. Thus, any optimization framework needs to apply some of the available cross-layer optimization techniques. IEEE 802.21 has defined a Media Independent Handover Function that provides cross-layer triggers to expedite a handover.
- It is always useful to have a handoff model that can predict the systems performance based on the schedule and the available systems resources. When the systems parameters and resource availability are varied, the performance of the system will also vary. Service providers can use such a handoff model to determine what types of protocol and optimization techniques are needed in a specific scenario.
- The scheduling of handoff primitives is largely determined by the systems resources and the data dependency among the events. Since the scheduling of handoff primitives affects the systems performance, it can be changed to meet performance requirements at the cost of added systems resources.
- The scheduling of handoff operations can also affect the trade-off between the resources expended (e.g., battery, CPU, and bandwidth) and systems performance (e.g., delay and packet loss). Thus, the types of optimization that should be used are largely determined by the extent of the trade-off that can be allowed against resources.
- In the case of multi-interface mobility, a make-before-break mechanism helps to reduce the delay and packet loss at the cost of additional resources,[1] since both of the interfaces remain active during handoff. The extent of overlap of the operations is determined by the amount of resources that can be expended during handoff.

---

[1] Several types of resources are defined in Appendix B.

- Proactive operations appear to be more attractive for providing the desired handoff performance (e.g., delay and packet loss) compared with sequential and parallel operations. However, there is a trade-off between the amount of resources and the performance when there are multiple target networks, since the mobile needs to complete proactive handoff-related operations with multiple target networks to increase the probability of a successful handover.

12.2 Summary of Contributions

This book contributes to the general theory of optimized handover. Some of the key contributions include identification of the basic properties of the mobility event, formulation of a mobility systems model, the design of optimization techniques based on some fundamental design principles for optimization, and evaluation of the associated optimization techniques by means of analysis, model-based simulation, and experiments. These contributions can be summarized in the following three main areas.

First, this book has addressed the need for a formal systems model that can characterize a mobility event and the associated mobility optimization methodologies. This provides a systematic and formal approach to the mobility event that works independently of the type of mobility protocol. After a thorough analysis of the abstract operations associated with several mobility protocols, we determined that these basic operations form a set of discrete events that can be modeled as a discrete-event dynamic system. We have used deterministic timed-transition Petri nets to model the mobility event and analyzed its behavioral properties and systems performance. This model has the ability to predict the systems performance based on the availability of systems resources and the mobility pattern. Analysis of this model and the optimization methodologies will help to define a set of principles and guidelines for designing any new mobility protocol, as well as to evaluate the effectiveness of any specific mobility protocol in a deployment scenario.

We have developed a Petri net model that can analyze behavioral properties such as deadlocks and validate the systems performance of any type of handoff optimization supporting intratechnology, intertechnology, simultaneous, and multilayer mobility. This model can also perform a trade-off analysis between the handoff performance of these optimization techniques and systems resources. The model-based approach provides the ability to define various handoff schedules under resource constraints and can determine the extent of the parallelism and proactive operations that are possible among the handoff components.

Second, we have developed a series of optimization techniques (e.g., reactive, proactive, and parallel) for different handoff components and have carried out extensive experiments to validate these optimization techniques. We have applied these optimization techniques to different mobility scenarios, such as simultaneous mobility, multilayer mobility, multicast mobility, and multi-interface mobility, and compared the results with the nonoptimized version. This series of experiments has provided a systematic methodology that can be carried out in a repetitive manner and be applied to optimize different handoff components.

Third, we have described a hierarchical scope-based multicast architecture to support multicast streaming using proxies in the access network. This proposed architecture introduces a novel local advertisement insertion technique, and program management between local and global programs. We have developed a few optimization techniques to support fast handoff for multicast traffic in a hierarchical scope-based environment. These techniques are based on proxy-based proactive triggering and application layer triggering to expedite multicast stream delivery by reducing the join latency.

12.3   Future Work

This book has laid the foundations for a systematic approach to mobility events that can be analyzed with a formal model using Petri nets. However, this model-based analysis of the mobility event could be enhanced to make it more useful to the wireless and mobility deployment communities. The following is a list of future work items that we believe could be pursued beyond what is covered in this book:

1. Although we have used this model to validate a few mobility optimization techniques, the model could be enhanced to study the behavioral properties and systems performance of any type of mobility protocol, such as transport layer protocols and mobility in other types of networks such as ad hoc networks.
2. Using the current model, we were able to study and detect behavioral properties such as system deadlocks, investigate the anomalies of specific schedules, and then compare various schedules, such as proactive, reactive, and concurrent schedules. The model could be enhanced so that one could use it in an automated fashion to generate a specific schedule for the handoff operations given a set of resource constraints and performance objectives, and a dependency graph. Automatic generation of schedules for handoff operations to provide the desired quality of service with the available resources will help one to use the right set of protocols.
3. Using a systematic analysis of the mobility functions, one can design a customized mobility protocol suitable for one's own set of requirements. Currently, any mobility event depends upon a set of protocols, each performing its own desired functions (e.g., DHCP for IP address acquisition and server discovery). However, each of these protocols adds additional overhead when used individually in a decoupled manner. Using our model, one can design a comprehensive mobility protocol that will define its own set of protocols for each of the desired functions instead of using the existing ones.
4. The current Petri net model has been used to look at the resource parameters of the mobile only. This model could be enhanced to predict performance based on the resource parameters of all of the network elements that are involved in the mobility event. Other system elements may include the layer 2 point of attachment (e.g., an access point), the layer 3 point of attachment (e.g., a router), and servers in the network. Distributed resource metrics would be useful for any wireless service providers that might wish to have an optimized service deployment.
5. The formalization of key techniques, the models of systems dependencies, and the ability to calculate or predict optimization metrics provide a foundation for the automated discovery and implementation of mobility optimization. We envision the specification of the functional components of mobility protocols as defined in Chapter 3 and then having tools that search for application- or context-specific optimizations, such as caching, proactive, or cross-layer techniques.

A

RDF Schema for Application Layer Discovery

In this appendix, we illustrate a sample RDF schema (Figure A.1) that can be used for information discovery with IEEE 802.21. For the sake of brevity, we provide only a subset of the schema. We show examples that include a combination of basic and extended sets of classes and their associated properties. For example, a network class will have properties of type L2 and L3, and an L2 class will have properties such as network-id, operator, location, and neighbor information.

A.1 Schema Primitives

In Figure A.2, we present sample primitives in ASN.1 format that can be transported as part of an RDF schema.

Mobility Protocols and Handover Optimization: Design, Evaluation and Application, First Edition.
Ashutosh Dutta and Henning Schulzrinne.
© 2014 John Wiley & Sons, Ltd. Published 2014 by John Wiley & Sons, Ltd.

```
<?xml version="1.0"?>
<!DOCTYPE rdf:RDF [
 <!ENTITY rdf 'http://www.w3.org/1999/02/22-rdf-syntax-ns#'>
 <!ENTITY rdfs 'http://www.w3.org/2000/01/rdf-schema#'>
 <!ENTITY mihbase 'URL_TO_BE_ASSIGNED'>
 ]>

<rdf:RDF xmlns:rdf="&rdf;" xmlns:rdfs="&rdfs;"
xmlns:mihbase="&mihbase;"
 xml:base="&mihbase;">


<rdfs:Class rdf:ID="Network">
 <rdfs:subClassOf rdf:resource="&rdfs;Resource"/>
<rdfs:comment>

The network class has two properties, namely l2 for layer 2 information
and l3 for higher-layer information. Any property can be added to this
class in an extended schema.
 <rdfs:comment/>
 </rdfs:Class>
<rdf:Property rdf:ID="l2">
 <rdfs:domain rdf:resource="#Network"/>
 <rdfs:range rdf:resource="#L2"/>
 <rdfs:comment>
This property is of type L2 class.
 <rdfs:comment/>
 </rdf:Property>


<rdf:Property rdf:ID="l3">
 <rdfs:domain rdf:resource="#Network"/>
 <rdfs:range rdf:resource="#L3"/>
 <rdfs:comment>
This property is of type L3 class.
 <rdfs:comment/>
 </rdf:Property>


<rdfs:Class rdf:ID="L2">
 <rdfs:subClassOf rdf:resource="&rdfs;Resource"/>
 <rdfs:comment>
The L2 class has properties that are specific to the link layer. These
properties include network-id, operator, location, and
neighbor-information properties. Any property can be added to this
class in an extended schema.
 <rdfs:comment/>
 </rdfs:Class>
<rdf:Property rdf:ID="operator">
 <rdfs:domain rdf:resource="#L2"/>
 <rdfs:range rdf:resource="&rdfs;Literal"/>
 <rdfs:comment>
This property contains a name of the operator. It could be the same as
the network-id property.
 <rdfs:comment/>
 </rdf:Property>
<rdf:Property rdf:ID="network-id">
 <rdfs:domain rdf:resource="#L2"/>
 <rdfs:range rdf:resource="&rdfs;Literal"/>
 <rdfs:comment>
This property contains an identifier of the network. It may contain an
SSID.
<rdfs:comment/>
 </rdf:Property>
 </rdf:RDF>
```

Figure A.1   Sample RDF schema for information service

```
Network ::= ENUMERATED{L2info, L3info, Location}
L2info ::= ENUMERATED {802.11, 802.16, GSM, GPRS, W-CDMA, cdma2000}
L3info ::= ENUMERATED {IPv4, IPv6}
Location ::= SEQUENCE {
                        Geo-location ::= String
                        Civic-addr ::= String
                      }
802.11 ::= SEQUENCE {
                      Standards ::= BITMAP{802.11a, 802.11b, 802.11g}
                      SSID_Network_Name ::= String(SIZE(1..32))
                      BSSID ::= NumericString(SIZE(6))
                      Channel ::= INTEGER
                      Phy ::= ENUMERATED{CCK, DSSS, OFDM}
                      Data_Rates ::= INTEGER
                      Network_Service_Provider_Code ::= String
                      Network_Service_Provider_Name ::= String
                      Network_Service_Provider_Tariff ::= String
                      Cipher Suites ::= BITMAP {WEP, TKIP, AES-CCMP}
           Authenticated_Key_Management_Suites::=BITMAP{WEP,Psk,802.1x}
              KeyManagementProtocol::=ENUMERATED{11i4WayHandshake}
           Quality_of_Service ::= ENUMERATED {802.11e}
           Cost ::= INTEGER
           Roaming_List ::= String
           Mobility ::= ENUMERATED {802.11r, 802.11u, 802.21, PreAuth}
                    }
IPv4 ::= SEQUENCE{
              Router_Address ::= String
                    DHCP_Server_Address ::= String
                    DomainName ::= String
                    Subnet ::= String
                    SIP_Server_Address ::= String
       KeyManagementProtocol ::= ENUMERATED{IKEv1, IKEv2}
                    Authentication ::= ENUMERATEED{PANA, UAM}
                    PacketCiphering ::= ENUMERATED{IPsec}
                    Internet_Service_Provider_Code ::= String
                    Internet_Service_Provider_Name ::= String
                    Internet_Service_Provider_Tariff ::= ???
                    Mobility ::= ENUMERATED {MIPv4, CT, CARD,
         Preauth}
Quality_of_Service ::= ENUMERATED{...}
                    VPN_Gateway_Address ::= String
                    NAT_Address ::= String
                    }
MIPv4 ::= SEQUENCE{
                    HomeAgent_Address ::= String
                    ForeignAgent_Address ::= String }
PANA ::= SEQUENCE{
                    PAA_Address ::= String
                    EP_Address ::= String }
```

Figure A.2 RDF schema of ASN.1 primitives

B

Definitions of Mobility-Related Terms

It is useful to define some mobility-related terms that are often used in the context of handover and optimization. In this appendix, we introduce a few of the relevant terms, including some from RFC 3753 (Manner and Kojo 2004) and ITU-T X.200 (ITU-T 2004), that are useful for handover optimization and have been used in various chapters of this book.

Definition B.1 A *mobile node* (MN) is a node that is capable of changing its point of attachment to the network across layers, namely layer 2 or layer 3.

A mobile node may be either a mobile host (with no forwarding functionality) or a mobile router (with forwarding functionality). A mobile node can have multiple interfaces.

Definition B.2 An *interface identifier* is a unique identifier that is assigned to a specific interface of a node, by which it can be addressed for data transfer.

An interface identifier can be a layer 2 or layer 3 identifier depending upon the specific configuration. For example, a MAC address is defined to be a layer 2 identifier, whereas an IP address is a layer 3 identifier.

Definition B.3 A *device identifier* is a unique identifier that is fixed and permanent and is assigned to a device during its manufacture.

A typical device identifier could be the MAC address of an IP-based device or the electronic serial number (ESN) of a cellular device. If multiple interfaces are assigned to a device, each interface will have a unique MAC address.

Definition B.4 A *network point of attachment* (PoA) is defined as the remote endpoint of the link that connects a mobile node to the network associated with a unique link layer or layer 3 identifier. These can be defined as the L2 PoA or L3 PoA, respectively.

Mobility Protocols and Handover Optimization: Design, Evaluation and Application, First Edition.
Ashutosh Dutta and Henning Schulzrinne.
© 2014 John Wiley & Sons, Ltd. Published 2014 by John Wiley & Sons, Ltd.

A point of attachment can be a layer 2 or layer 3 point of attachment. For example, an 802.11 access point or CDMA base station can be defined as a layer 2 point of attachment, whereas a router is considered to be a layer 3 point of attachment. The old point of attachment (oPoA) is defined as the point of attachment of the mobile node prior to the link switch. The new point of attachment (nPoA) is the new point of attachment of the mobile node that results from the link switch event.

Definition B.5 *Path identification* defines the unique connection path between a mobile and the point of attachment to the network.

A path identifier is defined as a pair of interface identifiers, where one of them is associated with the mobile and the other is associated with the point of attachment to the network.

Definition B.6 *Terminal identification* defines the unique communication interface associated with a fixed or mobile terminal that can be used to establish a communication between two end points and maintain it.

A terminal identifier can be the MAC address or IP address of a network interface associated with any terminal or IMEI (International Mobile Equipment Identity) of a mobile end point. In some cases, a single terminal may have multiple network interfaces, leading to primary and secondary terminal identification to take care of failover or load-balancing features.

Definition B.7 *User identification* defines the uniqueness of a user, irrespective of the devices the user is using or one's attachment point or location in a mobility environment.

A user identifier can be in the form of the URI (Unique Resource Identifier) most associated with the user's home domain or home PLMN. A dynamic mapping of a user's URI to the terminal identifier of the user's device enables the establishment of a connection between two users and assists in maintaining connectivity as users move from network to network.

Definition B.8 A *radio cell* is a geographical area within which an access point provides radio coverage, that is, where radio communication between a mobile node and the specific access point is possible.

A mobile node uses a radio cell to communicate with the point of attachment to the network.

Definition B.9 A *subnet* is a logical group of connected network nodes, where two hosts can communicate through a layer 2 connection and do not require a layer 3 entity.

In IP networks, nodes in a subnet share a common network mask (in IPv4) or a network prefix (in IPv6). All hosts within a subnet can be reached in one hop, implying that all hosts in a subnet are connected to the same logical layer 3 link.

Definition B.10 An *access router* resides on the edge of an access network and is connected to one or more access points.

An access router offers IP connectivity to mobile nodes, acting as a default router to the mobile nodes it is currently serving. An access router may include functionality beyond the simple forwarding service offered by ordinary IP routers, such as buffering, tunneling, and caching.

Definition B.11 An *access network* (AN) is an IP network which includes one or more access network routers.

An access network consists of multiple subnets that can be connected to the same access router or multiple routers.

Definition B.12 An *administrative domain* is a collection of networks under the same administrative control and grouped together for administrative purposes.

Definition B.13 The *serving access router* (SAR) is the access router that currently offers connectivity to a mobile node.

This is usually the point of departure for the mobile node as it makes its way towards a new access router (at which time the serving access router takes over the role of the previous access router). There may be several serving access routers serving a mobile node at the same time.

Definition B.14 The *next access router* (NAR) is the the access router that offers connectivity to a mobile node after a handover.

Definition B.15 The *previous access router* (PAR) is the access router that offered connectivity to a mobile node prior to a handover. This is the serving access router that will cease or has ceased to offer connectivity to the mobile node.

Definition B.16 A *candidate access router* (CAR) is an access router to which a mobile node may do a handoff.

Definition B.17 The *anchor MSC* is the MSC from which a handover has been initiated.

During a handover in a GSM network, the anchor MSC stays in the data path to help reduce the packet loss in the case of handoff forward and handoff backward.

Definition B.18 The *target MSC* is the MSC towards which a handover should take place.

The target MSC directs the base station to assign a channel to a mobile that a user intends to move to that base station's radio coverage area.

Definition B.19 *Handover* is the process by which an active mobile node changes its point of attachment to the network, or by which such a change is attempted.

The access network may provide features to minimize the interruptions in progress; this is termed "optimized handover." The term "handover" is often used interchangeably with "handoff."

Definition B.20 *Layer 2 handover* is a type of handover where the MN changes between access points (or some other aspect of the radio channel) connected to the same access router's interface.

This type of handover is transparent to the routing in the IP layer.

Definition B.21 *Layer 3 handover* is a type of handover where the mobile node changes between access points (or some other aspect of the radio channel) connected to different subnetwork interfaces.

The subnetwork interface may be connected to two different interfaces on the same router or two different routers altogether. During this type of handover, the mobile node moves to a new subnet.

Definition B.22 *Roaming* involves formal agreements between operators that allow a mobile to obtain connectivity from a foreign network.

Roaming includes the functionality by which users can send their identity to the local access network so that inter-access-network agreements can be activated and services and applications in the MN's home network can be made available to the user in the visited network.

Definition B.23 *Systems resources* are resources on the mobile host that are shared by different elementary operations during a handoff event.

These resources could include mobile-node resources such as CPU cycles and battery power, and network resources such as the number of bytes that can be transmitted per unit time on a specific access channel (e.g., per-user bandwidth).

Definition B.24 *Network resource parameters* are the parameters that are needed to perform the various handoff-related operations.

These resource parameters could include wireless parameters, such as channel number, frequency, authentication algorithm, and authentication server. A mobile node uses these parameters to carry out handoff-related operations.

Definition B.25 A *care-of address* (CoA) is an IP address associated with a mobile node while visiting a foreign link; the subnet prefix of this IP address is a foreign subnet prefix.

A packet addressed to a mobile node which arrives at the mobile node's home network when the mobile node is away from home and has registered a care-of address will be forwarded to that address by the home agent in the home network.

Definition B.26 A *home address* (HoA) is an IP address assigned to a mobile node, used as the permanent address of the mobile node.

This address is within the mobile node's home link. Standard IP routing mechanisms will deliver packets destined for a mobile node's home address to its home link.

Definition B.27 *Binding* can be defined as an association where N entities support connectionless mode by maintaining a binding with the appropriate N SAPs to deliver connectionless data to $(N + 1)$ entities (ITU-T X.200).

For example, in the case of mobility, a binding can be an association between a temporary care-of address obtained in the visited node and the permanent home address of the mobile node so that the data from the correspondent node can be routed to the mobile node.

Definition B.28 *Encapsulation* is the process of adding control information as a header to any protocol data frame; this is often for routing purposes. All of the control overhead and data of that protocol is considered as data after the encapsulation.

The IP–IP encapsulation often used in Mobile IP and the ESP encapsulation used in IPSec are examples of encapsulation.

Definition B.29 A *home agent* is a router on a mobile node's home link with which the mobile node has registered its current care-of address.

When a mobile node is away from home, the home agent intercepts packets on the home link destined for the mobile node's home address, encapsulates them, and tunnels them to the mobile node's registered care-of address.

Definition B.30 A *foreign agent* is a node on a mobile node's visited network that intercepts the traffic from the home agent and delivers it to the mobile node connected to the same visited node.

When a mobile node is connected to a visited network, it can use a foreign agent as the care-of address so that packets destined for the mobile node can be captured by the foreign agent and delivered to the mobile node in the visited network.

Definition B.31 An *encapsulation agent* encapsulates the data and sends it to a decapsulation agent, which removes the header and delivers the data to the target node.

A home agent is an encapsulation agent that encapsulates data traffic using source and destination IP headers and sends it to the foreign agent that delivers the packet to the target node.

Definition B.32 A *decapsulation agent* communicates with an encapsulation agent and decapsulates the data by stripping the header.

A foreign agent is a type of decapsulation agent that decapsulates an IP–IP tunnel and delivers to the mobile host.

Definition B.33 A *binding update* (BU) message indicates a mobile node's current mobility binding, and in particular its care-of address.

A mobile can send its binding update either directly to the other communicating node or to the home agent to associate its home address with the care-of address.

Definition B.34 *Tunneling* is a process of setting up a point-to-point virtual link between two end points by adding an encapsulation header to the data so that it can be routed properly over a tunnel.

A home agent uses a tunneling mechanism to send data from the home network to a foreign agent or to a mobile node in a visited network.

Definition B.35 A *bidirectional tunnel* is a tunnel between two communicating nodes where the data can be sent over the tunnel in either direction.

In the case of bidirectional tunnel mode operation for Mobile IP, both the home agent and the foreign agent act like encapsulation and decapsulation agents.

Definition B.36 *Horizontal handover* involves mobile nodes moving between access points of the same type (in terms of coverage, data rate, and mobility), such as UMTS to UMTS or 802.11 to 802.11.

A horizontal handover is also called an intratechnology handover.

Definition B.37 *Vertical handover* involves mobile nodes moving between access points of different type, such as UMTS to 802.11 or vice versa.

A vertical handover is also called an intertechnology handover.

Definition B.38 *Mobile-initiated handover* requires that the mobile node is the one that makes the initial decision to initiate the handover.

Definition B.39 *Network-initiated handover* requires that the network makes the initial decision to initiate the handover.

Definition B.40 *Handover latency* is the difference between the time a mobile node was last able to send and/or receive an IP packet by way of the PAR (previous access router) and the time the mobile node is able to send and/or receive an IP packet through the NAR (next access router).

Definition B.41 *Smooth handover* is a type of handover where there is no session interruption when the mobile is in transition from one base station to another.

Smooth handover is aimed primarily at minimizing packet loss, with no explicit concern for additional delays in packet forwarding.

Definition B.42 *Fast handover* is aimed primarily at minimizing handover latency, with no explicit interest in packet loss.

However, minimizing handover latency typically results in reduction of packet loss.

Definition B.43 *Macromobility* is defined as occurring when a mobile moves between two networks that belong to two different subnets. These subnets may belong to the same administrative domain or different ones.

Layer 3 mobility support and associated address registration procedures are needed when a mobile node is subject to macromobility. Inter-access-network handovers typically involve macromobility protocols. Mobile IP can be seen as a means to provide a macromobility solution.

Definition B.44 *Micromobility* usually means movement of a terminal within a mobility domain, where a mobility domain may be confined to a subnet or collection of subnets within an administrative domain.

Micromobility protocols exploit the locality of movement by confining movement-related changes and signaling to the access network without the need to interact with Mobile IP. Some well-known examples of IP micromobility architectures include HAWAII (Handoff-Aware Wireless Access Internet Infrastructure), Cellular IP, and HMIP (Hierarchical Mobile IP).

Definition B.45 During a *make-before-break handover,* the mobile node prepares the new connection path before the old one is broken.

Thus, if a mobile has multiple interfaces, it can communicate with both the old AR and the new AR at the same time using either of the interfaces. However, only one interface is used for transmitting data, while the other is engaged in handover preparation. Alternatively, many of the handover-related operations for the second interface can be performed by the first interface. When the mobile has a single interface, a virtual interface is used to enable make-before-break handoff.

Definition B.46 During a *break-before-make handover,* the mobile node breaks the existing connection before the new connection is made.

In a break-before-make handover, there is an appreciable amount of handover delay, as the second interface comes up only after the first interface is disconnected. If the handover involves a single interface, the same interface is configured again in the new network.

Definition B.47 *Intradomain handover* refers to a handover scenario where a mobile node's movement is confined to a domain.

A "domain" can be defined as an administrative domain, a DNS-based domain, or a mobility domain anchored by a mobility agent. Several types of domain are defined by 1989.

Definition B.48 *Interdomain handover* refers to a handover scenario where a mobile node moves between two domains.

These two domains can be two different administrative domains, each with its own anchor mobility agent.

Definition B.49 *Route optimization* is a process of minimizing the data path between communicating nodes by having a direct path for data forwarding.

When a mobile node changes its network identifier, data is rerouted to a new point of attachment and may take a longer route. Route optimization helps to maintain a direct path between the communicating nodes.

Definition B.50 *Network-assisted mobile and network-controlled* (NAMONC) handoff allows a mobile node to be involved in an anticipated IP-layer handoff procedure.

The mobile node is therefore assisted by the network in performing an anticipated L3 handoff before it completes the L2 handoff.

Definition B.51 *Network-initiated mobile-terminated* (NIMOT) handoff allows the network to initiate a handoff and register proactively on behalf of the mobile.

This handoff method enables a rapid establishment of service at the new point of attachment so that the effect of the handoff on real-time applications is minimized.

Definition B.52 *Network-controlled handoff* (NCHO) involves the network in handling the necessary RSS measurements and handoff decisions.

NCHO is used in first-generation cellular systems such as Advanced Mobile Phone System (AMPS), where the mobile telephone switching office (MTSO) is responsible for the overall handoff decision.

Definition B.53 *Mobile-assisted handoff* (MAHO) involves the mobile in making RSS measurements and sending them periodically to the BS.

MAHO is used in the Global System for Mobile Communications (GSM), where, based on the received measurements, the BS or the MSC decides when to hand off.

Definition B.54 *Mobile-controlled handoff* (MCHO) extends the role of the MS by giving overall control to it.

Both the MS and the BS make the necessary measurements and the BS sends them to the MS. Then the MS decides when to hand off, based on the information gained from the BS and itself.

Definition B.55 During a *connected state*, a mobile is in the process of actively receiving data from a correspondent node.

After all the handoff operations are complete, the mobile node goes into a connected mode.

Definition B.56 During a *disconnected state*, a mobile does not receive any data from a correspondent node.

A mobile node is said to be in disconnected mode when it is in the process of handing off to a new point of attachment in a network.

Definition B.57 *Join latency* is defined as the elapsed time between a host joining a group and the router sending a multicast packet towards the mobile.

A mobile node can send an unsolicited join request to the router to trigger a multicast flow, failing which it can wait to respond to the router's query.

Definition B.58 *Leave latency* is defined as the time between the moment the last host leaves a group and when the routing protocol is notified that there are no more multicast members.

To reduce the leave latency, the last mobile node to leave the group sends a group leave message to all router multicast addresses so that the router prunes the multicast tree. In the absence of a specific group leave report from the mobile node, for example in the case of IGMPv1 (Internet Group Management Protocol), the router needs to wait until it has not received a response from any of the clients.

References

3GPP (2005) 3GPP System to Wireless Local Area Network (WLAN) Interworking: Systems Description, TS23.234 v6.10.0 2005.

3GPP (2008) Feasibility Study on Multimedia Session Continuity, 3GPP TR23893.

3GPP (2009) Architecture Enhancements for Non-3GPP Accesses, TS23.402 v9.5.0.

Aboba B and Simon D (1999) PPP EAP TLS Authentication Protocol. RFC 2716, Internet Engineering Task Force.

Aboba B, Blunk L, Vollbrecht J, Carlson J, and Levkowetz E (2004) Extensible Authentication Protocol (EAP). RFC 3748, Internet Engineering Task Force.

Aboba B, Carlson J, and Cheshire S (2006a) Detecting Network Attachment in IPv4 (DNAv4). RFC 4436, Internet Engineering Task Force.

Aboba B, Simon D, Eronen P, and Levkowetz H (2006b) Extensible Authentication Protocol (EAP) Key Management Framework. IETF Draft (work in progress).

Acharya A, Bakre A, and Nath B (1995) IP Multicast Extensions for Mobile Internetworking. Technical Report LCSR-TR-243, Department of Computer Science, Rutgers University, New Brunswick, NJ.

Almeroth K (2000) A long-term analysis of growth and usage patterns in the multicast backbone (MBone). Proceedings of the Conference on Computer Communications (IEEE Infocom), Tel Aviv, Israel.

Almeroth K and Ammar M (1997) Multicast group behavior in the Internet's multicast backbone (MBone). *IEEE Communications Magazine* **35**(6), 124–129.

Almesberger W, Boudec JY, and Oechslin P (1997) Application REQuested IP over ATM (AREQUIPA). RFC 2170, Internet Engineering Task Force.

Amadio R and Prasad S (1998) Modelling IP mobility, in *CONCUR'98: Concurrency Theory: 9th International Conference* (eds D Sangiorgi and R de Simone). Springer, pp. 301–316.

Antoniadis P, Courcoubetis C, and Strulo B (2005) Incentives for content availability in memory-less peer-to-peer file sharing systems. *SIGecom Exchange* **5**(4), 11–20.

Arabshian K and Schulzrinne H (2004) Gloserv: Global service discovery architecture. First Annual International Conference on Mobile and Ubiquitous Systems: Networking and Services (Mobiquitous).

Arkko J, Aboba B, Kohonen J, and Bari F (2008) Network Discovery and Selection Problem. RFC 5113, Internet Engineering Task Force.

Baker M, Zhao X, Cheshire S, and Stone J (1996) Supporting mobility in MosquitoNet. Proceedings of the 1996 USENIX Technical Conference, pp. 127–140.

Ballardie T, Francis P, and Crowcroft J (1993) Core based trees (CBT), in *SIGCOMM Symposium on Communications Architectures and Protocols* (ed. D Sidhu). ACM, San Francisco, pp. 85–95. Also in *Computer Communication Review* **23**(4), Oct. 1992.

Bargh MS, Hulsebosch B, Eertink H, Prasad A, Wang H and Schoo P (2004) Fast authentication methods for handovers between IEEE 802.11 wireless LANs. Second ACM International Workshop on Wireless Mobile Applications and Services on WLAN Hotspots, Philadelphia, PA,.

Baset S and Schulzrinne H (2006) An analysis of the Skype peer-to-peer Internet telephony protocol, in *IEEE INFOCOM 2006 – IEEE International Conference on Computer Communications*. IEEE, pp. 2695–2705.

Bauer M, Schefczik P, Soellner M, and Speltacker W (2003) Evolution of the UTRAN architecture, in *4th International Conference on 3G Mobile Communication Technologies, 2003 (3G 2003)*, IEE Conf. Publ. No. 494. IEE, pp. 244–248.

Baugher M, McGrew D, Naslund M, Carrara E, and Norrman K (2004) The Secure Real-Time Transport Protocol (SRTP). RFC 3711, Internet Engineering Task Force.

Belhe U and Kusiak A (1993) Performance analysis of design process using timed Petri nets. *Concurrent Engineering* 1(3), 147.

Berners-Lee T, Fielding R, and Masinter L (1998) Uniform Resource Identifiers (URI): Generic Syntax. RFC 2396, Internet Engineering Task Force.

Boussinot F and de Simone R (1991) The ESTEREL language. *Proceedings of the IEEE* 79(9), 1293–1304.

Brik V, Mishra A, and Banerjee S (2005) Eliminating handoff latencies in 802.11 WLANs using multiple radios: Applications, experience, and evaluation. ACM SIGCOMM IMC, Berkeley, CA.

Buragohain C, Agrawal D, and Suri S (2003) A game theoretic framework for incentives in P2P systems, in *Proceedings of the Third International Conference on Peer-to-Peer Computing (P2P2003)*. IEEE Computer Society, Washington, DC, p. 48.

Buttyán L, Holczer T, and Schaffer P (2005) Spontaneous cooperation in multi-domain sensor networks, in *Security and Privacy in Ad-hoc and Sensor Networks, Second European Workshop, ESAS 2005, Visegrad, Hungary, July 13–14, 2005: Revised Selected Papers* (eds R Molva, G. Tsudik, and D Westhoff), Lecture Notes in Computer Science 3813. Springer, pp. 42–53.

Cabellos-Aparicio A, Núñez-Martínez J, Julian-Bertomeu H, Jakab L, Serral-Gracià R, and Domingo-Pascual J (2005) Evaluation of the fast handover implementation for mobile IPv6 in a real testbed, in *Operations and Management in IP-Based Networks* (eds P Dini, S Jürgen, T Magedanz, and ERM Madeira). Springer, pp. 181–190.

Cain B, Deering S, Kouvelas I, Fenner B, and Thyagarajan A (2002) Internet Group Management Protocol, Version 3. RFC 3376, Internet Engineering Task Force.

Calhoun P, Hiller T, Kempf J, McCann P, Pairla C, Singh A, and Thalanany S (2000) Foreign Agent Assisted Hand-off. Internet-Draft, Internet Engineering Task Force. Work in progress.

Calhoun P, Loughney J, Guttman E, Zorn G, and Arkko J (2003a) Diameter Base Protocol. RFC 3588, Internet Engineering Task Force.

Calhoun P, Montenegro G, and Perkins C (2003b) Mobile IPv4 Regional Registration. Internet-Draft draft-ietf-mobileip-reg-tunnel-08, Internet Engineering Task Force. Work in progress.

Camarillo G, Eriksson G, Holler J, and Schulzrinne H (2002) Grouping of Media Lines in the Session Description Protocol (SDP). RFC 3388, Internet Engineering Task Force.

Campbell AT, Gomez J, Kim SH, Valko AG, and Wan CY (2000) Design, implementation, and evaluation of cellular IP. *IEEE Personal Communications* 7(4), 42–49.

Cao X and Ho Y (1990) Models of discrete event dynamic systems. *IEEE Control Systems Magazine* 10(4), 69–76.

Carli M, Neri A, and Picci AR (2001) Mobile IP and cellular IP integration for inter access networks hand-off, in *Conference Record of the International Conference on Communications (ICC), Helsinki*. IEEE, pp. 2467–2471.

Carpenter B (2000) Internet Transparency. RFC 2775, Internet Engineering Task Force.

Carson M and Santay D (2003) NIST Net: A Linux-based network emulation tool. *ACM SIGCOMM Computer Communication Review* 33(3), 111–126.

Chandra R, Bahl V, and Bahl P (2004) MultiNet: Connecting to multiple IEEE 802.11 networks using a single wireless card. Proceedings of the IEEE INFOCOM Conference.

Chandra R, Padmanabhan VN, and Zhang M (2006) WiFiProfiler: Cooperative diagnosis in wireless LANs, in *MobiSys* 2006, pp. 205–219. ACM Press, New York.

Chandra R, Padhye J, Ravindranath L, and Wolman A (2007) Beacon-stuffing: Wi-Fi without associations. Eighth IEEE Workshop on Mobile Computing Systems and Applications, 2007 (Hot Mobile 2007), pp. 53–57.

Chen Y and Mary Q (2003) Soft Handover Issues in Radio Resource Management for 3G WCDMA Networks. Department of Electronic Engineering, Queen Mary University of London.

Cheshire S, Aboba B, and Guttman E (2005) Dynamic Configuration of IPv4 Link-Local Addresses. RFC 3927, Internet Engineering Task Force.

Chiba T, Dutta A, and Schulzrinne H (2007) Trombone routing mitigation techniques for IMS/MMD networks. Proceedings of IEEE WCNC, Hong Kong.

Chiba T, Yokota H, Dutta A, Chee D, and Schulzrinne H (2008) Route optimization for Proxy Mobile IPv6 in IMS network. Proceedings of the 2008 International Conference on Signal Processing and Communication Systems.

Choi J (2005) Goals of Detecting Network Attachment in IPv6. RFC 4135, Internet Engineering Task Force.

Clancy T and Tschofenig H (2009) Extensible Authentication Protocol – Generalized Pre-Shared Key (EAP-GPSK) Method. RFC 5433, Internet Engineering Task Force.

Clancy T, Nakhjiri M, Narayanan V, and Dondeti L (2008) Handover Key Management and Re-Authentication Problem Statement. RFC 5169 (Informational), Internet Engineering Task Force.

Coffman E, Elphick M, and Shoshani A (1971) System deadlocks. *Computing Surveys* **3**(2), 67–78.

Combs G et al. (2004) Ethereal: A network protocol analyzer. Software on line, http://www.wireshark.org (accessed November 12, 2013).

Conta A and Deering S (1998) Generic Packet Tunneling in IPv6 Specification. RFC 2473, Internet Engineering Task Force.

Cruz Filho F, Maciel P, Barros E, and de Informatica C (2000) Using Petri Nets for Data Dependency Analysis, in *Proceedings of IEEE International Conference on Systems, Man, and Cybernetics, Nashville, TN*, vol. 4. IEEE, pp. 2998–3003.

Cuervo F, Greene N, Rayhan A, Huitema C, Rosen B, and Segers J (2000) MEGACO Protocol Version 1.0. RFC 3015, Internet Engineering Task Force.

Cuninghame-Green R (1991) Minimax algebra and applications. *Fuzzy Sets and Systems* **41**(3), 251–267.

Das S, Misra A, Agrawal P, and Das S (2000) TeleMIP: Telecommunications-enhanced mobile IP architecture for fast intradomain mobility. *IEEE Personal Communications* **7**(4), 50–58.

Das S, Dutta A, McAuley A, Misra A, and Das S (2002) IDMP: An intradomain mobility management protocol for next-generation wireless networks. *IEEE Wireless Communications* **9**(3), 38.

Das S, Tauil M, Cheng Y, Dutta A, Baker D, Yajnik M, Famolari D, et al. (2009) IEEE 802.21: Media independent handover: Features, applicability, and realization. *IEEE Communications Magazine*, p. 113.

Day M, Rosenberg J, Sugano H (2000) A Model for Presence and Instant Messaging. RFC 2778, Internet Engineering Task Force.

de Laat C, Gross G, Gommans L, Vollbrecht J, and Spence D (2000) Generic AAA Architecture. RFC 2903, Internet Engineering Task Force.

Deering S and Hinden R (1998) Internet Protocol, Version 6 (IPv6) Specification. RFC 2460, Internet Engineering Task Force.

Deering SE, Estrin D, Farinacci D, Jacobson V, Liu CM, and Wei L (1994) An architecture for wide-area multicast routing. SIGCOMM Symposium on Communications Architectures and Protocols, London, pp. 126–135.

del Prado Pavon J and Choi S (2003) Link adaptation strategy for IEEE 802.11 WLAN via received signal strength measurement, in *IEEE International Conference on Communications*. IEEE, pp. 1108–1113.

Devarapalli V, Wakikawa R, Petrescu A, and Thubert P (2005) Network Mobility (NEMO) Basic Support Protocol. RFC 3963, Internet Engineering Task Force.

Dierks T and Allen C (1999) The TLS Protocol Version 1.0. RFC 2246, Internet Engineering Task Force.

Dreibholz T, Jungmaier A, and Tuxen M (2003) A new scheme for IP-based Internet-mobility, in *Conference on Local Computer Networks*, vol. 28. IEEE, pp. 99–108.

Droms R (1997) Dynamic Host Configuration Protocol. RFC 2131, Internet Engineering Task Force.

Droms R (1999) Procedure for Defining New DHCP Options. RFC 2489, Internet Engineering Task Force.

Droms R (ed.), Bound J, Volz B, Lemon T, Perkins C, and Carney M (2003) Dynamic Host Configuration Protocol for IPv6 (DHCPv6). RFC 3315, Internet Engineering Task Force.

Duong H, Dadej A, and Gordon S (2004) Proactive context transfer in WLAN-based access networks, in *Proceedings of the 2nd ACM International Workshop on Wireless Mobile Applications and Services on WLAN Hotspots*. ACM, New York, pp. 61–70.

Dutta A and Schulzrinne H (2001) A streaming architecture for next generation Internet, in *Conference Record of the International Conference on Communications (ICC), Helsinki*. IEEE, p. 7,

Dutta A and Schulzrinne H (2004) MarconiNet: overlay mobile content distribution network. *IEEE Communications Magazine* **42**(2), 64–75.

Dutta A, Vakil F, Chen JC, Tauil M, Baba S, and Schulzrinne H (2001) Application layer mobility management scheme for wireless Internet. 3G Wireless, San Francisco, p. 7.

Dutta A, Altintas O, Chen W, and Schulzrinne H (2002a) Mobility approaches for all IP wireless networks. SCI, Orlando, FL.

Dutta A, Burns J, Wong K, Jain R, Young K, Schulzrinne H, and McAuley A (2002b). Realization of integrated mobility management protocol for ad-hoc networks. MILCOM, vol. 1, pp. 448–454.

Dutta A, Wong D, Burns J, Jain R, Young K, McAuley A, and Schulzrinne H (2002c) Realization of integrated mobility management for ad-hoc networks. IEEE Milcom, Anaheim, CA.

Dutta A, Chennikara J, Chen W, Altintas O, and Schulzrinne H (2003a) Multicasting streaming media to mobile users. *IEEE Communications Magazine* **41**(10), 81–89.

Dutta A, Madhani S, Chen W, Altintas O, and Schulzrinne H (2003b) MobiCom poster: Optimized fast-handoff schemes for application layer mobility management. *ACM SIGMOBILE Mobile Computing and Communications Review* **7**(1), 17–19.

Dutta A, Agrawal P, Das S, Elaoud M, Famolari D, Madhani S, McAuley A, Li P, Tauil M, and Schulzrinne H (2004a) Realizing mobile wireless Internet telephony and streaming multimedia testbed. *Computer Communications* **27**(8), 725–738.

Dutta A, Das S, Li P, McAuley A, Ohba Y, Baba S, and Schulzrinne H (2004b) Secured mobile multimedia communication for wireless internet. International Conference on Network Sensing and Control, Taipei, Taiwan.

Dutta A, Madhani S, Chen W, Altintas, O and Schulzrinne H (2004c) Fast-handoff schemes for application layer mobility management. 15th IEEE International Symposium on Personal, Indoor and Mobile Radio Communications, vol. 3.

Dutta A, Zhang T, Madhani S, Taniuchi K, Fujimoto K, Schulzrinne H, Ohba Y, and Katsube Y (2004d) Secure universal mobility for wireless internet. Second ACM International Workshop on Wireless Mobile Applications and Services on WLAN Hotspots, Philadelphia, PA.

Dutta A, Burns J, Jain R, Wong D, Young K, and Schulzrinne H (2005a) Implementation and performance evaluation of application layer MIP-LR. 2005 International Conference on Wireless Networks, Communications and Mobile Computing, Maui, HI, vol. 2.

Dutta A, Das S, Famolari D, Ohba Y, Taniuchi K, Kodama T, and Shulzrinne H (2005b) Seamless handover across heterogeneous networks – an IEEE 802.21 centric approach. Proceedings of IWS-WPMC.

Dutta A, Kim B, Zhang T, Technologies T, Baba S, Taniuchi K, Ohba Y, and Schulzrinne H (2005c) Experimental analysis of multi interface mobility management with SIP and MIP. IEEE Wireless Conference, Maui, HI.

Dutta A, Zhang T, Madhani S, Taniuchi K, Fujimoto K, Katsube Y, Ohba Y, and Schulzrinne H (2005d) Secure universal mobility for wireless internet. *ACM SIGMOBILE Mobile Computing and Communications Review* **9**(3), 45–57.

Dutta A, Zhang T, Ohba Y, Taniuchi K, and Schulzrinne H (2005e) MPA assisted proactive handoff scheme. ACM Mobiquitous, SIGMOBILE, pp. 155–165.

Dutta A, Das S, Famolari D, Ohba Y, Taniuchi K, Fajardo V, Kodama T, and Schulzrinne H (2006a) Secured seamless convergence across heterogeneous access networks. IEEE World Telecommunication Congress, Budapest.

Dutta A, Madhani S, Chen W, and Schulzrinne H (2006b) GPS assisted fast-handoff mechanism for real-time communication. IEEE Sarnoff Symposium, Princeton, NJ.

Dutta A, Madhani S, Zhang T, Ohba Y, Taniuchi K, and Schulzrinne H (2006c) Network discovery mechanisms for fast-handoff. Third International Conference on BROADNETS, San Jose, CA.

Dutta A, Schulzrinne H, Chiba T, Yokota H, and Das S (2006d) Comparative analysis of network layer and application layer IP mobility protocols for IPv6 networks. Proceedings of WPMC 2006, San Diego, CA.

Dutta A, van den Berg E, Famolari D, Fajardo V, Ohba Y, Taniuchi K, and Schulzrinne H (2006e) Dynamic buffering scheme for mobile handoff. IEEE PIMRC, Helsinki.

Dutta A, Chakravarty S, Taniuchi K, Fajardo V, Ohba Y, Famolari D, and Schulzrinne H (2007a) An experimental study of location assisted proactive handover. IEEE GLOBECOM, Washington, DC.

Dutta A, Das S, Famolari D, Ohba Y, and Schulzrinne H (2007b) Seamless proactive handover across heterogeneous access networks. *Wireless Personal Communication* **43**(3), 837–855.

Dutta A, Lyles B, Schulzrinne H, Chiba T, Yokota H, and Idoue A (2007c) Generalized modeling framework for handoff analysis. Annual IEEE International Symposium on Personal, Indoor and Mobile Radio Communications (PIMRC'07), Athens.

Dutta A, Manousakis K, Das S, and Lin F (2007d) Mobility testbed for 3GPP2-based multimedia domain networks. *IEEE Communications Magazine* **45**(7), 118–126.

Dutta A, Famolari D, Das S, Ohba Y, Fajardo V, Taniuchi K, Lopez R, and Schulzrinne H (2008) Media-independent pre-authentication supporting secure interdomain handover optimization. *IEEE Wireless Communications* **15**(2), 55–64.

Dutta A, Lyles B, Schulzrinne H, and Wang J (2009) Systems modeling for IP-based handoff using timed Petri nets. IEEE International Conference on Systems Sciences (HICSS 2009), Big Island, HI.

Dutta A, Ohba Y, Fajardo V, Taniuchi K, and Schulzrinne H (2010) A Framework of Media-Independent Pre-Authentication (MPA). Internet-Draft draft-irtf-mobopts-mpa-framework-07, Internet Engineering Task Force. Work in progress.

Dutta A, Fajardo V, Ohba Y, Taniuchi K, and Schulzrinne H (2011) A Framework of Media-Independent Pre-Authentication (MPA) for Inter-Domain Handover Optimization. RFC 6252 (Informational), Internet Engineering Task Force.

Egevang K and Francis P (1994) The IP Network Address Translator (NAT). RFC 1631, Internet Engineering Task Force.

Eriksson H (1994) MBone: The multicast backbone. *Communications of the ACM*, pp. 54–60.

Ernst T (2007) Network Mobility Support Goals and Requirements. RFC 4886, Internet Engineering Task Force.

Eronen ES (2006) IKEv2 Mobility and Multihoming Protocol. RFC 4555, Internet Engineering Task Force.

Farinacci D, Tweedly A, and Speakman T (1996/1997) Cisco Group Management Protocol (CGMP), ftp://ftpeng.cisco.com/ipmulticast/specs/cgmp.txt.

Fathi H, Kobara K, Chakraborty S, Imai H, and Prasad R (2005) On the impact of security on the latency in WLAN 802.11. IEEE Global Telecommunications Conference, 2005 (GLOBECOM'05), St Louis, MO.

Feeney L and Nilsson M (2001) Investigating the energy consumption of a wireless network interface in an ad hoc networking environment, in *Proceedings of IEEE INFOCOM*, vol. 3. IEEE, pp. 1548–1557.

Feldman M, Lai K, Stoica I, and Chuang J (2004) Robust incentive techniques for peer-to-peer networks, in EC'04. ACM Press, New York, pp. 102–111.

Fenner W (1997) Internet Group Management Protocol, Version 2. RFC 2236, Internet Engineering Task Force.

Ferguson DA (1994) Measurement of mundane TV behaviors: Remote control device flipping frequency. *Journal of Broadcasting and Electronic Media* **38**, 35–47.

Finlayson R (2003) The UDP Multicast Tunneling Protocol. Internet-Draft draft-finlayson-umtp-09, Internet Engineering Task Force. Work in progress.

Flament M, Gessler F, Lagergren F, Queseth O, Stridh R, Unbehaun M, Wu JLC, and Zander J (1999) An approach to 4th generation wireless infrastructures, scenarios and research issues. VTC, Houston, TX.

Floyd R (1962) Algorithm 97: Shortest path. *Communications of the ACM* **5**(6), 345.

Fogelstroem E, Jonsson A, and Perkins C (2007) Mobile IPv4 Regional Registration. RFC 4857, Internet Engineering Task Force.

Forsberg D, Ohba Y, Patil B, Tschofenig H, and Yegin A (2008) Protocol for Carrying Authentication for Network Access (PANA). RFC 5191, Internet Engineering Task Force.

Forte A and Schulzrinne H (2007) Cooperation between stations in wireless networks, in *IEEE International Conference on Network Protocols 2007 (ICNP 2007)*, Beijing. IEEE, pp. 31–40.

Forte A, Shin S, and Schulzrinne H (2006a) Passive Duplicate Address Detection for Dynamic Host Configuration Protocol (DHCP). Technical Report cucs-011-06, Computer Science Department, Columbia University.

Forte AG, Shin S, and Schulzrinne H (2006b) IEEE 802.11 in the Large: Observations at an IETF Meeting. Technical report, Columbia University.

Forte AG, Shin S, and Schulzrinne H (2006c) Improving L3 handoff delay in IEEE 802.11 wireless networks. WICON '06, ACM.

FreeRADIUS Project (2006) FreeRADIUS.

Fretzagias C and Papadopouli M (2004) Cooperative location-sensing for wireless networks, in *PERCOM '04*. IEEE Computer Society, Washington, DC, p. 121.

Gast M (2005) 802.11 Wireless Networks: The Definitive Guide. O'Reilly Media, Inc.

Geier J (2002) Understanding 802.11 frame types. *Wi-Fi Planet*, http://www.wi-fiplanet.com/tutorials/article.php/1447501 (accessed November 12, 2013).

Georgides M (2004) Context transfer support for IP-based mobility management. CCSR Awards for Research Excellence.

Greis M (2001) 3GPP TS23. 107 v5. 0.0. Quality of Service, Concept and Architecture. Report, 3GPP.

Gundavelli S, Leung K, Devarapali V, Chowdhury K, and Patil B (2008) Proxy Mobile IPv6. RFC 5213, Internet Engineering Task Force.

Guttman E, Perkins C, Veizades J, and Day M (1999) Service Location Protocol, Version 2. RFC 2608, Internet Engineering Task Force.

Gwon Y, Fu G, and Jain R (2003) Fast handoffs in wireless LAN networks using mobile initiated tunneling handoff protocol for IPv4 MITHv4, in *IEEE Wireless Communications and Networking*. IEEE, pp. 1248–1253.

Han Y et al. (2003) Advance Duplicate Address Detection. Internet-Draft, Internet Engineering Task Force. Work in progress.

Handley M (1996) The SDR session directory: An MBone conference scheduling and booking system, http://www-mice.cs.ucl.ac.uk/multimedia/software/sdr/ (accessed November 12, 2013).

Handley M and Jacobson V (1998) SDP: Session Description Protocol. RFC 2327, Internet Engineering Task Force.

Handley M, Perkins C, and Whelan E (2000) Session Announcement Protocol. RFC 2974, Internet Engineering Task Force.

Hansen T, Hardie T, and Masinter L (2006) Guidelines and Registration Procedures for New URI Schemes. RFC 4395, Internet Engineering Task Force.

Hares S and Katz D (1989) Administrative Domains and Routing Domains: A Model for Routing in the Internet. RFC 1136, Internet Engineering Task Force.

Harkins D and Carrel D (1998) The Internet Key Exchange (IKE). RFC 2409, Internet Engineering Task Force.

Hempstead M, Tripathi N, Mauro P, Wei G, and Brooks D (2005) An ultra low power system architecture for sensor network applications. *ACM SIGARCH Computer Architecture News* **33**(2), 208–219.

Henderson T, Kotz D, and Abyzov I (2004) The changing usage of a mature campus-wide wireless network, in *MobiCom '04*. ACM Press, New York, pp. 187–201.

Herrin G (2000) Linux IP networking: A guide to the implementation and modification of the Linux protocol stack. *Connections* **3**, 2.

Hirel C, Tun B, and Trivedi K (2000) SPNP version 6.0. 11th International Conference on Computer Performance Evaluation: Modelling Tools and Techniques (TOOLS), pp. 354–357.

Holbrook H and Cain B (2006) Source-Specific Multicast for IP. RFC 4607, Internet Engineering Task Force.

Holliday M and Vernon M (1987) A generalized timed Petri net model for performance analysis. *IEEE Transactions on Software Engineering* **13**, 12.

Hormozi A and Dube L (1999) Establishing project control: Schedule, cost, and quality. *SAM Advanced Management Journal* **64**(4), 32–38.

Housely R and Aboba B (2007) Guidance for Authentication, Authorization, and Accounting (AAA) Key Management. RFC 4962, Internet Engineering Task Force.

Hsieh P, Dutta A, and Schulzrinne H (2003a) Application Layer Mobility Proxy for real-time communication. 3G Wireless Conference.

Hsieh R, Zhou ZG, and Seneviratne A (2003b) S-MIP: A seamless handoff architecture for Mobile IP. Proceedings of the IEEE INFOCOM conference.

Hunter TE and Nosratinia A (2004) Diversity through coded cooperation *IEEE Transactions on Wireless Communications* **5**, 283–289.

IEEE (1999) IEEE Standard for Information Technology – Telecommunications and Information Exchange between Systems – Local and Metropolitan Area Networks – Specific Requirements – Part 11: Wireless LAN Medium Access Control (MAC) and Physical Layer (PHY) Specifications, 802.11-1999.

IEEE (2003a) IEEE Recommended Practice for Multi-vendor Access Point Interoperability via an Inter-Access Point Protocol Across Distribution Systems Supporting IEEE 802.11 Operation.

IEEE (2003b) IEEE Draft Amendment to Standard for Information Technology – Telecommunications and Information Exchange between Systems – Specification for Radio Resource Measurement.

IEEE (2004) Amendment to IEEE Std 802.11, 1999 Edition (Reaff 2003). Medium Access Control (MAC) Security Enhancements.

IEEE (2005a) IEEE Standard for Information Technology, Telecommunications and Information Exchange between Systems, Amendment 8: Medium Access Control (MAC) Quality of Service Enhancements.

IEEE (2005b) IEEE Draft Standard for Local and Metropolitan Area Networks: Media Independent Handover Services.

IEEE (2006a) IEEE Draft Amendment to Standard for Information Technology – Telecommunications and Information Exchange between Systems – LAN/MAN Amendment 2: Fast BSS Transition.

IEEE (2006b) 802.1X – Port Based Network Access Control, http://www.ieee802.org/1/pages/802.1x.html (accessed November 11, 2013).

IEEE Computer Society LAN MAN Standards Committee et al. (1997) Wireless LAN Medium Access Control (MAC) and Physical Layer (PHY) Specifications. IEEE.

Internet System Consortium (2005) DHCP Client Version 3.

ITU-T (2004) Open Systems Interconnection – Model and Notation. Recommendation X.200, ITU-T.

Jaimes-Romero F, Munoz-Rodriguez D, Molina C, and Tawfik H (1997) Modeling resource management in cellular systems using Petri nets. *IEEE Transactions on Vehicular Technology* **46**(2), 298–312.

Jain A (2003) Handoff Delay for 802.11b Wireless LAN. Technical report, University of Kentucky.

Jain R, Raleigh T, Graff C, Bereschinsky M, and Patel M (1998) Mobile Internet access and QoS guarantees using mobile IP and RSVP with location registers, in *IEEE International Conference on Communications*, Atlanta, GA, vol. 3. IEEE, pp. 1690–1695.

Jain R, Raleigh T, Yang D, Chang LF, Graff CJ, Bereschinsky M, and Patel M (1999) Enhancing survivability of mobile Internet access using mobile IP with location registers. Proceedings of the Conference on Computer Communications (IEEE Infocom), New York.

Jardosh AP, Ramachandran KN, Almeroth KC, and Belding-Royer EM (2005) Understanding link-layer behavior in highly congested IEEE 802.11b wireless networks, in E-WIND '05. ACM Press, New York, pp. 11–16.

Jayaraman P, Lopez R, and Ohba Y (2008) Protocol for Carrying Authentication for Network Access (PANA) Framework. RFC 5193, Internet Engineering Task Force.

Jiang W and Schulzrinne H (2000) Modeling of packet loss and delay and their effect on real-time multimedia service quality. Proceedings of NOSSDAV.

Johner H (1998) Understanding LDAP. IBM Corporation.

Johnson D, Perkins C, and Arkko J (2004) Mobility Support in IPv6. RFC 3775, Internet Engineering Task Force.

Johnson D, Perkins C, and Arkko J (2009) Mobility Support in IPv6. Internet-Draft draft-ietf -mext-rfc3775bis-05.txt, Internet Engineering Task Force. Work in progress.

Jones C, Sivalingam K, Agrawal P, and Chen J 2001 A survey of energy efficient network protocols for wireless networks. *Wireless Networks* **7**(4), 343–358.

Kaufman C (ed.) (2005) Internet Key Exchange (IKEv2) Protocol. RFC 4306, Internet Engineering Task Force.

Kaur S, Madan B, and Ganeshan S (1999) Multicast support for mobile IP using modified IGMP, in *IEEE WCNC*. IEEE, pp. 948–952.

Kempf J (2007) Goals for Network Based Localized Mobility Management. RFC 4830, Internet Engineering Task Force.

Kent S and Atkinson R (1998a) IP Encapsulating Security Payload (ESP). RFC 2406, Internet Engineering Task Force.

Kent S and Atkinson R (1998b) Security Architecture for the Internet Protocol. RFC 2401, Internet Engineering Task Force.

Kent S and Seo K (2005) Security Architecture for the Internet Protocol. RFC 4301, Internet Engineering Task Force.

Kershaw M et al. (2005) Kismet: 802.11 Layer 2 Wireless Network Sniffer, http://www.kismetwireless.net (accessed November 10, 2013).

Khalaf-Bitar R and Rubin I (2009) Throughput-capacity and bit-per-joule performance of IEEE 802.11 based wireless mesh networks. IEEE Ad Hoc Networking Workshop, pp. 34–41.

Khalil M, Akhtar H, Qaddoura E, Perkins C, and Cerpa A 1999 Buffer Management for Mobile IP. Internet-Draft draft-mkhalil-mobileip-buffer-00.txt, Internet Engineering Task Force.

Kim B and Han K (2004) Multicast handoff agent mechanism for all-IP mobile network. *Mobile Networks and Applications* **9**(3), 185–191.

Kim K, Ha J, Hyun E, and Kim S (2001) New approach for mobile multicast based on SSM, in *Proceedings of Ninth IEEE International Conference on Networks*. IEEE, pp. 405–408.

Kim P, Lee M, Park S, and Kim Y (2004) A new mechanism for SIP over Mobile IPv6, in *Computational Science and Its Applications (ICCSA 2004)* (eds A Laganá, ML Gavrilova, V Kumar, Y Mun, CJK Tan, and O Gervasi), Lecture Notes in Computer Science 3045. Springer, pp. 975–984.

Klas G and Lepold R (1992) TOMSPIN – A tool for modeling with stochastic Petri nets. Proceedings of Computer Systems and Software Engineering (CompEuro'92), pp. 618–623.

Koh S et al. (2003) Use of SCTP for Seamless Handover. Internet-Draft, Internet Engineering Task Force. Work in progress.

Koodli R (2005) Fast handovers for mobile IPv6. RFC 4068, Internet Engineering Task Force.

Koodli R (2008) Mobile IPv6 Fast Handovers. RFC 5268, Internet Engineering Task Force. Obsoleted by RFC 5568.

Kosugi K and Davies J (1973) Basic Joseki. Ishi Press, Tokyo.

Koutsonikolas D, Das SM, Hu YC, Lu YH, and Lee CG (2006) CoCoA: Coordinated Cooperative Localization for Mobile Multi-Robot Ad Hoc Networks. *ICDCSW* **0**, 9.

Kravets R and Krishnan P (2000) Application-driven power management for mobile communication. *Wireless Networks* **6**(4), 263–277.

Kravets R, Carter C, and Magalhaes L (2001) A cooperative approach to user mobility. *ACM SIGCOMM Computer Communication Review* **31**(5), 57–69.

Krishnamurthi G, Chalmers R, and Perkins C (2001) Buffer Management for Smooth HandOvers in Mobile IPv6. Internet-Draft, Internet Engineering Task Force. Work in progress.

Kwon H, Jung KR, Park A, and Ryou JC (2005) Consideration of UMTS–WLAN seamless handover, in *Seventh IEEE International Symposium on Multimedia*. IEEE, p. 6.

Lassila O, Swick R, et al. (1999) Resource Description Framework (RDF) Model and Syntax Specification. World Wide Web Consortium (W3C).

Lawler E (2001) *Combinatorial Optimization: Networks and Matroids*. Dover Publications.

Lee H, Lee S, and Cho D (2003) Mobility management based on the integration of mobile IP and session initiation protocol in next generation mobile data networks. 58th IEEE Vehicular Technology Conference, 2003 (VTC 2003-Fall), vol. 3.

Leech M, Ganis M, Lee Y, Kuris R, Koblas D, and Jones L (1996) SOCKS Protocol Version 5. RFC 1928, Internet Engineering Task Force.

Lemon T and Sommerfield B (2006) Node-specific Client Identifiers for Dynamic Host Configuration Protocol Version Four (DHCPv4). RFC 4361, Internet Engineering Task Force.

Lennox J and Schulzrinne H (1999) Transporting User Control Information in SIP REGISTER Payloads. Internet-Draft, Internet Engineering Task Force. Work in progress.

Leung K, Dommety G, Narayanan V, and Petrescu A (2008) Network Mobility (NEMO) Extensions for Mobile IPv4. RFC 5177, Internet Engineering Task Force. Updated by RFC 6626.

Levon J and Elie P (2005) Oprofile: A System Profiler for Linux, http://oprofile.sourceforge.net (accessed November 12, 2013).

Liebsch M, Singh A, Chaskar H, and Funato D (2005) Candidate Access Router Discovery (CARD). RFC 4066, Internet Engineering Task Force.

Lin C and Wang KM (2000) Mobile multicast support in IP networks. Proceedings of the Conference on Computer Communications (IEEE Infocom), Tel Aviv, Israel.

Liu P, Tao Z, and Panwar S (2005) A cooperative MAC protocol for wireless local area networks, in *ICC '05*. IEEE Computer Society, pp. 2962–2968.

Lopez R, Dutta A, Ohba Y, and Schulzrinne H (2007) Network-layer assisted mechanism to optimize authentication delay during handoff in 802.11 networks. ACM Mobiquitous, Philadelphia, PA.

Loughney J, Nakhjiri M, Perkins C, and Koodli R (2005) Context Transfer Protocol. RFC 4067, Internet Engineering Task Force.

Maciel P, Cruz Filho F, and Barros E (2001) A Petri net based method for resource estimation: an approach considering data-dependency, causal and temporal precedences. 14th Symposium on Integrated Circuits and Systems Design, 2001, Mirenopolis, Brazil, pp. 78–84.

MADWiFi Driver (2013) http://wireless.kernel.org/en/users/Drivers/madwifi (accessed November 12, 2013).

Malinen J (2004) HostAP driver for Intersil Prism2/2.5/3.

Malinen J (2005a) HostAP: Wireless driver for Intersil Prism2/2.5/3.

Malinen J (2005b) Linux WPA/WPA2/IEEE 802.1x supplicant.

Malki K (2004) Low Latency Handoffs in Mobile IPv4. Internet-Draft draft-ietf-mobileip-lowlatency-handoffs -v4-08, Internet Engineering Task Force. Work in progress.

Malki KE (2007) Low-Latency Handoffs in Mobile IPv4. RFC 4881, Internet Engineering Task Force.

Maltz DA and Bhagwat P (1998) MSOCKS: An architecture for transport layer mobility, in *Proceedings of the Conference on Computer Communications (IEEE Infocom), San Francisco*. IEEE, p. 1037.

Manner J and Kojo M (2004) Mobility Related Terminology. RFC 3753, Internet Engineering Task Force.

Marsan M, Meo M, Gribaudo M, and Sereno M (2001) On Petri net-based modeling paradigms for the performance analysis of wireless internet access. International Workshop on Petrinet and Performance Models, IEEE, Aachen, Germany.

Matcovschi M, Mahulea C, and Pastravanu O (2003) Petri net toolbox for MATLAB. 11th IEEE Mediterranean Conference on Control and Automation MED'03, Rhodes, Greece.

Maughan D, Schertler M, Schneider M, and Turner J (1998) Internet Security Association and Key Management Protocol (ISAKMP). RFC 2408, Internet Engineering Task Force.

McAuley A, Bommaiah E, Misra A, Talpade R, Thomson S, and Young KC (1999) Mobile multicast proxy. IEEE Milcom, Atlantic City, NJ.

McAuley A, Misra A, Wong L, and Manousakis K (2001) Experience with autoconfiguring a network with IP addresses, in MILCOM, vol. 1. IEEE, pp. 272–276.

McBride B (2002) Jena: A semantic Web toolkit. *IEEE Internet Computing* 6(6), 55–59.

McBride B (2004) The resource description framework (RDF) and its vocabulary description language RDFS, in *Handbook on Ontologies* (eds S Staab and R Studer). Springer, pp. 51–66.

McGregor G (1992) The PPP Internet Protocol Control Protocol (IPCP). RFC 1332, Internet Engineering Task Force.

McGuinness D, Van Harmelen F, et al. (2004) OWL Web Ontology Language Overview. W3C Recommendation 10, 2004–03.

Miller B and Pascoe R (2000) Salutation Service Discovery in Pervasive Computing Environments. IBM Pervasive Computing White Paper.

Mishra A, Shin M, and Arbaugh W (2003a) An empirical analysis of the IEEE 802.11 MAC layer handoff process. *ACM Computer Communication Review* **33**(2), 93–102.

Mishra A, Shin M, and Arbaugh W (2003b) An empirical analysis of the IEEE 802.11 MAC layer handoff process. *SIGCOMM Computer Communication Review* **33**(2), 93–102.

Mishra A, Shin M, Petroni Jr N, Clancy T, and Arbaugh W (2004) Proactive key distribution using neighbor graphs. *IEEE Wireless Communications* **11**(1), 26–36.

Misra A, Das S, Dutta A, McAuley A, and Das S (2002) IDMP based fast-handoff and paging in IP based 4G mobile networks. *IEEE Communications Magazine* **40**(3), 138–145.

Miu A and Bahl P (2001) Dynamic host configuration for managing mobility between public and private networks. 3rd Usenix Internet Technical Symposium, San Francisco, CA.

Molina-Ramirez C, Munoz-Rodriguez D, and Lopez-Mellado E (1994) Modelling and Analysis of Telecommunication Cellular Systems Using Petri Nets. Technical report, IEEE.

Montavont J, Montavont N, and Noel T (2005) Enhanced schemes for L2 handover in IEEE 802.11 networks and their evaluations. 16th International Symposium on Personal Indoor and Mobile Radio Communications, Berlin.

Montenegro G (ed.) (2001) Reverse Tunneling for Mobile IP, revised. RFC 3024, Internet Engineering Task Force.

Moore N (2004) Edge Handovers for Mobile IPv6. Internet-Draft draft-moore-mobopts-edge-handovers-00, Internet Engineering Task Force. Work in progress.

Moore N (2006) Optimistic Duplicate Address Detection DAD for IPv6. RFC 4429, Internet Engineering Task Force.

Moskowitz R and Nikander P (2006) Host Identity Protocol (HIP) architecture. RFC 4423, Internet Engineering Task Force.

Mostafa H and Cicak P (2006) Hands on Roaming Duration: Petri-Nets Modeling of a Wireless Mobile-IP Procedure in Cisco Platform. International Conference on Networking and Services, 2006 (ICNS'06), pp. 28–28.

Moy J (1993) Multicast routing extensions for OSPF, in *International Networking Conference (INET)*. Internet Society, San Francisco, pp. BCC-1–BCC-7.

Murata T (1985) Use of resource-time product concept to derive a performance measure of timed Petri nets. Proceedings 1985 Midwest Symposium on Circuits and Systems, Louisville, KY.

Murata T (1989) Petri nets: Properties, analysis and applications. *Proceedings of IEEE*, p. 541.

Mysore J and Bhargavan V (1997) A new multicasting-based architecture for Internet host mobility, in *Third Annual ACM/IEEE International Conference on Mobile Computing and Networking, Budapest*. ACM/IEEE, pp. 161–172.

Nakajima N, Dutta A, Das S, and Schulzrinne H (2003) Handoff delay analysis and measurement for SIP based mobility in IPv6. ICC 2003 – Personal Communication Systems and Wireless LANs, Anchorage, AK.

Narayanan S (2006) Detecting Network Attachment in IPv6 Networks (DNAv6). IETF Draft, Internet Engineering Task Force. Work in progress.

Narten T, Nordmark E, and Simpson W (1998) Neighbor Discovery for IP Version 6 (IPv6). RFC 2461, Internet Engineering Task Force.

Niemi A, Arkko J, and Torvinen V (2002) Hypertext Transfer Protocol (HTTP) Digest Authentication Using Authentication and Key Agreement (AKA). RFC 3310, Internet Engineering Task Force.

nsnam (2005) nsnam Web pages, http://www.isi.edu/nsnam (accessed November 11, 2013).

O'Hara B (2004) *IEEE 802.11 Handbook: A Designer's Companion*. IEEE.

Ohba Y and Yegin A (2010) Pre-Authentication Support for the Protocol for Carrying Authentication for Network Access (PANA). RFC 5873, Internet Engineering Task Force.

Ohba Y, Wu Q, and Zorn G (2009) Extensible Authentication Protocol (EAP) Early Authentication Problem Statement. Internet-Draft draft-ietf-hokey-preauth-ps, Internet Engineering Task Force. Work in progress.

Ohba Y, Wu Q, and Zorn G (2010) Extensible Authentication Protocol (EAP) Early Authentication Problem Statement. RFC 5836, Internet Engineering Task Force.

Ote DV, Paskalis S, Kaloxylos A, and Merakos L (2003) A SIP-based method for intra-domain handoffs, in VTC-Fall '03. IEEE Computer Society, pp. 2068–2072.

Ott J, Wenger S, Sato N, Burmeister C, and Rey J (2006) Extended RTYP Profile for Real-Time Transport Control Protocol (RTCP)-Based Feedback (RTP/AVPF). RFC 4585, Internet Engineering Task Force.

Pack S and Choi Y (2002) Fast inter-AP handoff using predictive authentication scheme in a public wireless LAN, in *Networks: Proceedings of the Joint International Conference on Wireless LANs and Home Networks (ICWLHN 2002) and Networking (ICN 2002), Atlanta, USA, 26-29 August 2002*. World Scientific, p. 15.

Pack S and Choi Y (2003) Pre-authenticated fast handoff in a public wireless LAN based on IEEE 802.1 x model. Proceedings of IFIP TC6/WG6, vol. 8, pp. 175–182.

Papadopouli M and Schulzrinne H (2001) Effects of power conservation, wireless coverage and cooperation on data dissemination among mobile devices, in MobiHoc '01. ACM Press, New York, pp. 117–127.

Park J (2002) Mobile Multicast Routing Protocol: TBMOM (Timer-Based Mobile Multicast). RMT Workshop 2002.

Park S, Kim H, Park C, Kim J, and Ko S (2004) Selective channel scanning for fast handoff in wireless LAN using neighbor graph, in Personal Wireless Communications. Springer, pp. 629–629.

Park S, Kim P, and Volz B (2005) Rapid Commit Option for the Dynamic Host Configuration Protocol Version 4. RFC 4039, Internet Engineering Task Force.

Patsidou E and Kantor J (1991) Application of minimax algebra to the study of multipurpose batch plants. *Computers & Chemical Engineering* **15**(1), 35–46.

Pering T, Agarwal Y, Gupta R, and Want R (2006) Coolspots: Reducing the power consumption of wireless mobile devices with multiple radio interfaces, in *Proceedings of the 4th International Conference on Mobile Systems, Applications and Services*. ACM, p. 232.

Perkins C (1996a) IP Encapsulation within IP. RFC 2003, Internet Engineering Task Force.

Perkins C (1996b) Minimal Encapsulation within IP. RFC 2004, Internet Engineering Task Force.

Perkins C (2002a) IP Mobility Support for IPv4. RFC 3220, Internet Engineering Task Force.

Perkins C (2002b) IP Mobility Support for IPv4. RFC 3344, Internet Engineering Task Force.

Perkins C (2002c) Mobile IP. *IEEE Communications Magazine* **40**(5), 66–82.

Perkins C and Hodson O (1998) Options for Repair of Streaming Media. RFC 2354, Internet Engineering Task Force.

Perkins CE and Johnson DB (1998) Route Optimization for Mobile IP. *Cluster Computing* **1**(2), 161–176.

Perkins CE and Wang KY (1999) Optimized smooth handoffs in mobile IP, in Fourth IEEE *Symposium on Computers and Communications (ISCC '99)*. IEEE, pp. 340–346.

Peterson J (1981) *Petri Net Theory and the Modeling of Systems*. Prentice Hall, Englewood Cliffs, NJ, p. 290.

Peterson J and Jennigs C (2006) Enhancements for Authenticated Identity Management in the Session Initiation Protocol (SIP). RFC 4474, Internet Engineering Task Force.

Plummer D (1982) Ethernet Address Resolution Protocol: Or Converting Network Protocol Addresses to 48.bit Ethernet Address for Transmission on Ethernet Hardware. RFC 826, Internet Engineering Task Force.

Politis C, Chew K, Akhtar N, Georgiades M, Tafazolli R, and Dagiuklas T (2004) Hybrid multilayer mobility management with AAA context transfer capabilities for all-IP networks. *IEEE Wireless Communications* **11**(4), 76–88. [See also *IEEE Personal Communications*.]

Polk J, Schnizlein J, and Linsner M (2004) Dynamic Host Configuration Protocol Option for Coordinate-Based Location Configuration Information. RFC 3825, Internet Engineering Task Force.

Pollini G (1996) Trends in handover design. *IEEE Communications Magazine* **34**(3), 82–90.

Potlapally N, Ravi S, Raghunathan A, and Jha N (2003) Analyzing the energy consumption of security protocols, in *Proceedings of the 2003 International Symposium on Low Power Electronics and Design*. ACM, New York, pp. 30–35.

Quinn B and Almeroth K (2001) IP Multicast Applications: Challenges and Solutions. RFC 3170, Internet Engineering Task Force.

Rahnema M (1993) Overview of the GSM system and protocol architecture. *IEEE Communications Magazine* **31**(4), 92–100.

Ramamoorthy C and Ho G (1980) Performance evaluation of asynchronous concurrent systems using Petri nets. *IEEE Transactions on Software Engineering* **6**(5), 440–449.

Ramani I and Savage S (2005) Syncscan: Practical fast handoff for 802.11 infrastructure networks. Proceedings of the IEEE INFOCOM Conference.

Ramchandani C (1974) Analysis of asynchronous concurrent systems by timed Petri nets. Ph.D dissertation, Massachusetts Institute of Technology, Cambridge, MA.

Ramjee R, LaPorta TF, Salgarelli L, Thuel S, Varadhan K, and Li LE (2000) IP-based access network infrastructure for next-generation wireless networks. *IEEE Personal Communications Magazine* **7**(4), 34–41.

Rigney C, Rubens AC, Simpson WA, and Willens S (2000) Remote Authentication Dial In User Service (RADIUS). RFC 2865, Internet Engineering Task Force.

Rodriguez P, Chakravorty R, Chesterfield J, Pratt I, and Banerjee S (2004) MAR: A commuter router infrastructure for the mobile internet. Proceedings of the 2nd International ACM Conference on Mobile Systems, Applications, and Services, Boston, MA, pp. 217–230,

Romdhani I, Kellil M, Lach H, Bouabdallah A, and Bettahar H (2004) IP mobile multicast: Challenges and solutions. *IEEE Communications Surveys & Tutorials* **6**(1), 18–41.

Rosenberg J and Schulzrinne H (1999) An RTP Payload Format for Generic Forward Error Correction. RFC 2733, Internet Engineering Task Force.

Rosenberg J, Schulzrinne H, Camarillo G, Johnston A, Peterson J, Sparks R, Handley M, and Schooler E (2002) SIP: Session Initiation Protocol. RFC 3261, Internet Engineering Task Force.

Rosenberg J, Peterson J, Schulzrinne H, and Camarillo G (2004) Best Current Practices for Third Party Call Control (3PCC) in the Session Initiation Protocol (SIP). RFC 3725, Internet Engineering Task Force.

Roshan P and Leary J (2003) *802.11 Wireless LAN Fundamentals*. Cisco Press.

Ruckforth T and Linder J (2004) AAA Context Transfer for Fast Authenticated Inter-domain Handover. Swisscom SA, March.

Ryuther N (2004) A tool for PetriNet simulation. Petri Nets World, http://www.informatik.uni-hamburg.de/TGI/PetriNets (accessed November 11, 2013).

Salawu N and Elizabeth N (2009) Energy optimization mechanism for mobile terminals using vertical handoff between WLAN and CDMA2000 networks. *Leonardo Electronic Journal of Practices and Technologies* **15** 51–58.

Sanders W (1995) UltraSAN Users Manual, Version 3.0. Center for Reliable and High-Performance Computing, University of Illinois.

Sasse MA, Hardman VJ, Kouvelas I, Perkins CE, Hodson O, Watson AI, Handley M, and Crowcroft J (1995) RAT (Robust-Audio Tool). University College London.

Schosser S, Böhm K, Schmidt R, and Vogt B (2006) Incentives engineering for structured P2P systems – a feasibility demonstration using economic experiments, in EC '06. ACM Press, New York, pp. 280–289.

Schulzrinne H (1996) Personal mobility for multimedia services in the Internet, in *Interactive Distributed Multimedia Systems and Services* (eds B Butscher, E Moeller, and H Pusch), Lecture Notes in Computer Science 1045. Springer, pp. 143–162.

Schulzrinne H (2001) SIP Registration. Internet-Draft, Internet Engineering Task Force. Work in progress.

Schulzrinne H and Hancock R (2008) GIMPS: General Internet Messaging Protocol for Signaling. Internet-Draft draft-ietf-nsis-ntlp-15, Internet Engineering Task Force. Work in progress.

Schulzrinne H and Wedlund E (2000a) Application-layer mobility using SIP. *Mobile Computing and Communications Review (MC2R)* **4**(3), 47–57.

Schulzrinne H, Rao A, and Lanphier R (1998) Real Time Streaming Protocol (RTSP). RFC 2326, Internet Engineering Task Force.

Schulzrinne H, Casner S, Frederick R, and Jacobson V (2003) RTP: A Transport Protocol for Real-Time Applications. RFC 3550, Internet Engineering Task Force.

Seaborne A (2004) RDQL – A Query Language for RDF. W3C Member Submission, 9, 29-1.

Shacham R, Schulzrinne H, Thakolsri S, and Kellerer W (2007) Ubiquitous device personalization and use: The next generation of IP multimedia communications. *ACM Transactions on Multimedia Computing, Communications, and Applications (TOMCCAP)* 3(2), 12.

Shacham R, Schulzrinne H, Thakolsri S, and Kellerer W (2008) Session Initiation Protocol (SIP) Session Mobility. Internet-Draft draft-shacham-sipping-session-mobility-05, Internet Engineering Task Force. Work in progress.

Shin H, Suh Y, and Kwon D (2000) Multicast routing protocol by multicast agent in mobile networks, in *Proceedings of the 2000 International Conference on Parallel Processing*. IEEE Computer Society, p. 271.

Shin M, Mishra A, and Arbaugh W (2004a) Improving the latency of 802.11 hand-offs using neighbor graphs, in *Proceedings of the 2nd International Conference on Mobile Systems, Applications, and Services*. ACM, pp. 70–83.

Shin S, Forte AG, Rawat AS, and Schulzrinne H (2004b) Reducing MAC layer handoff latency in IEEE 802.11 wireless LANs, in *ACM International Workshop on Mobility Management and Wireless Access (MobiWac '04)*. ACM Press, New York, pp. 19–26.

Shnayder V, Hempstead M, Chen B, Allen G, and Welsh M (2004) Simulating the power consumption of large-scale sensor network applications, in *Proceedings of the 2nd International Conference on Embedded Networked Sensor Systems*. ACM, New York, pp. 188–200.

Simpson W (1994) The Point-to-Point Protocol (PPP). RFC 1661, Internet Engineering Task Force.

Snoeren AC and Balakrishnan H (2000) An end-to-end approach to host mobility, in *ACM/IEEE International Conference on Mobile Computing and Networking (MobiCom), Boston, MA*. ACM/IEEE, pp. 155–166.

Soliman H, Castelluccia C, el Malki K, and Bellier L (2006) Hierarchical Mobile IPv6 Mobility Management HMIPv6. RFC 4140, Internet Engineering Task Force.

Sparks R (2003) The Session Initiation Protocol (SIP) Refer Method. RFC 3515, Internet Engineering Task Force.

Stallings W (2004) *IEEE 802.11: Wireless LANs from a to n*. IEEE Computer Society.

Steele R, Li J, and Gould P (2001) *GSM, cdmaOne and 3G Systems*. John Wiley and Sons, Inc., New York.

Stefanov A and Erkip E (2005) Cooperative space–time coding for wireless networks. *IEEE Transactions on Communications*, pp. 1804–1809.

Stemm M and Katz R (1998) Vertical handoffs in wireless overlay networks. *Mobile Networks and Applications* 3(4), 335–350.

Stewart R, Xie Q, Morneault K, Sharp C, Schwarzbauer H, Taylor T, Rytina I, and Kalla M (2000) Stream Control Transmission Protocol. RFC 2960, Internet Engineering Task Force.

Tachikawa K (2002) *W-CDMA Mobile Communications System*. John Wiley and Sons, Ltd, Chichester.

Tan CL and Pink S (2000a) MobiCast: A multicast scheme for wireless networks. *Mobile Networks and Applications* 5(4), 259–271.

Tan C, Lye K, and Pink S (1999) A fast handoff scheme for wireless networks, in *Proceedings of the 2nd ACM International Workshop on Wireless Mobile Multimedia*. ACM, p. 90.

Telecommunication Industry Association (2013) TR-45 Mobile and Personal Communications Systems Standards, www.tiaonline.org (accessed November 12, 2013).

Teraoka F, Gogo K, Mitsuya K, Shibui R, and Mitani K (2008) Unified Layer 2 (L2) Abstractions for Layer 3 (L3)-Driven Fast Handover. RFC 5184, Internet Engineering Task Force.

Thajchayapong S and Peha J (2006) Mobility patterns in microcellular wireless networks. *IEEE Transactions on Mobile Computing*, pp. 52–63.

Thaler D (2004) Border Gateway Multicast Protocol. RFC 3913, Internet Engineering Task Force.

Thaler D, Talwar M, Aggarwal A, Vicisano L, and Pusateri T (2007) Automatic IP Multicast without Explicit Tunnels (AMT). Internet-Draft draft-ietf-mboned-auto-multicast-08.txt, Internet Engineering Task Force. Work in progress.

Thaler D, Vicisano L, et al. (2002) IPv4 Automatic Multicast without Explicit Tunnels (AMT). Internet-Draft, Internet Engineering Task Force. Work in progress.

Third Generation Partnership Project 2 and Telecommunications Industry Association (2007) Voice Call Continuity (VCC), 3GPP2-TIA PN-3-0231 (TIA-1093)/X. P0042.

Thomson S and Narten T (1998) IPv6 Stateless Address Autoconfiguration. RFC 2462, Internet Engineering Task Force.

Tilak S and Abu-Ghazaleh N (2001) A concurrent migration extension to an end-to-end host mobility architecture. *ACM SIGMOBILE Mobile Computing and Communications Review* **5**(3), 26–31.

Time O (2000) ITU-T Recommendation G. 114, ITU-T.

Tripathi N, Reed J, and VanLandingham HF (1998) Handoff in cellular systems. *IEEE Personal Communications* **5**(6), 26–37.

Tseng C, Yen L, Chang H, and Hsu K (2005) Topology-aided cross-layer fast handoff designs for IEEE 802.11/mobile IP environments. *IEEE Communications Magazine* **43**(12), 156–163.

Tuominen AJ and Petander HL (2001) MIPL mobile IPv6 for Linux in HUT campus network mediapoli. Proceedings of Ottawa Linux Symposium, 2001, Ottawa.

Tutsch D and Sokol J (2001) Petri net based performance evaluation of USAIAs bandwidth partitioning for the wireless cell level. Proceedings of the 9th International Workshop on Petri Nets and Performance Models (PNPM'01).

Vakil F, Famolari D, Baba S, and Famolari D (2001) Virtual soft hand-off in IP-centric wireless CDMA networks, in *3G Wireless, Conference Proceedings, San Francisco*. Delson, pp. 704–709.

Valkó A (1999) Cellular IP: A new approach to Internet host mobility. *ACM SIGCOMM Computer Communication Review* **29**(1), 50–65.

Van Brussel H, Peng Y, and Valckenaers P (1993) Modelling flexible manufacturing systems based on Petri nets. *CIRP Annals* **42**(1), 479–484.

Varshney U and Chatterji S (1999) Architectural issues in IP multicasting over wireless networks. IEEE Conference on Wireless Communication and Networking, New Orleans.

Vatn JO and Maguire GC (1998) The effect of using co-located care-of addresses on macro handover latency. 14th Nordic Tele-traffic Seminar, Technical University of Denmark, Lyngby, Denmark.

Velayos H and Karlsson G (2004) Techniques to reduce IEEE 802.11b handoff time. Wireless Networking Symposium, Paris.

Viswanadham N, Narahari Y, and Johnson T (1990) Deadlock prevention and deadlock avoidance in flexible manufacturing systems using Petri net models. *IEEE Transactions on Robotics and Automation* **6**(6), 713–723.

Vogt C (2006) A Comprehensive Delay Analysis for Reactive and Proactive Handoffs with Mobile IPv6 Route Optimization. Institute of Telematics, Universität Karlsruhe (TH), Karlsruhe, Germany, TM-2006-1.

Wahl M, Howes T, and Kille S (1997) Lightweight Directory Access Protocol (v3). RFC 2251, Internet Engineering Task Force.

Waitzman D, Partridge C, and Deering S (1988) Distance Vector Multicast Routing Protocol. RFC 1075, Internet Engineering Task Force.

Waldo J (1999) The Jini architecture for network-centric computing. *Communications of the ACM* **42**(7), 76–82.

Wang J (1998) *Timed Petri Nets: Theory and Application*. Kluwer Academic.

Wang J, Sun L, Jiang X, and Wu Z (2002) IGMP snooping: a VLAN-based multicast protocol, in *5th IEEE International Conference on High Speed Networks and Multimedia Communications*. IEEE, pp. 335–340.

Wedlund E and Schulzrinne H (1999) Mobility support using SIP. 2nd ACM/IEEE International Conference on Wireless and Mobile Multimedia (WoWMoM), Seattle, WA.

Williamson B (2000) *Developing IP Multicast Networks: The Definitive Guide to Designing and Deploying CISCO IP Multicast Networks*, vol. 1. Cisco Press, San Francisco.

Williamson C, Harrison T, Mackrell WL, and Bunt RB (1998) Performance evaluation of the MoM mobile multicast protocol. *ACM Mobile Networks and Applications (MONET) Journal* **3**(2), 189–201.

Wong K (2002) Architecture alternatives for integrating Cellular IP and Mobile IP, in *21st IEEE International Conference on Performance, Computing, and Communications, 2002, Phoenix, AZ*. IEEE, pp. 197–204.

Wong K and Dutta A (2005) Simultaneous mobility in MIPv6, in *2005 IEEE International Conference on Electro Information Technology*. IEEE, p. 5.

Wong D and Lim T (1997) Soft handoffs in CDMA mobile systems. *IEEE Personal Communications* **4**(6), 6–17. [See also *IEEE Wireless Communications*.]

Wong K, Wei HY, Dutta A, Young K, and Schulzrinne H (2002) IP micro-mobility management using host-based routing, in *Wireless IP and Building the Mobile Internet* (eds S Dixit and R Prasad). Artech House.

Wong D, Dutta A, Burns J, Young K, and Schulzrinne H (2003a) A multilayered mobility management scheme for auto-configured wireless IP networks. *IEEE Wireless Communications Magazine* **10**(5), 62–69.

Wong D, Dutta A, Schulzrinne H, and Young K (2003b) Managing simultaneous mobility of IP hosts. IEEE International Military Communications Conference (MILCOM 2003), Boston, MA.

Wong KD, Dutta A, Schulzrinne H, and Young K (2007) Simultaneous mobility: Analytical framework, theorems and solutions. *Wireless Communications and Mobile Computing*, **7**, 623–642.

Wongrujira K and Seneviratne A (2005) Monetary incentive with reputation for virtual market-place based P2P, in *CoNEXT'05*. ACM Press, New York, pp. 135–145.

Wu C, Cheng A, Lee S, Ho J, and Lee D (2002) Bi-directional route optimization in mobile IP over wireless LAN, in *Proceedings of 56th IEEE Vehicular Technology Conference, 2002 (VTC 2002-Fall)*. IEEE Computer Society, pp. 1168–1172.

Wu JLC (1999) An IP mobility support architecture for 4GW wireless infrastructure. Proceedings of the 1999 Personal Computing and Communication Workshop, pp. 67–71.

Xylomenos G and Polyzos GC (1997) IP multicast for mobile hosts. *IEEE Communications Magazine* **35**(1), 54–58.

Yang S and Park S (2001) An efficient multicast routing scheme for mobile hosts in IPv6-based networks. *Journal of the Institute of Electronics Engineers of Korea* **38**(8), 11–18.

Yemini Y (1983) A bang-bang principle for real-time transport protocols. *ACM SIGCOMM Computer Communication Review* **13**(2), 262–268.

Yokota H, Idoue A, Hasegawa T, and Kato T (2002) Link layer assisted mobile IP fast handoff method over wireless LAN networks. In *Proceedings of the 8th Annual International Conference on Mobile Computing and Networking (MobiCom '02)*. ACM, New York, pp. 131–139.

Zaid M (1994) Personal mobility in PCS. *IEEE Personal Communications* **1**(4), 12. [See also *IEEE Wireless Communications*.]

Zeadally S, Siddiqui F, DeepakMavatoor N, and Randhavva P (2004) SIP and mobile IP integration to support seamless mobility. 15th IEEE International Symposium on Personal, Indoor and Mobile Radio Communications, 2004 (PIMRC 2004), vol. 3.

Zhang T, Madhani S, Dutta A, VanDenberg E, Ohba Y, Taniuchi K, and Mohanty S (2005) Implementation and evaluation of autonomous collaborative discovery of neighboring networks. 3rd International Conference on Information Technology: Research and Education, 2005 (ITRE 2005), pp. 12–17.

Zhou M and DiCesare F (1991) Parallel and sequential mutual exclusions for Petri net modeling of manufacturing systems with shared resources. *IEEE Transactions on Robotics and Automation* **7**(4), 515–527.

Zhou M and Robbi A (1994) Applications of Petri net methodology to manufacturing systems. In Computer Control of Flexible Manufacturing Systems (eds SB Joshi and JS Smith). Chapman and Hall, pp. 207–230.

Zhou M and Venkatesh K (1999) Modeling, Simulation, and Control of Flexible Manufacturing Systems: A Petri Net Approach. World Scientific.

Zhuang S, Lai K, Stoica I, Katz R, and Shenker S (2005) Host mobility using an Internet indirection infrastructure. *Wireless Networks* **11**(6), 741–756.

Zimmermann A, German R, Freiheit J, and Hommel G (1999a) Timenet 3.0 tool description. International Conference on Petri Nets and Performance Models (PNPM'99), Zaragoza, Spain.

Zimmermann A, Rodriguez D, and Silva M (1999b) Modelling and optimisation of manufacturing systems: Petri nets and simulated annealing. Proceedings of the 1999 European Control Conference (ECC99), Karlsruhe, Germany.

Zuberek W (1980) Timed Petri nets and preliminary performance evaluation, in *Proceedings of the 7th Annual Symposium on Computer Architecture*. ACM, New York, pp. 88–96.

Zuberek W (1991) Timed Petri nets definitions, properties, and applications. *Microelectronics and Reliability* **31**(4), 627–644.

Zuberek W (2000) Hierarchical analysis of manufacturing systems using Petri nets. IEEE International Conference on Systems, Man, and Cybernetics, 2000, Nashville, TN, vol. 4.

Zuberek WM and Kubiak W (1999) Timed Petri nets in modeling and analysis of simple schedules for manufacturing cells. *Computers and Mathematics with Applications* **37**(11–12), 191–206.

Index

Mobility Protocols and Handover Optimization: Design, Evaluation and Application, First Edition.
Ashutosh Dutta and Henning Schulzrinne.
© 2014 John Wiley & Sons, Ltd. Published 2014 by John Wiley & Sons, Ltd.